"Was du ererbt von deinen Vätern hast, erwirb es, um es zu besitzen."
("That which you have inherited from your fathers,
acquire it so as to make it your own.")

—Goethe in Faust

In memory of my father, Krishnarao Shivamoggi

Bhimsen K. Shivamoggi

Perturbation Methods
for Differential Equations

Springer Science+Business Media, LLC

Bhimsen K. Shivamoggi
Department of Mathematics
University of Central Florida
Orlando, FL 32816-1364
U.S.A.

Library of Congress Cataloging-in-Publication Data

Shivamoggi, Bhimsen K.
  Perturbation methods for differential equations / Bhimsen K. Shivamoggi.
    p. cm.
  Includes bibliographical references and index.
    ISBN 978-0-8176-4189-4   ISBN 978-1-4612-0047-5 (eBook)
  DOI 10.1007/978-1-4612-0047-5
    1. Perturbation (Mathematics) 2. Differential equations–Numerical solutions. 3.
  Differential equations, Partial–Numerical solutions. I. Title.

  QA871 .S44 2002
  515'.35–dc 21   2002066657
                                                                        CIP

AMS Subject Classifications: 34E05, 34E10, 34E13, 34E15, 34E20, 35B20, 35B25

Printed on acid-free paper.
©2003 Springer Science+Business Media New York
Originally published by Birkhäuser Boston in 2003

ISBN 978-0-8176-4189-4   SPIN 10764753

Typeset by the author.

9 8 7 6 5 4 3 2 1

Table of Contents

Preface

In nonlinear problems, essentially new phenomena occur which have no place in the corresponding linear problems. Therefore, in the study of nonlinear problems the major purpose is not so much to introduce methods that improve the accuracy of linear methods, but to focus attention on those features of the nonlinearities that result in distinctively new phenomena. Among the latter are –

* existence of solutions of periodic problems for all frequencies rather than only a set of characteristic values,
* dependence of amplitude on frequency,
* removal of resonance infinities,
* appearance of jump phenomena,
* onset of chaotic motions.

On the other hand, mathematical problems associated with nonlinearities are so complex that a comprehensive theory of nonlinear phenomena is out of the question.[1] Consequently, one practical approach is to settle for something less than complete generality. Thus, one gives up the study of global behavior of solutions of a nonlinear problem and seeks nonlinear solutions in the neighborhood of (or as perturbations about) a known linear solution. This is the basic idea behind a perturbative solution of a nonlinear problem.

---

[1] In an alternative approach via numerical solution of nonlinear problems, computers have contributed enormously by providing detailed information. But, this approach lacks organization and does not afford complete insight into the underlying complex physical behavior. Besides, for problems of singular-perturbation type (see below), determination of the accuracy of a numerical solution is extremely difficult without first obtaining some idea about the salient features of the solution afforded by the singular perturbation approach.

The method of perturbation expansion is carried out with respect to a small parameter ε, which may be a non-dimensionalized amplitude of a typical perturbation. The coefficients in these expansions are obtained as solutions of a sequence of linear problems. The lowest-order terms are governed by problems that typically result from a linearization of the original problem and are known. The higher-order quantities are produced as solutions of linear inhomogeneous differential equations where the inhomogeneities involve only the previously determined lower-order quantities.

One has a regular perturbation problem if a straightforward perturbation expansion is uniformly valid. However, if a straightforward expansion is not uniformly valid, and if this difficulty is compounded in higher approximations, then one has a singular perturbation problem. Over the course of more than a century, starting with problems of dynamical astronomy, a host of singular perturbation procedures have been developed to resolve these non-uniformities. A number of excellent accounts of these procedures exist in the literature emphasizing several different points of view.

This book has grown out of my lecture notes for an interdisciplinary graduate level course on perturbation methods which I have taught for the past several years to a mix of students in applied mathematics, physics, and engineering. I have endeavored to describe the standard procedures (which together cover a wide area of applicability) along with the underlying basic concepts in a unified way. To facilitate an interdisciplinary readership, I have adopted a straightforward intuitive approach and paid more attention to the procedures and the underlying ideas than to mathematical rigor. Thus, I have side-stepped the theoretical aspects such as error estimates and soft-pedaled the formal proofs.

The book starts with an introduction to the idea of asymptotic approximations to the evaluation of transcendental functions and the solution of differential equations and a discussion of the regular perturbation approach. It then embarks on a systematic and unified discussion of the resolution of the non-uniformities produced, as a rule, by the regular perturbation method. We discuss several singular perturbation methods, primarily

* the method of strained parameters,
* the method of averaging,
* the method of matched asymptotic expansions,
* the method of multiple scales,
* the quantum-field-theoretic renormalization method.

These procedures are illustrated via simple but useful worked examples and are applied to ordinary and partial differential equations arising in various problems of solid mechanics, fluid dynamics, and plasma physics. I have tried to provide brief background material for several mathematical as well as application topics to make the discussion reasonably self-contained.

Several exercises are provided at the end of each chapter along with answers to selected problems (given at the end of the book) to consolidate the ideas and procedures and to occasionally indicate developments from the research literature.

The selection of the application topics reflects, to some extent, my own research interests. In the same vein, the list of references given at the end of the book contains primarily books and papers I have used in developing the graduate lecture material (and this book) which may also provide the reader with some useful leads into the more advanced aspects of the subject. Perturbation methods has a vast literature and the list of references given in this book cannot pretend to be fully representative.

Acknowledgements

In the preparation of this manuscript (and the graduate lecture material), I have drawn enormously from the ideas, procedures, and treatments developed by several eminent authors – J.D. Cole, K.O. Friedrichs, M.H. Holmes, J.B. Keller, J. Kevorkian, P.A. Lagerstrom, A.H. Nayfeh, R.A. Struble, M.D. Van Dyke, and others. I have acknowledged the various sources, in this connection, in the References, given at the end of this book and wish to apologize to any others I may have inadvertently missed. I wish to thank my students and colleagues who rendered valuable assistance as the course material was developed. I wish to express my thanks to my colleagues, Professors Larry Andrews, John Cannon, David Rollins, and Mike Taylor for several valuable discussions on many aspects of the subject. Most of the manuscript was read by Professors John Cannon and David Rollins who made valuable suggestions. I wish to thank also the referees for their valuable suggestions.

My sincere thanks are due to Jackie Callahan for the exemplary patience, dedication, and skill with which she did the whole typing job. My thanks are due to Ronee Trantham for her excellent job in doing the figures. My thanks are also due to Birkhäuser for their splendid cooperation with this whole project and working with them was really a delight. Last, but by no means least, my immense thanks are due to my wife, Jayashree, and my daughters, Vasudha and

Rohini for their enormous understanding and cooperation with this endeavor (as they have done with my previous book projects). My father Krishnarao could never attend college but was always immensely enthusiastic about higher learning and knowledge. This book is dedicated to his memory.

Orlando, Florida   *Bhimsen K. Shivamoggi*

Chapter 1

Asymptotic Series and Expansions

1.1. Introduction

The method of perturbations can be used to develop approximate solutions to differential equations, which have nonlinearities or variable coefficients so that an exact solution cannot be constructed. Questions of convergence of the perturbation expansion in terms of a suitable small parameter are abandoned if such an expansion can provide a useful representation of the exact solution. However, the perturbation expansions are usually not uniformly valid and various techniques have been developed to render them uniformly valid.

A perturbative solution of a nonlinear problem becomes viable if it is "close" to another problem we already know how to solve. We study the solution of the simpler problem and then try to express the solution of the more difficult problem in terms of the simpler one modified by a small correction. This approach is sequential, in the sense that once an improved approximation has been found, the process may be repeated to obtain a still better approximation. The latter step involves developing a formal perturbation expansion with respect to a small parameter ε. The accuracy of this approximation gets better for smaller values of ε. For example, ε may be the nondimensionalized amplitude of a typical perturbation. The coefficients in these expansions are obtained as solutions of a sequence of linear problems. The quantities of relevance are essentially developed in a Taylor series, which is taken to be uniformly convergent in ε in any compact part of the respective domain of analyticity. The justification to expect the solution to exhibit such an analytic dependence on the parameter ε follows if the parameter ε enters the differential equations and the boundary conditions in an analytic way. The lowest-order terms in these series are given by quite simplified forms of the original problem, so that often these terms are well known. The higher-order terms are then produced as solutions of linear inhomogeneous differential equations where the inhomogeneities involve only the previously determined linear-order quantities.

1.2. Taylor Series Expansions

Suppose a function $f(x)$ is infinitely differentiable at $x = x_0$, then we may express it in a power series of $(x - x_0)$ as

$$f(x) = a_0 + a_1 (x - x_0) + a_2 (x - x_0)^2 + \dots \tag{1.2.1}$$

where,

$$a_n \equiv \frac{1}{n!} f^{(n)}(x_0). \tag{1.2.2}$$

Example 1: Taylor series expansions of some standard functions are –

$$\left. \begin{array}{l} \sin x = x - \dfrac{x^3}{3!} + \dfrac{x^5}{5!} - \dots \\[2mm] \cos x = 1 - \dfrac{x^2}{2!} + \dfrac{x^4}{4!} - \dots \\[2mm] e^x = 1 + \dfrac{x}{1!} + \dfrac{x^2}{2!} + \dots \\[2mm] \ell n(1+x) = x - \dfrac{x^2}{2} + \dfrac{x^3}{3} - \dots \end{array} \right\}.$$

1.3. Gauge Functions

Consider the limit of a function $f(\varepsilon)$ as $\varepsilon \to 0$. This limit may depend on whether $\varepsilon \to 0^+$ or $\varepsilon \to 0^-$.

Example 2: $$\lim_{\varepsilon \to 0^+} e^{-\frac{1}{\varepsilon}} = 0, \qquad \lim_{\varepsilon \to 0^-} e^{-\frac{1}{\varepsilon}} = \infty.$$

Suppose $\varepsilon > 0$. Suppose $f(\varepsilon)$ has no essential singularity at $\varepsilon = 0$ (e.g., $\lim_{\varepsilon \to 0} \sin\left(\frac{1}{\varepsilon}\right)$), then one has three possibilities –

$$\lim_{\varepsilon \to 0} |f(\varepsilon)| = \begin{cases} 0 \\ A, \quad 0 < A < \infty. \\ \infty \end{cases} \tag{1.3.1}$$

In order to determine the rate at which $f(\varepsilon)$ tends to 0 or ∞, we compare these rates with the rates at which known functions, called gauge functions, tend to 0 or ∞.

The simplest set of gauge functions are

$$\left. \begin{array}{l} 1, \varepsilon, \varepsilon^2, \ldots \\ \varepsilon^{-1}, \varepsilon^{-2}, \varepsilon^{-3}, \ldots \end{array} \right\}$$

which satisfy, for ε near 0,

$$\left. \begin{array}{l} 1 > \varepsilon > \varepsilon^2 > \ldots \\ \varepsilon^{-1} < \varepsilon^{-2} < \varepsilon^{-3} < \ldots \end{array} \right\}. \tag{1.3.2}$$

Example 3: We have

$$\left. \begin{array}{l} \sin \varepsilon = \varepsilon - \dfrac{\varepsilon^3}{3!} + \dfrac{\varepsilon^5}{5!} - \ldots \\ \cos \varepsilon = 1 - \dfrac{\varepsilon^2}{2!} + \dfrac{\varepsilon^4}{4!} - \ldots \end{array} \right\},$$

from which

$$\left. \begin{array}{l} \lim_{\varepsilon \to 0} \dfrac{\sin \varepsilon}{\varepsilon} = \lim_{\varepsilon \to 0} \left(1 - \dfrac{\varepsilon^2}{3!} + \dfrac{\varepsilon^4}{4!} - \ldots \right) = 1 \\ \lim_{\varepsilon \to 0} \dfrac{1 - \cos \varepsilon}{\varepsilon^2} = \lim_{\varepsilon \to 0} \left(\dfrac{1}{2!} - \dfrac{\varepsilon^2}{4!} + \ldots \right) = \dfrac{1}{2!} \end{array} \right\}.$$

Thus,

$$\left.\begin{array}{l} \sin\varepsilon \to 0 \text{ at the same rate as } \varepsilon \to 0 \\ 1-\cos\varepsilon \to 0 \text{ at the same rate as } \varepsilon^2 \to 0 \end{array}\right\}.$$

On the other hand, in order to determine the rate at which $e^{-1/\varepsilon} \to 0$ as $\varepsilon \to 0$, we try to expand $e^{-1/\varepsilon}$ in a Taylor series for small ε.

Letting $f(\varepsilon) = e^{-1/\varepsilon}$, we have

$$f'(\varepsilon) = \frac{1}{\varepsilon^2} e^{-1/\varepsilon}, \quad f''(\varepsilon) = \left(\frac{1}{\varepsilon^4} - \frac{2}{\varepsilon^3}\right) e^{-1/\varepsilon},$$

etc.

Thus,

$$\left.\begin{array}{l} f'(0) = \lim_{\varepsilon\to 0} \dfrac{1}{\varepsilon^2} e^{-\frac{1}{\varepsilon}} = \lim_{x\to\infty} \dfrac{x^2}{e^x} = 0 \\[3mm] f''(0) = \lim_{\varepsilon\to 0} \left(\dfrac{1}{\varepsilon^4} - \dfrac{2}{\varepsilon^3}\right) e^{-\frac{1}{\varepsilon}} \\[3mm] \qquad = \lim_{x\to\infty} \dfrac{x^4 - 2x^3}{e^x} = 0 \end{array}\right\}$$

etc.

where $x \equiv 1/\varepsilon$.

Therefore, the function $e^{-\frac{1}{\varepsilon}}$ cannot be represented by a power series in ε. In fact, $e^{-\frac{1}{\varepsilon}} \to 0$ faster than any power of ε, because

$$\lim_{\varepsilon\to 0} \frac{e^{-\frac{1}{\varepsilon}}}{\varepsilon^n} = \lim_{x\to\infty} \frac{x^n}{e^x} = 0.$$

1.4. Asymptotic Series and Expansions

Definition: Let $f(\varepsilon)$ and $g(\varepsilon)$ be defined in some neighborhood of $\varepsilon = 0$.

We write

$$f(\varepsilon) = o\big(g(\varepsilon)\big) \quad as \quad \varepsilon \to 0 \tag{1.4.1}$$

if

$$\lim_{\varepsilon \to 0} \left| \frac{f(\varepsilon)}{g(\varepsilon)} \right| = 0$$

and we write

$$f(\varepsilon) = O(g(\varepsilon)) \quad as \quad \varepsilon \to 0 \qquad (1.4.2)$$

if there exists a positive bounded constant M such that

$$|f(\varepsilon)| \le M |g(\varepsilon)| \quad or \quad \left| \frac{f(\varepsilon)}{g(\varepsilon)} \right| \text{ is bounded}$$

for all ε *in some neighborhood of* $\varepsilon = 0$.

Example 4: $e^{-\frac{1}{\varepsilon}} = o(\varepsilon^\alpha)$, $\forall \alpha$. Thus, $e^{-\frac{1}{\varepsilon}}$ is transcendentally small with respect to ε^α.

Example 5: $\varepsilon^2 \ln \varepsilon = o(\varepsilon)$ as $\varepsilon \to 0^+$.

By L'Hospital's rule

$$\lim_{\varepsilon \to 0^+} \frac{\varepsilon^2 \ln \varepsilon}{\varepsilon} = \lim_{\varepsilon \to 0^+} \frac{\ln \varepsilon}{(1/\varepsilon)} = \lim_{\varepsilon \to 0^+} \frac{1/\varepsilon}{(-1/\varepsilon^2)} = 0.$$

Thus, $\varepsilon^2 \ln \varepsilon = o(\varepsilon)$ as $\varepsilon \to 0^+$.

Example 6: $\sin \varepsilon = O(\varepsilon)$ as $\varepsilon \to 0^+$.

By the Mean Value Theorem of calculus, there exists a number c between 0 and ε such that

$$\frac{\sin \varepsilon - \sin 0}{\varepsilon - 0} = \cos c.$$

Thus,

$$|\sin \varepsilon| = |\varepsilon \cos c| \le |\varepsilon|$$

since $|\cos c| \le 1$. Therefore,

$$\sin \varepsilon = O(\varepsilon) \quad as \quad \varepsilon \to 0^+.$$

The above definition may be extended to functions of ε and another variable t lying in an interval I.

Definition: *Let* $f(t,\varepsilon)$ *and* $g(t,\varepsilon)$ *be defined for all* $t \in I$ *and all* ε *in a neighborhood of* $\varepsilon = 0$. *We write*

$$f(t,\varepsilon) = o\big(g(t,\varepsilon)\big) \ \text{ as } \ \varepsilon \to 0 \ \text{ uniformly on } I \tag{1.4.3}$$

if

$$\lim_{\varepsilon \to 0} \left| \frac{f(t,\varepsilon)}{g(t,\varepsilon)} \right| = 0, \quad \forall \ t \in I.$$

If there exists a positive bounded function $M(t)$ *on* I *such that*

$$\left| f(t,\varepsilon) \right| \le M(t) \left| g(t,\varepsilon) \right|$$

for all $t \in I$ *and for all* ε *in some neighborhood of zero, then we write*

$$f(t,\varepsilon) = O\big(g(t,\varepsilon)\big) \ \text{ as } \ \varepsilon \to 0, \ \text{uniformly on } I. \tag{1.4.4}$$

A measure of decreasing orders of magnitude is provided by an asymptotic sequence of functions.

Definition: *A sequence* $\left\{ \varphi_n(\varepsilon) \right\}_{n=1}^{\infty}$ *is an asymptotic sequence if*

$$\varphi_{n+1}(\varepsilon) = o\big(\varphi_n(\varepsilon)\big) \quad \text{as} \quad \varepsilon \to 0.$$

A sum of terms in an asymptotic sequence is an asymptotic expansion. Various linear operations and multiplications can be performed with asymptotic expansions.

If a function possesses an asymptotic approximation in terms of an asymptotic sequence, namely, an asymptotic expansion, then the latter is unique for the given asymptotic sequence. Besides, with an asymptotic expansion, the error of approximation can be understood as well as controlled.

Example 7: $\displaystyle\sum_{n=1}^{\infty} a_n \, \varphi_n(\varepsilon)$ is an asymptotic expansion of $f(\varepsilon)$ as $\varepsilon \to 0$ if

$$f(\varepsilon) - \sum_{n=0}^{N} a_n \, \varphi_n(\varepsilon) = o(\varphi_N(\varepsilon)) \quad \text{or} \quad O(\varphi_{N+1}(\varepsilon)) \quad \text{as} \quad \varepsilon \to 0.$$

Note that

$$a_m = \lim_{\varepsilon \to 0} \frac{f(\varepsilon) - \sum_{n=1}^{m-1} a_n \, \varphi_n(\varepsilon)}{\varphi_m(\varepsilon)}$$

which gives a_m uniquely.

In a physical problem, the coefficients in an asymptotic expansion will depend on space or time variables other than ε. The series is said to be uniformly valid (in space and time) if the error is small uniformly in these variables.

Definition: An asymptotic expansion

$$f(x,\varepsilon) \sim \sum_{m=0}^{\infty} \delta_m(\varepsilon) \, a_m(x) \quad as \quad \varepsilon \to 0 \tag{1.4.5}$$

is uniformly valid if

$$f(x,\varepsilon) - \sum_{m=0}^{N} \delta_m(\varepsilon) \, a_m(x) = R_N(x,\varepsilon) \tag{1.4.6}$$

where $R_N(x,\varepsilon) = O[\delta_{N+1}(\varepsilon)]$ *or* $o[\delta_N(\varepsilon)]$ *uniformly, for all x; otherwise it is nonuniform.*

Example 8: $\sin(x+\varepsilon) = \sin x + \varepsilon \, \cos x - \dfrac{\varepsilon^2}{2!} \sin x + \dots$ is a uniformly valid expansion as $\varepsilon \to 0$.

Example 9: $\sqrt{x+\varepsilon} = \sqrt{x} \, (1+\varepsilon/x)^{1/2}$

$$= \sqrt{x} \left(1 + \frac{\varepsilon}{2x} - \frac{\varepsilon^2}{8x^2} + \dots \right).$$

This is a nonuniform expansion because this converges only if $|\varepsilon/x| \leq 1$.

Example 10: $y(t,\varepsilon) = e^{-\varepsilon t}$, $t > 0$, $\varepsilon \ll 1$.

The first three terms in the Taylor expansion in powers of ε provide an approximation

$$y_{approx}(t,\varepsilon) = 1 - \varepsilon t + \frac{1}{2}\varepsilon^2 t^2.$$

The error is

$$E(t,\varepsilon) = e^{-\varepsilon t} - 1 + \varepsilon t - \frac{1}{2}\varepsilon^2 t^2 = -\frac{1}{3!}\varepsilon^3 t^3 + \dots.$$

For a fixed t, the error can be made as small as desired by choosing ε small enough. However, for a fixed ε, no matter how small, the approximation breaks down, as t becomes large. Thus, this approximation is not uniformly valid on $I = [0,\infty)$. In other words, we cannot write $E(t,\varepsilon) = O(\varepsilon^3)$, as $\varepsilon \to 0$, uniformly on $[0,\infty)$.

Example 11: Consider

$$\frac{dy}{dx} + y = \frac{1}{x}, \quad x \gg 1.$$

This has the solution given by

$$y = e^{-x} \int_{-\infty}^{x} \frac{e^z}{z}\, dz$$

$$= \frac{1}{x} + e^{-x} \int_{-\infty}^{x} \frac{e^z}{z^2}\, dz$$

$$= \frac{1}{x} + \frac{1}{x^2} + 2e^{-x} \int_{-\infty}^{x} \frac{e^z}{z^3}\, dz$$

$$= \frac{0!}{x} + \frac{1!}{x^2} + \frac{2!}{x^3} + \dots + \frac{(n-1)!}{x^n} + n!\, e^{-x} \int_{-\infty}^{x} \frac{e^z}{z^{n+1}}\, dz.$$

Note that this series diverges for all x, as $n \to \infty$. However, for $x < 0$, the residue after n terms is

$$|R_n| = \left| n! \, e^{-x} \int_{-\infty}^{x} \frac{e^z}{z^{n+1}} \, dz \right|$$

$$\leq n! \left| \frac{1}{x^{n+1}} \right| e^{-x} \int_{-\infty}^{x} e^z \, dz = \frac{n!}{|x^{n+1}|}.$$

Thus, the error committed in truncating the series after n terms is in magnitude less than the first neglected term. Observe that, as $|x| \to \infty$, with n fixed, $|R_n| \to 0$; so, for a fixed large n, the first n terms in the series can represent y with an error which can be made arbitrarily small by taking x sufficiently large, the divergence of the series notwithstanding.

An infinite asymptotic series may either converge for some range of ε or diverge for all ε. Mathematical convergence depends on the behavior of terms of indefinitely high order. However, in developing an approximation in practice, one can calculate only the first few terms and hope that they rapidly approach the true solution. This requirement may sometimes be better met with a divergent series than with a convergent series which requires many terms to give an accurate approximation. The error incurred by the nth sum is usually smallest when $\varepsilon \sim (1/n)$.

Example 12: The expansion of the Bessel function

$$J_0(\varepsilon) = \sum_{k=0}^{\infty} \frac{(-1)^k \, \varepsilon^{2k}}{2^{2k} \, (k!)^2}$$

has an infinite radius of convergence, but many terms are needed for accurate results unless ε is small. On the other hand, the asymptotic expansion

$$J_0\left(\frac{1}{\varepsilon}\right) \sim \sqrt{\frac{2\varepsilon}{\pi}} \left[\left(1 - \frac{1^2 \cdot 3^2}{2! \, 8^2} \varepsilon^2 + \frac{1^2 \cdot 3^2 \cdot 5^2 \cdot 7^2}{4! \, 8^4} \varepsilon^4 + \cdots \right) \right.$$

$$\times \cos\left(\frac{1}{\varepsilon} - \frac{\pi}{4} \right)$$

$$\left. + \left(\frac{1^2}{1! \, 8^1} \varepsilon^1 - \frac{1^2 \cdot 3^2 \cdot 5^2}{3! \, 8^3} \varepsilon^3 + \cdots \right) \sin\left(\frac{1}{\varepsilon} - \frac{\pi}{4} \right) \right]$$

is divergent for all ε, no matter how small, but a few terms given good accuracy for moderately small ε.

1.5. Asymptotic Solutions of Differential Equations

Frequently, even though the solution of a given differential equation is not a simple combination of elementary functions, an asymptotic representation of that solution can be constructed in terms of elementary functions.

Example 13: Consider the differential equation –

$$u'' + u - \frac{\beta^2 - 1/4}{x^2} u = 0.$$

This equation results by putting $y = \dfrac{u(x)}{\sqrt{x}}$ in Bessel's equation

$$x^2 y'' + x y' + \left(x^2 - \beta^2 \right) y = 0.$$

For $x \gg 1$, this equation gives

$$u'' + u \approx 0,$$

from which

$$u \approx e^{ix}.$$

We may therefore seek the following solution –

$$u \sim e^{ix} \sum_{n=0}^{\infty} a_n x^{-n}, \qquad a_0 = 1.$$

Following the usual procedure for a power-series solution, we obtain

$$u \sim e^{ix} \left[1 + \frac{\left(\beta^2 - 1/4 \right)}{2x} - \frac{\left(\beta^2 - 1/4 \right)\left(\beta^2 - 9/4 \right)}{8x^2} + \cdots \right].$$

Note: If $\beta = \pm 1/2$, we have

$$u = e^{ix}$$

as is to be anticipated readily from the given equation.

Example 14: Consider the differential equation –

$$u'' - \lambda x u = 0, \quad \lambda \gg 1.$$

Look for a solution of the form

$$u = e^{\int^{x} g \, dt}.$$

Substituting this in the given equation, we obtain

$$g' + g^2 = \lambda x.$$

Let

$$g[x, \lambda] \sim \lambda^{1/2} \sum_{n=0}^{\infty} g_n(x) \, \lambda^{-n/2}$$

which leads to

$$\left[\left(g_0^2 - x\right) + \lambda^{-1/2} \left(2g_0 g_1 + g_0'\right) + \frac{1}{\lambda} \left(g_1' + 2g_0 g_2 + g_1^2\right) + \cdots \right] = 0.$$

Equating coefficients of the various powers of $1/\lambda$ to zero, we obtain

$$\left. \begin{array}{l} g_0^2 = x \\[2mm] g_1 = -\dfrac{\left(\ell n \, g_0\right)'}{2} \\[4mm] g_2 = -\dfrac{g_1^2 + g_1'}{2 g_0} \\[4mm] \text{etc.} \end{array} \right\}$$

Thus,

$$u(x, \lambda) \sim x^{-1/4} \, e^{\pm 2/3 \, \lambda^{1/2} \, x^{3/2}}.$$

This solution breaks down near $x = 0$. The domain of nonuniformity corresponds to $g' \sim g^2 \sim \lambda x$ or (on using $g \sim x^{1/2}$) $x \lambda^{1/3} \sim 0(1)$.

1.6. Exercises

1. Show that $\dfrac{\sqrt{\varepsilon}}{1 - \cos \varepsilon} = O\left(\varepsilon^{-3/2}\right)$ as $\varepsilon \to 0^+$.

2. Show that $\ln \varepsilon = o\left(\varepsilon^{-p}\right)$ as $\varepsilon \to 0^+$ for all $p > 0$.

3. Show that $e^{\tan \varepsilon} = O(1)$ as $\varepsilon \to 0$.

4. Find an asymptotic expansion for large w for $f(w) = \int\limits_0^\infty \dfrac{w}{w + x} e^{-x} dx$.

5. Find an asymptotic expansion for large x for $F(x) = \int\limits_x^\infty \dfrac{e^{-t}}{t} dt$.

6. Find an asymptotic expansion for large x for $E_n(x) = \int\limits_1^\infty \dfrac{e^{-xt}}{t^n} dt$.

7. Find an asymptotic solution for large x for the equation $\dfrac{dy}{dx} = e^{-2xy}$.

Chapter 2

Regular Perturbation Methods

2.1. Introduction

The regular perturbation method works only for exceptionally special problems, and fails in general.

The procedure consists of –

(i)   substituting the power series

$$y(x;\varepsilon) \sim \sum_{n=0}^{\infty} \varepsilon^n y_n(x)$$

into the differential equation and the boundary/initial conditions,

(ii) expanding all quantities in a power series in ε,

(iii) collecting terms with same powers of ε and equating them to zero,

(iv) solving this hierarchy of boundary/initial value problems sequentially, i.e., if $y_0, y_1, \ldots, y_{k-1}$ are known, then y_k is determined by an equation of the form

$$\mathscr{L} y_k = f_k(y_0, y_1, \ldots, y_{k-1}),$$

where $\mathscr{L}$ is the linearized operator of the reduced problem evaluated at the known solution y_0; the possibility that the above problem has a solution locally whenever $\mathscr{L}$ is invertible is indicated by the Implicit Function Theorem.

Theorem 1: Consider an implicit equation

$$f(x,\varepsilon) = 0.$$

If there exists $x = x_0$ *such that*

$$f(x_0, \varepsilon) = 0$$

and if $f_x(x_0, 0)$ *is an invertible linear map, then there exists a unique solution of the given equation in the neighborhood of* $\varepsilon = 0$ *given by*

$$x = g(\varepsilon).$$

2.2. Algebraic Equations

Many of the essential concepts of perturbation methods can be discussed in the simpler context of algebraic equations before differential equations are considered.

Example 1: Consider the algebraic equation (Simmonds and Mann, 1986)

$$z^2 - 2z + \varepsilon = 0, \quad \varepsilon \ll 1.$$

This equation has two roots given by

$$\left. \begin{aligned} z_1 &= 1 - \sqrt{1-\varepsilon} \\ z_2 &= 1 + \sqrt{1-\varepsilon} \end{aligned} \right\}.$$

Expanding $\sqrt{1-\varepsilon}$ as a Taylor series

$$\sqrt{1-\varepsilon} = 1 - \frac{1}{2}\varepsilon - \frac{1}{8}\varepsilon^2 - \cdots$$

we obtain

$$\left. \begin{aligned} z_1 &\sim \frac{1}{2}\varepsilon + \frac{1}{8}\varepsilon^2 + \cdots \\ z_2 &\sim 2 - \frac{1}{2}\varepsilon - \frac{1}{8}\varepsilon^2 - \cdots \end{aligned} \right\}.$$

Let us now obtain these results using a regular perturbation expansion,

$$z(\varepsilon) \sim a_0 + \varepsilon a_1 + \varepsilon^2 a_2 + \cdots + a_N \varepsilon^N + O(\varepsilon^{N+1}).$$

On substituting this into the above equation, we obtain

$$\left[a_0 + \varepsilon\, a_1 + \varepsilon^2\, a_2 + \cdots + \varepsilon^N a_N + O\!\left(\varepsilon^{N+1}\right) \right]^2$$
$$-2 \left[a_0 + \varepsilon\, a_1 + \varepsilon^2\, a_2 + \cdots + \varepsilon^N a_N + O\!\left(\varepsilon^{N+1}\right) \right] + \varepsilon = 0$$

from which

$$\left[a_0^2 + \varepsilon \cdot 2\, a_0 a_1 + \varepsilon^2 \left(a_1^2 + 2 a_0 a_2 \right) + O\!\left(\varepsilon^3\right) \right]$$
$$-2 \left[a_0 + \varepsilon\, a_1 + \varepsilon^2\, a_2 + O\!\left(\varepsilon^3\right) \right] + \varepsilon = 0.$$

Rearranging this, we obtain

$$\left(a_0^2 - 2 a_0 \right) + \varepsilon \left(2 a_0 a_1 - 2 a_1 + 1 \right) + \varepsilon^2 \left(a_1^2 + 2 a_0 a_2 - 2 a_2 \right) + O\!\left(\varepsilon^3\right) = 0.$$

Equating coefficients of like powers of ε to zero, we obtain an infinite system of equations

$$\left. \begin{array}{l} a_0^2 - 2 a_0 = 0 \\ 2 a_0 a_1 - 2 a_1 + 1 = 0 \\ a_1^2 + 2 a_0 a_2 - 2 a_2 = 0 \\ \text{etc.} \end{array} \right\}$$

which is to be solved recursively.

Thus,

$$a_0 = 0 \ \text{ or } \ 2.$$

For $a_0 = 0$, we obtain

$$a_1 = \frac{1}{2}, \ a_2 = \frac{1}{8}, \ \text{etc.}$$

which gives the first root

$$z_1 \sim \frac{1}{2}\,\varepsilon + \frac{1}{8}\,\varepsilon^2 + \cdots.$$

For $a_0 = 2$, we obtain

$$a_1 = -\frac{1}{2}, \ a_2 = -\frac{1}{8}, \ \text{etc.}$$

which gives the second root

$$z_2 \sim 2 - \frac{1}{2}\,\varepsilon - \frac{1}{8}\,\varepsilon^2 - \cdots.$$

Example 2: Consider the algebraic equation (Simmonds and Mann, 1986)

$$\varepsilon z^2 - 2z + 1 = 0, \quad \varepsilon \ll 1.$$

Note that, in the limit $\varepsilon \to 0$, the degree of the equation drops from 2 to 1. So, this problem cannot be treated adequately by a regular perturbation expansion. This is a singular-perturbation problem.

This equation has two roots given by

$$\left. \begin{aligned} z_1 &= \frac{1 - \sqrt{1-\varepsilon}}{\varepsilon} \\ z_2 &= \frac{1 + \sqrt{1-\varepsilon}}{\varepsilon} \end{aligned} \right\}.$$

Expanding $\sqrt{1-\varepsilon}$ as a Taylor series

$$\sqrt{1-\varepsilon} = 1 - \frac{1}{2}\varepsilon - \frac{1}{8}\varepsilon^2 - \cdots$$

we obtain

$$\left. \begin{aligned} z_1 &= \frac{1}{2} + \frac{1}{8}\varepsilon + \cdots \\ z_2 &= \frac{2}{\varepsilon} - \frac{1}{2} - \frac{1}{8}\varepsilon + \cdots \end{aligned} \right\}.$$

Let us now obtain these results using a regular perturbation expansion,

$$z(\varepsilon) \sim a_0 + \varepsilon a_1 + \varepsilon^2 a_2 + O(\varepsilon^3)$$

(this presupposes that εz^2 is small compared with $-2z + 1$ in the given equation). On substituting this into the above equation, we obtain

$$\varepsilon \left[a_0 + \varepsilon a_1 + \varepsilon^2 a_2 + O(\varepsilon^3) \right]^2 - 2\left[a_0 + \varepsilon a_1 + \varepsilon^2 a_2 + O(\varepsilon^3) \right] + 1 = 0$$

from which,

$$\varepsilon \left[a_0^2 + \varepsilon \cdot 2a_0 a_1 + O(\varepsilon^2) \right] - 2\left[a_0 + \varepsilon a_1 + \varepsilon^2 a_2 + O(\varepsilon^3) \right] + 1 = 0.$$

Rearranging this, we obtain

$$\left(-2a_0 + 1 \right) + \varepsilon \left(a_0^2 - 2a_1 \right) + O(\varepsilon^2) = 0.$$

Equating coefficients of like powers of ε to zero, we obtain

$$\left.\begin{array}{r} -2a_0 + 1 = 0 \\ a_0^2 - 2a_1 = 0 \\ \text{etc.} \end{array}\right\}.$$

Thus,

$$a_0 = \frac{1}{2}, \quad a_1 = \frac{1}{8}, \quad \text{etc.}$$

These give only the first root

$$z_1 = \frac{1}{2} + \frac{1}{8}\varepsilon + \dots.$$

Now, in order to find the second root z_2, note that we may write

$$\varepsilon z^2 - 2z + 1 = \varepsilon(z - z_1)(z - z_2)$$
$$= \varepsilon z^2 - \varepsilon(z_1 + z_2)z + \varepsilon z_1 z_2$$

from which

$$\varepsilon z_1 z_2 = 1 \quad \text{or} \quad z_2 \sim \frac{1}{\varepsilon}$$

so that, for the root z_2, εz^2 is not small compared with $-2z + 1$, in the limit $\varepsilon \to 0$, so the previous regular perturbation expansion does not set up the right dominant balance for z_2, and is, therefore, not suited for recovering z_2.

This suggests that, in order to obtain z_2, we should set

$$\varepsilon z(\varepsilon) = w(\varepsilon), \quad \text{with } w(0) \neq 0,$$

so that the given equation becomes

$$w^2 - 2w + \varepsilon = 0.$$

Observe that the singular problem posed by the second root has been rendered regular by a change of variable (z to w).

The roots of this equation can now be found using a regular perturbation expansion

$$w(\varepsilon) \sim b_0 + \varepsilon b_1 + \varepsilon^2 b_2 + O(\varepsilon^3).$$

Then we obtain

$$(b_0^2 - 2b_0) + \varepsilon(2b_0 b_1 - 2b_1 + 1) + O(\varepsilon^2) = 0$$

from which

$$\left.\begin{aligned}
b_0^2 - 2b_0 &= 0 \\
2b_0 b_1 - 2b_1 + 1 &= 0 \\
\text{etc.}
\end{aligned}\right\}.$$

We then obtain

$$b_0 = 2, \quad b_1 = -1/2, \quad \text{etc.}$$

Note that the other root for b_0, viz., $b_0 = 0$, has to be discarded since $w(0) \neq 0$.

Thus, we have

$$w(\varepsilon) = 2 - \frac{1}{2}\varepsilon + O(\varepsilon^2)$$

from which, we get for the second root,

$$z_2 = \frac{2}{\varepsilon} - \frac{1}{2} + O(\varepsilon).$$

Example 3: Consider the algebraic equation (Nayfeh, 1981)

$$(x-1)(x-\tau) = -\varepsilon x.$$

When $\varepsilon = 0$, this equation becomes

$$(x_0 - 1)(x_0 - \tau) = 0$$

from which,

$$x_0 = 1 \quad \text{or} \quad x_0 = \tau.$$

Look for a regular perturbation expansion,

$$x \sim x_0 + \varepsilon x_1 + \varepsilon^2 x_2 + \dots.$$

Substituting into the given equation, and expanding we obtain

$$(x_0 - 1)(x_0 - \tau) + \varepsilon\left[(2x_0 - 1 - \tau)\, x_1 + x_0\right]$$
$$+ \varepsilon^2\left[(2x_0 - 1 - \tau)\, x_2 + x_1^2 + x_1\right] + \dots = 0.$$

Equating the coefficients of each power of ε to zero, we obtain

$$\left.\begin{array}{l} (x_0 - 1)(x_0 - \tau) = 0 \\ (2x_0 - 1 - \tau)\, x_1 + x_0 = 0 \\ (2x_0 - 1 - \tau)\, x_2 + x_1^2 + x_1 = 0 \\ \text{etc.} \end{array}\right\}.$$

Thus,

$$\left.\begin{array}{l} x_0 = 1 \ \text{ or } \ x_0 = \tau \\[2mm] x_1 = -\dfrac{1}{1-\tau} \ \text{ or } \ \dfrac{\tau}{1-\tau} \\[4mm] x_2 = -\dfrac{\tau}{(1-\tau)^3} \ \text{ or } \ \dfrac{\tau}{(1-\tau)^3} \\[4mm] \text{etc.} \end{array}\right\}.$$

Therefore,

$$\left.\begin{array}{l} x \sim 1 - \dfrac{\varepsilon}{1-\tau} - \dfrac{\varepsilon^2 \tau}{(1-\tau)^3} + \cdots \\[4mm] x \sim \tau + \dfrac{\varepsilon \tau}{1-\tau} + \dfrac{\varepsilon^2 \tau}{(1-\tau)^3} + \cdots \end{array}\right\}.$$

Observe that these expansions break down as $\tau \to 1$, and the singularity gets worse at higher orders. The region of non-uniformity can be determined by examining the conditions under which the successive terms in these expansions are of the same order. This happens when

$$1 - \tau = O(\varepsilon^{1/2}).$$

In order to obtain a uniform expansion in this region, introduce a detuning parameter σ defined by

$$1 - \tau = \varepsilon^{1/2}\, \sigma$$

and look for a solution of the form

$$x \sim x_0 + \varepsilon^{1/2}\, x_1 + \cdots.$$

We then obtain

$$\left[(x_0 - 1) + \varepsilon^{1/2}\, x_1 + \cdots\right]\left[(x_0 - 1) + \varepsilon^{1/2}\, (x_1 + \sigma) + \cdots\right]$$
$$= -\varepsilon\left[x_0 + \varepsilon^{1/2}\, x_1 + \cdots\right],$$

from which

$$\left. \begin{array}{l} \left(x_0 - 1\right)^2 = 0 \\ x_1^2 + \sigma x_1 + 1 = 0, \text{ etc.} \end{array} \right\}.$$

Thus,

$$\left. \begin{array}{l} x_0 = 1 \text{ or } 1 \\ x_1 = \dfrac{1}{2}\left(-\sigma \pm \sqrt{\sigma^2 - 4}\right), \text{ etc.} \end{array} \right\}.$$

Therefore,

$$\left. \begin{array}{l} x = 1 - \dfrac{1}{2}\,\varepsilon^{1/2}\left(\sigma + \sqrt{\sigma^2 - 4}\right) + \cdots \\ x = 1 - \dfrac{1}{2}\,\varepsilon^{1/2}\left(\sigma - \sqrt{\sigma^2 - 4}\right) + \cdots \end{array} \right\}$$

which are regular at $\sigma = 0$ or $\tau = 1$!

Example 4: Consider the roots of Bessel's function $J_0(x)$ for large x.

Note the asymptotic expression

$$J_0(x) \sim \sqrt{\frac{2}{\pi x}}\left[\cos\left(x - \frac{\pi}{4}\right) + \frac{1}{8x}\sin\left(x - \frac{\pi}{4}\right)\right], \quad x \text{ large.}$$

The roots of

$$J_0(x) = 0$$

are then given by

$$\cot\left(x - \frac{\pi}{4}\right) = -\frac{1}{8x}, \, x \text{ large.}$$

So, to first approximation, we have

$$\cot\left(x - \frac{\pi}{4}\right) \approx 0,$$

from which

$$x - \frac{\pi}{4} \approx \left(n + \frac{1}{2}\right)\pi$$

or

$$x \approx \left(n+\frac{3}{4}\right)\pi, \ n \text{ large.}$$

To improve on this approximation, let us put

$$x = \left(n+\frac{3}{4}\right)\pi + \delta.$$

This leads to

$$\cot\left[\left(n+\frac{1}{2}\right)\pi + \delta\right] = -\frac{1}{2\pi(4n+3)+8\delta} + \cdots$$

or

$$\frac{\cot\left(n+\frac{1}{2}\right)\pi \cdot \cot\delta - 1}{\cot\left(n+\frac{1}{2}\right)\pi + \cot\delta} = -\frac{1}{2\pi(4n+3)}\frac{1}{1+\dfrac{4}{\pi(4n+3)}\delta} + \cdots$$

or

$$-\tan\delta = -\delta + \cdots = -\frac{1}{2\pi(4n+3)}\left(1 - \frac{4}{\pi(4n+3)}\delta + \cdots\right),$$

from which

$$\delta \approx \frac{1}{2\pi(4n+3)}.$$

Therefore,

$$x \approx \left(n+\frac{3}{4}\right)\pi + \frac{1}{2\pi(4n+3)}.$$

Root No.	1	2	3	4	5
Perturbation	2.40308	5.52004	8.65372	11.79153	14.93092
Tabulated	2.40482	5.52008	8.65373	11.79153	14.93092

Table 2.1. Comparison of Approximate and Tabulated Roots
of Bessel's Function of order zero (from Nayfeh, 1981).

2.3. Ordinary Differential Equations

We will now consider applications of regular perturbation methods to ordinary differential equations.

Example 5: Consider the motion of an object projected radially upward from the surface of the earth (Holmes, 1995). Letting $x(\varepsilon)$ denote the height of the object, measured from the surface, we have from Newton's Second Law, the following equation of motion –

$$\frac{d^2 x}{dt^2} = -\frac{g R^2}{(x + R)^2}, \quad t > 0$$

where R is the radius of the earth and g is the gravitational constant.

Let the initial conditions be

$$t = 0 : x = 0, \quad \frac{dx}{dt} = V_0.$$

Non-dimensionalizing as follows –

$$\tau = \frac{t}{(V_0/g)}, \quad y = \frac{x}{(V_0^2/g)}.$$

The above initial-value problem becomes

$$\left. \begin{array}{c} \dfrac{d^2 y}{d\tau^2} = -\dfrac{1}{(1 + \varepsilon y)^2}, \quad \tau > 0 \\[3mm] \tau = 0 : y = 0, \quad \dfrac{dy}{d\tau} = 1 \end{array} \right\},$$

where

$$\varepsilon \equiv \frac{V_0^2}{Rg} \ll 1.$$

Look for a solution of the form –

$$y \sim y_0(\tau) + \varepsilon y_1(\tau) + 0(\varepsilon^2).$$

Substituting into the initial-value problem and equating the coefficients of like powers of ε, we obtain

$$O(1) : y_0'' = -1$$
$$\left.\begin{array}{l} \tau = 0 : y_0 = 0, \ y_0' = 1 \end{array}\right\}$$
$$O(\varepsilon) : y_1'' = 2y_0$$
$$\left.\begin{array}{l} \tau = 0 : y_1 = 0, \ y_1' = 0 \end{array}\right\}$$

etc.

Here, primes denote differentiation with respect to τ.

Solving these problems successively, we obtain

$$\left.\begin{array}{l} y_0 = -\dfrac{1}{2}\tau^2 + \tau \\[2mm] y_1 = \dfrac{1}{3}\tau^3 - \dfrac{1}{12}\tau^4 \\[2mm] \text{etc.} \end{array}\right\}.$$

Thus,

$$y(\tau) \sim \tau\left(1 - \frac{\tau}{2}\right) + \varepsilon\frac{1}{3}\tau^3\left(1 - \frac{\tau}{4}\right) + \cdots.$$

A comparison between the asymptotic expansion and the exact numerical solution is shown in Figure 1 – the two-term asymptotic solution is quite close to the exact numerical solution! Observe that y_1 contributes to increase the flight time of the object. This is due to the fact that the decrease of the gravitational force with height allows the object to stay up longer.

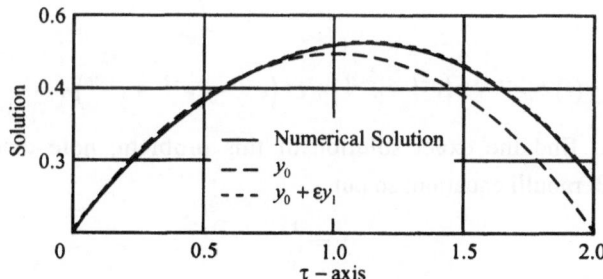

Figure 2.1. Comparison between the asymptotic solutions and the numerical solution in Example 5. In these calculations $\varepsilon = 10^{-1}$. There is little difference between the numerical solution and the two-term expansion (From Holmes, 1998).

Example 6: Consider the differential equation –

$$\left. \begin{array}{l} \dfrac{dy}{dt} = -y + \varepsilon y^2, \quad t > 0 \\[2mm] y(0) = 1 \end{array} \right\}.$$

Look for a solution of the form

$$y \sim y_0(t) + \varepsilon y_1(t) + \varepsilon^2 y_2(t).$$

Substituting into the above equations and equating coefficients of like powers of ε, we obtain

$$O(1): \left. \begin{array}{l} y_0 = -y_0 \\[1mm] y_0(0) = 1 \end{array} \right\}$$

$$O(\varepsilon): \left. \begin{array}{l} \dot{y}_1 = -y_1 + y_0^2 \\[1mm] y_1(0) = 0 \end{array} \right\}$$

$$O(\varepsilon^2): \left. \begin{array}{l} \dot{y}_2 = -y_2 + 2y_0 y_1 \\[1mm] y_2(0) = 0 \end{array} \right\}$$

etc.

Solving these problems successively, we obtain

$$\left. \begin{array}{l} y_0 = e^{-t} \\[1mm] y_1 = e^{-t} - e^{-2t} \\[1mm] y_2 = e^{-t} - 2e^{-2t} + e^{-3t} \\[1mm] \text{etc.} \end{array} \right\}.$$

Thus,

$$y(t) \sim e^{-t} + \varepsilon\left(e^{-t} - e^{-2t}\right) + \varepsilon^2\left(e^{-t} - 2e^{-2t} + e^{-3t}\right) + \cdots$$

In order to find the exact solution of this problem, note that the given equation is a Bernoulli equation; so put

$$y = \frac{1}{w}.$$

We then obtain

$$\left. \begin{array}{l} \dfrac{dw}{dt} - w = -\varepsilon \\[2mm] w(0) = 1 \end{array} \right\}.$$

This initial-value problem has the solution

$$w = \varepsilon + (1-\varepsilon)\, e^t.$$

In terms of y, this is

$$y_{\text{exact}} = \frac{1}{\varepsilon + (1-\varepsilon)e^t} = \frac{e^{-t}}{1+\varepsilon\left(e^{-t}-1\right)},$$

from which

$$\varepsilon = 1 : y_{\text{exact}} \equiv 1,$$

which shows the fact that the unperturbed problem (corresponding to $\varepsilon = 0$) describes motion which dies away, while the motion of the perturbed problem does not die away! (In fact, when $\varepsilon = 1$, we have $y(t) \equiv 1$.) Thus, the "small perturbation" ε alters the character of motion. The expansion of the exact solution in powers of ε gives

$$y(t) = e^{-t} + \varepsilon\left(e^{-t} - e^{-2t}\right) + \varepsilon^2\left(e^{-t} - 2e^{-2t} + e^{-3t}\right) + \cdots$$

which agrees with the perturbation solution.

Observe also that the convergence of the series in the approximate solution is uniform, as $\varepsilon \to 0$, for $0 \le t < \infty$ because, for a fixed $t > 0$, the error approaches zero as $\varepsilon \to 0$.

Example 7: Consider the initial-value problem for Duffing's equation –

$$\left.\begin{array}{c} \dfrac{d^2 u}{dt^2} + u + \varepsilon u^3 = 0, \quad t > 0 \\[2mm] u(0) = A, \quad \dot{u}(0) = 0 \end{array}\right\}.$$

Look for a solution of the form

$$u(t;\varepsilon) \sim u_0(t) + \varepsilon u_1(t) + \varepsilon^2 u_2(t) + \cdots.$$

We then obtain, to various orders in ε,

$$\left.\begin{array}{c} O(1): \ddot{u}_0 + u_0 = 0 \\[2mm] u_0(0) = A, \quad \dot{u}_0(0) = 0 \end{array}\right\}$$

$$\left.\begin{array}{c} O(\varepsilon): \ddot{u}_1 + u_1 = -u_0^3 \\[2mm] u_1(0) = 0, \quad \dot{u}_1(0) = 0 \end{array}\right\},$$

etc.

where the dots overhead denote differentiation with respect to t.

Solving the $O(1)$ problem, we obtain

$$u_0(t) = A\cos t.$$

Substituting this, the $O(\varepsilon)$ problem becomes

$$\left.\begin{aligned}\ddot{u}_1 + u_1 &= -A^3\cos^3 t = -\frac{A^3}{4}\left(3\cos t + \cos 3t\right)\\ u_1(0) &= 0, \quad \dot{u}_1(0) = 0\end{aligned}\right\}$$

from which,

$$u_1(t) = \frac{A^3}{32}\left(\cos 3t - \cos t\right) - \frac{3A^3}{8}\, t\sin t.$$

Thus,

$$u(t) \sim A\cos t + \varepsilon A^3\left[\frac{1}{32}\left(\cos 3t - \cos t\right) - \frac{3}{8}\, t\sin t\right] + \cdots.$$

Observe that the second term has a secular (*secular* is the Latin word used for century – this adjective comes from the astronomical context) term which causes the $O(\varepsilon)$ term to dominate the $O(1)$ term as t becomes larger than ε^{-1}. Note that it is not possible to alleviate this difficulty by including higher-order terms because the latter only worsen this difficulty.

In order to see the source of this difficulty, let us look at the exact solution of this problem.

The first integral of the given equation is given by

$$\frac{1}{2}\dot{u}^2 + \frac{1}{2}u^2 + \frac{1}{4}\varepsilon u^4 = h$$

where h is a constant of integration. On using the given initial conditions, we obtain

$$h = \frac{1}{2}A^2 + \frac{\varepsilon}{4}A^4.$$

Integrating the first integral further, we obtain

$$t = \pm\int_A^u \frac{du}{\sqrt{2h - u^2 - \dfrac{1}{2}\varepsilon u^4}},$$

which may be rewritten as

$$t = \pm \int_A^u \frac{du}{\sqrt{A^2 - u^2}\sqrt{1 + \frac{1}{2}\,\varepsilon\,A^2 + \frac{1}{2}\,\varepsilon u^2}}.$$

Putting

$$u = -A\cos\theta,$$

we obtain

$$t = \pm \int_\pi^\theta \frac{d\theta}{\sqrt{1 + \frac{1}{2}\,\varepsilon\,A^2 + \frac{1}{2}\,\varepsilon\,A^2\cos^2\theta}}$$

which may be rewritten as

$$t = \pm \frac{1}{\sqrt{1 + \varepsilon\,A^2}} \int_\pi^\theta \frac{d\theta}{\sqrt{1 - m^2\sin^2\theta}}$$

where

$$m^2 \equiv \frac{(\varepsilon/2)\,A^2}{1 + \varepsilon\,A^2}.$$

The period of the oscillation is then given by

$$T = \frac{2}{\sqrt{1 + \varepsilon\,A^2}} \int_0^\pi \frac{d\theta}{\sqrt{1 - m^2\sin^2\theta}}$$

$$= \frac{4}{\sqrt{1 + \varepsilon\,A^2}} \int_0^{\pi/2} \frac{d\theta}{\sqrt{1 - m^2\sin^2\theta}} = K(m^2)$$

where $K(m^2)$ is the complete elliptic integral of the first kind.

For small ε, T may be expanded in powers of ε,

$$T = \frac{4}{\sqrt{1 + \varepsilon\,A^2}} \int_0^{\pi/2} \left(1 + \frac{m^2}{2}\sin^2\theta + \cdots\right) d\theta$$

$$= \frac{4}{\sqrt{1 + \varepsilon\,A^2}} \left(\frac{\pi}{2} + \frac{m^2\,\pi}{8} + \cdots\right)$$

$$= 2\pi\left(1 - \frac{3}{8}\,\varepsilon\,A^2 + \cdots\right).$$

Thus, the source of the difficulty with the regular perturbation solution is that the correction terms of $O(\varepsilon)$, $O(\varepsilon^2)$, etc. do not correct for the difference between the exact period of oscillation T and the approximate period 2π of the leading-order term $A\cos t$. The cumulative effect of this error over several oscillations destroys the uniform validity of the regular perturbation solution.

Example 8: Consider the boundary-value problem –

$$\left.\begin{array}{l} \varepsilon y'' + y' + y = 0, \ \ 0 \le x \le 1 \\ y(0) = a, \ \ y(1) = b \end{array}\right\},$$

where primes denote differentiation with respect to x.

Look for a regular perturbation expansion of the form

$$y(x,\varepsilon) \sim y_0(x) + \varepsilon y_1(x) + \cdots.$$

We then obtain at various powers of ε:

$$\left.\begin{array}{l} y_0' + y_0 = 0 \\ y_1' + y_1 = -y_0'' \\ \text{etc.} \end{array}\right\}.$$

Note that the order of the differential equations for $y_0, y_1, \ldots$ is reduced. Consequently, these differential equations cannot take on both of the boundary conditions, and one of these boundary conditions, viz., $y(0) = a$, must be dropped[1]. Thus, we have the following boundary conditions –

$$\left.\begin{array}{l} y_0(1) = b \\ y_1(1) = 0 \\ \text{etc.} \end{array}\right\}.$$

We therefore have the following solutions –

$$\left.\begin{array}{l} y_0 = b\,e^{1-x} \\ y_1 = b(1-x)\,e^{1-x} \\ \text{etc.} \end{array}\right\}.$$

Thus,

$$y \sim b\left[1+\varepsilon(1-x)\right]e^{1-x} + O(\varepsilon^2).$$

---

[1] We will discuss in Chapter 5 why the boundary condition at $x = 0$ is the one that needs to be dropped for this problem.

Observe that this solution does not satisfy the other boundary condition $y(0) = a$, and so this solution is valid everywhere except near the point $x = 0$. However, for $\varepsilon \ll 1$, this solution is close to the exact solution everywhere except in a small interval at $x = 0$, where the exact solution changes rapidly in order to retrieve the boundary condition there which is about to be lost.

The source of the difficulty with the regular perturbation solution is that the correction terms of $O(\varepsilon)$, $O(\varepsilon^2)$, ... do not recover the boundary condition $y(0) = a$ that is missed by the lowest order solution y_0. The successive terms actually worsen the behavior near $x = a$!

2.4. Partial Differential Equations

We will now consider applications of regular perturbation methods to partial differential equations (see Appendix for a review of partial differential equations).

Example 9: Consider the nonlinear hyperbolic wave-propagation problem –

$$\left. \begin{array}{l} u_{tt} - u_{xx} = u_x u_{tt}, \quad x \geq 0, \quad t \geq 0 \\ u(0,t) = \varepsilon f(t), \quad f(t) = 0 \text{ for } t \leq 0, \\ u(x,0) = 0 \end{array} \right\}.$$

Let us expand the solution as follows –

$$u \sim \varepsilon u_1 + \varepsilon^2 u_2 + \cdots.$$

We then obtain from the given equation to various orders in ε:

$$\left. \begin{array}{l} O(\varepsilon) : u_{1tt} - u_{1xx} = 0 \\ \quad u_1(0,t) = f(t) \\ \quad u_1(x,0) = 0 \end{array} \right\}$$

$$O(\varepsilon^2): u_{2tt} - u_{2xx} = u_{1x} u_{1tt}$$
$$u_2(0,t) = 0$$
$$u_2(x,0) = 0$$

etc.

Solving the $O(\varepsilon)$ problem, we obtain

$$u_1(x,t) = f(t-x).$$

Let us introduce the following characteristics of the (linearized) problem –

$$\xi = t - x, \quad \eta = t + x.$$

The $O(\varepsilon^2)$ problem then becomes

$$\frac{\partial^2 u_2}{\partial \xi \partial \eta} = -\frac{1}{4} f'(\xi) f''(\xi),$$

from which we have

$$u_2(x,t) = -\frac{1}{8} \left[f'(\xi) \right]^2 \eta + G(\xi).$$

Using the condition $u_2(0,t) = 0$, we obtain

$$G(t) = \frac{1}{8} \left[f'(t) \right]^2 t.$$

Thus,

$$u_2(x,t) = \frac{1}{8} \left[f'(\xi) \right]^2 (\xi - \eta) = -\frac{1}{4} \left[f'(\xi) \right]^2 x.$$

Finally,

$$u(x,t) \sim \varepsilon f(\xi) - \varepsilon^2 \left\{ \frac{1}{4} \left[f'(\xi) \right]^2 x \right\} + \cdots.$$

Observe the nonuniformity in the above solution. Physically, the latter is due to the erroneous assumption of constant speed of propagation. The cumulative effects of variation in the speed of propagation produce errors in the far field.

Thus, two common sources of non-uniformities in asymptotic expansions are –

- infinite domains which allow long-term effects of small perturbations to accumulate,

- singularities in governing equations, which lead to localized regions of rapid changes.

2.5. Applications to Fluid Dynamics: Decay of a Line Vortex

Consider the decay of a vortex filament in an incompressible fluid (Shivamoggi, 1998). Let us use the cylindrical coordinates and take the z-axis along the axis of the filament. The vorticity component along the z-direction given by

$$\zeta = \frac{\partial u_\theta}{\partial r} + \frac{u_\theta}{r} \tag{2.5.1}$$

where u_θ is the azimuthal component of the fluid velocity, evolves according to

$$\frac{\partial \zeta}{\partial t} = \nu \left(\frac{\partial^2 \zeta}{\partial r^2} + \frac{1}{r} \frac{\partial \zeta}{\partial r} \right), \tag{2.5.2}$$

ν being the kinematic viscosity of the fluid.

Using (2.5.1), equation (2.5.2) becomes

$$\frac{\partial}{\partial t} (r u_\theta) = \nu \left[\frac{\partial^2}{\partial r^2} (r u_\theta) - \frac{1}{r} \frac{\partial}{\partial r} (r u_\theta) \right]. \tag{2.5.3}$$

Non-dimensionalize the various quantities using a reference length L and a reference time τ, so that equation (2.5.3) becomes

$$\frac{\partial}{\partial t} (r u_\theta) = \varepsilon \left[\frac{\partial^2}{\partial r^2} (r u_\theta) - \frac{1}{r} \frac{\partial}{\partial r} (r u_\theta) \right] \tag{2.5.4}$$

where

$$\varepsilon \equiv \frac{\nu \tau}{L}.$$

The boundary conditions are

$$\left. \begin{array}{l} r = 0 \;\; : u_\theta = 0 \\ r \Rightarrow \infty : u_\theta \Rightarrow 0 \end{array} \right\} \;\; \text{for } t > 0. \tag{2.5.5}$$

Let us assume $\varepsilon \ll 1$ and seek a solution of equation (2.5.4) in the form of a straightforward expansion –

$$u_{\theta \, approx} \sim \sum_{n=0}^{\infty} \varepsilon^n u_{\theta_n} (r, t). \tag{2.5.6}$$

Substituting in equation (2.5.4), and equating the coefficients of equal powers of ε, we obtain

$$O(1): \frac{\partial}{\partial t}\left(ru_{\theta_0}\right)=O, \tag{2.5.7}$$

$$O(\varepsilon): \frac{\partial\left(ru_{\theta_1}\right)}{\partial t}=\frac{\partial^2\left(ru_{\theta_0}\right)}{\partial r^2}-\frac{1}{r}\frac{\partial\left(ru_{\theta_0}\right)}{\partial r}, \tag{2.5.8}$$

etc.

From the nature of the equations (2.5.7) and (2.5.8), it is clear that their solutions cannot satisfy both of the boundary conditions (2.5.5), and one of them must be dropped.

Solving equations (2.5.7) and (2.5.8), and requiring the satisfaction of the boundary conditions $u_\theta \Rightarrow 0$ as $r \Rightarrow \infty$, one finds

$$u_{\theta\,approx} \sim \frac{A(r)}{r}+\varepsilon\left[A''(r)-\frac{A'(r)}{r}\right]t+.... \tag{2.5.9}$$

According to (2.5.9), $u_{\theta\,approx} \Rightarrow \infty$ at $r \Rightarrow 0$ so that the error in (2.5.6) is not uniform over $[0,\infty)$, and the expansion (2.5.5) breaks down at the axis. It is of interest to note that the higher-order terms in the expansion (2.5.5) introduce successively higher-order singularities at the axis. This simply implies that the problem (2.5.3) is of singular-perturbation type.

In order to understand further the nature of the nonuniformity, note the exact solution of equation (2.5.3),

$$u_{\theta\,exact} = \frac{\Gamma_0}{2\pi r}\left[1-\exp\left(-\frac{r^2}{4\varepsilon t}\right)\right] \tag{2.5.10a}$$

which is in agreement with the first term in (2.5.9) and satisfies the boundary condition $u_\theta(\infty) \Rightarrow 0$ (Γ_0 can be shown to be the circulation of the vortex filament at time $t = 0$). In order to understand what happens at the boundary $r = 0$, write (2.5.10a) in the form

$$u_{\theta\,exact} = \frac{\Gamma_0}{2\pi r}-\frac{\Gamma_0}{2\pi r}\exp\left(-\frac{r^2}{4\varepsilon t}\right). \tag{2.5.10b}$$

The second term in (2.5.10b) is not negligible even as $\varepsilon \Rightarrow 0$ since we are interested in the region $r \Rightarrow 0$. Besides, in this form, the order of the error is uniform in $[0,\infty)$. The behavior of $u_{\theta\,exact}$ is shown in Figure 2.2 together with the first term of $u_{\theta\,approx}$. Note that near $r = 0$, the motion is a rigid-body rotation. It can be seen that for small ε, $u_{\theta\,exact}$ agrees with $u_{\theta\,approx}$ except in the

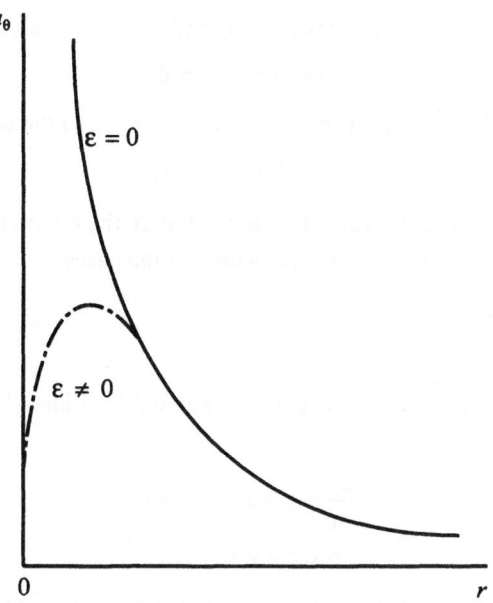

Figure 2.2. Velocity distribution in a vortex filament.

small region near the axis where it changes rapidly in order to satisfy the boundary condition there, which is about to be lost. Physically, this means that for small times there is a narrow vortex core near the axis, and the rest of the flow field is irrotational.

For $r \ll \sqrt{4vt}$, we have, from (2.5.10a),

$$u_{\theta\,exact} \approx \frac{\Gamma_0}{8\pi vt}\, r$$

which corresponds to an almost rigid rotation with angular velocity $\Gamma_0/8\pi vt$. The intensity of the vortex thus decreases with time as the "core" spreads radially outward.

2.6. Exercises

1. Using suitable perturbation expansions, find the two roots of the equation (Hinch, 1991)

$$(1-\varepsilon)\, x^2 - 2x + 1 = 0.$$

(Note: The perturbation expansion here is in non-integral powers of ε.)

2. Using suitable perturbation expansions, find the three roots of the equation
$$\varepsilon x^3 + x^2 - 1 = 0.$$

3. Using the quadratic-root formula, find the two roots of the equation
$$(x-1)(x-\tau) = -\varepsilon x.$$

   Expand these roots of powers of ε and compare them with the roots obtained by using regular perturbation expansions for the cases

   (i) $\tau \neq 1$,

   (ii) $\tau \approx 1$.

4. Using a regular perturbation expansion, solve the initial-value problem (Nayfeh, 1973)
$$\left.\begin{array}{l} \dfrac{dy}{dt} + y = \varepsilon y^2, \quad t > 0 \\[2mm] t = 0 : y = 1 \end{array}\right\}.$$

5. Using a regular perturbation expansion, solve the initial-value problem
$$\left.\begin{array}{l} \dfrac{dy}{dt} + 1 = \varepsilon y \\[2mm] t = 0 : y = \varepsilon \end{array}\right\}.$$

   Compare the solution with an expansion in powers of ε of the exact solution of this problem. Discuss the change in the nature of the solution produced by the perturbation $(\varepsilon \neq 0)$.

6. Using a regular perturbation expansion, solve the initial-value problem
$$\left.\begin{array}{l} \dfrac{d^2 y}{dt^2} + \left(1 - \varepsilon e^{-at}\right) y = 0, \quad a > 0 \\[2mm] t = 0 : y = 0, \quad y' = 1 \end{array}\right\}.$$

   Compare this solution in the limit $a \Rightarrow 0$ with an expansion in powers of ε of the exact solution.

7. Using a regular perturbation method to solve the initial-value problem
$$\left.\begin{array}{l} \dfrac{d^2 y}{dt^2} + \varepsilon y = f(t) \\[2mm] t = 0 : y = a, \quad y' = b \end{array}\right\}.$$

   Compare this solution with an expansion in powers of ε of the exact solution.

8. The asymptotic expansion of $J_v(x)$ for large x is

$$J_v(x) \sim \sqrt{\frac{2}{\pi x}} \left[\cos\left(x - \frac{v\pi}{2} - \frac{\pi}{4}\right) - \frac{4v^2 - 1}{8x} \sin\left(x - \frac{v\pi}{2} - \frac{\pi}{4}\right) \right].$$

Show that the roots of $J_v(x) = 0$ are given by

$$x = \left(n + \frac{3}{4} + \frac{v}{2}\right) \pi - \frac{4v^2 - 1}{2\pi(4n + 3 + 2v)} + \cdots.$$

2.7. Appendix

Review of Partial Differential Equations

A1. Transformation to Canonical Forms

Consider a second-order, linear partial differential equation (PDE)

$$AU_{xx} + BU_{xy} + CU_{yy} + DU_x + EU_y + FU = G \qquad (A.1)$$

where $A, B, C, \ldots$ are functions of x and y.

Consider a change of independent variables, given by

$$\left. \begin{array}{l} \xi = \xi(x, y) \\ \eta = \eta(x, y) \end{array} \right\} \qquad (A.2)$$

with the Jacobian

$$J = \begin{vmatrix} \xi_x & \xi_y \\ \eta_x & \eta_y \end{vmatrix} \neq 0 \qquad (A.3)$$

in the region under consideration; here $\xi(x, y)$ and $\eta(x, y)$ are assumed to be twice continuously differentiable.

Note the relations –

$$\left.\begin{aligned}
u_x &= u_\xi \, \xi_x + u_\eta \, \eta_x \\
u_y &= u_\xi \, \xi_y + u_\eta \, \eta_y \\
u_{xx} &= u_{\xi\xi} \, \xi_x^2 + 2u_{\xi\eta} \, \xi_x \, \eta_x + u_{\eta\eta} \, \eta_x^2 + u_\xi \, \xi_{xx} + u_\eta \, \eta_{xx} \\
u_{xy} &= u_{\xi\xi} \, \xi_x \, \xi_y + u_{\xi\eta}\left(\xi_x \, \eta_y + \xi_y \, \eta_x\right) + u_{\eta\eta} \, \eta_x \, \eta_y + u_\xi \, \xi_{xy} + u_\eta \, \eta_{xy} \\
u_{yy} &= u_{\xi\xi} \, \xi_y^2 + 2u_{\xi\eta} \, \xi_y \, \eta_y + u_{\eta\eta} \, \eta_y^2 + u_\xi \, \xi_{yy} + u_\eta \, \eta_{yy}
\end{aligned}\right\}. \qquad \text{(A.4)}$$

Using the relations (A.4), equation (A.1) becomes

$$A^* u_{\xi\xi} + B^* u_{\xi\eta} + C^* u_{\eta\eta} + D^* u_\xi + E^* u_\eta + F^* = G^* \qquad \text{(A.5)}$$

where,

$$\left.\begin{aligned}
A^* &= A\xi_x^2 + B\xi_x\xi_y + C\xi_y^2 \\
B^* &= 2A\xi_x \, \eta_x + B\left(\xi_x \, \eta_y + \xi_y \, \eta_x\right) + 2C\xi_y \, \eta_y \\
C^* &= A\eta_x^2 + B\eta_x\eta_y + C\eta_y^2 \\
D^* &= A\xi_{xx} + B\xi_{xy} + C\xi_{yy} + D\xi_x + E\xi_y \\
E^* &= A\eta_{xx} + B\eta_{xy} + C\eta_{yy} + D\eta_x + E\eta_y \\
F^* &= F \\
G^* &= G
\end{aligned}\right\}. \qquad \text{(A.6)}$$

Noting that

$$B^{*2} - 4A^*C^* = J^2\left(B^2 - 4AC\right) \qquad \text{(A.7)}$$

we see that the nature of equation (A.1) is invariant under the transformation (A.2), provided $J \neq 0$.

For the purpose of classification, write equation (A.1) in the form

$$A u_{xx} + B u_{xy} + C u_{yy} = H \qquad \text{(A.8)}$$

which transforms to

$$A^* u_{\xi\xi} + B^* u_{\xi\eta} + C^* u_{\eta\eta} = H^*. \qquad \text{(A.9)}$$

Suppose now that ξ and η are such that

$$\left.\begin{aligned}
A^* &= A\xi_x^2 + B\xi_x\xi_y + C\xi_y^2 = 0 \\
C^* &= A\eta_x^2 + B\eta_x\eta_y + C\eta_y^2 = 0
\end{aligned}\right\}. \qquad \text{(A.10)}$$

Along the curves $\xi = const$ and $\eta = const$, we have, respectively,

$$d\xi = \xi_x dx + \xi_y dy = 0 \\ d\eta = \eta_x dx + \eta_y dy = 0 \Big\}$$ (A.11a)

or

$$\frac{dy}{dx} = -\frac{\xi_x}{\xi_y} \\ \frac{dy}{dx} = -\frac{\eta_x}{\eta_y} \Bigg\}.$$ (A.11b)

Thus, we have, along the curves $\xi = const$ and $\eta = const$,

$$A\left(\frac{dy}{dx}\right)^2 - B\left(\frac{dy}{dx}\right) + C = 0,$$ (A.12)

from which

$$\frac{dy}{dx} = \frac{B + \sqrt{B^2 - 4AC}}{2A} \\ \frac{dy}{dx} = \frac{B - \sqrt{B^2 - 4AC}}{2A} \Bigg\}.$$ (A.13)

If the solutions of these equations are

$$\phi_1(x,y) = c_1 \\ \phi_2(x,y) = c_2 \Big\}$$ (A.14)

which represent the characteristic curves, then

$$\xi = \phi_1(x,y) \\ \eta = \phi_2(x,y) \Big\}$$ (A.15)

are the relevant transformations.

If $B^2 - 4AC > 0$, then we have two real/distinct families of characteristics, and equation (A.9) becomes

$$B^* u_{\xi\eta} = H^*, \quad B^* \neq 0.$$ (A.16)

If one makes a further transformation

$$\alpha = \xi + \eta \\ \beta = \xi - \eta \Big\}$$ (A.17)

then equation (A.16) becomes

$$u_{\alpha\alpha} - u_{\beta\beta} = H^*$$ (A.18)

which is the canonical form of the hyperbolic PDEs.

If, on the other hand, $B^2 - 4AC < 0$, then the characteristics are not real; nonetheless, equation (A.9) becomes again

$$B^* u_{\xi\eta} = H^*, \quad B^* \neq 0 \tag{A.19}$$

where ξ and η are now complex conjugate functions. So, if one makes a further transformation

$$\left. \begin{array}{l} \alpha = \dfrac{1}{2} \left(\xi + \eta \right) \\[2mm] \beta = \dfrac{1}{2i} \left(\xi - \eta \right) \end{array} \right\} \tag{A.20}$$

where α and β are real functions, then equation (A.16) becomes

$$u_{\alpha\alpha} + u_{\beta\beta} = H^{**} \tag{A.21}$$

which is the canonical form of the elliptic PDE's.

Finally, if $B^2 - 4AC = 0$, then two real families of characteristics (A.14) degenerate into one family.

Suppose that we choose ξ such that

$$A^* \equiv A\xi_x^2 + B\xi_x\xi_y + C\xi_y^2 = \left(\sqrt{A}\ \xi_x + \sqrt{C}\ \xi_y \right)^2 = 0. \tag{A.22}$$

We then have, from (A.6),

$$\begin{aligned} B^* &= 2A\xi_x\eta_x + B\left(\xi_x\eta_y + \xi_y\eta_x \right) + 2C\xi_y\eta_y \\[2mm] &= 2\left(\sqrt{A}\ \xi_x + \sqrt{C}\ \xi_y \right)\left(\sqrt{A}\ \eta_x + \sqrt{C}\ \eta_y \right) = 0 \end{aligned} \tag{A.23}$$

for any η such that $J \neq 0$.

Equation (A.9) then becomes

$$C^* u_{\eta\eta} = H^*, \quad C^* \neq 0 \tag{A.24}$$

which is the canonical form of the parabolic PDEs.

One may choose

$$\eta = y \tag{A.25}$$

for simplicity.

Example A.1: $y^2 u_{xx} - x^2 u_{yy} = 0$.

Here,

$$B^2 - 4AC = 4x^2 y^2 > 0.$$

So, this equation is hyperbolic everywhere except on the lines $x = 0$ or $y = 0$.

The characteristics are given by

$$\left. \begin{aligned} \frac{dy}{dx} &= \frac{x}{y} \\ \frac{dy}{dx} &= -\frac{x}{y} \end{aligned} \right\}$$

or

$$y^2 - x^2 = c_1 \quad \text{and} \quad y^2 + x^2 = c_2.$$

Therefore, consider a canonical transformation generated by

$$\xi = y^2 - x^2, \quad \eta = y^2 + x^2.$$

The given PDE then becomes

$$u_{\xi\eta} = \frac{\eta}{2\left(\xi^2 - \eta^2\right)} u_\xi - \frac{\xi}{2\left(\xi^2 - \eta^2\right)} u_\eta.$$

Chapter 3

The Method of Strained Coordinates/Parameters

3.1. Introduction

This technique is devised to prevent the appearance of secular terms in perturbative solutions of equations such as

$$\ddot{u} + \omega_0^2 u = \varepsilon f(u, \dot{u}), \quad \varepsilon \ll 1. \tag{3.1.1}$$

The secular terms in this context represent non-uniformities associated with infinite domains.

This technique takes note of the fact that the nonlinearities alter the frequency of the system. This is accounted for by introducing a new independent variable $\tau = \omega t$ and expanding u and ω in powers of ε as (Poincaré, 1892)

$$\left. \begin{array}{l} u \sim u_0(\tau) + \varepsilon u_1(\tau) + \varepsilon^2 u_2(\tau) + \cdots \\ \omega = \omega_0 + \varepsilon \omega_1 + \varepsilon^2 \omega_2 + \cdots \end{array} \right\}. \tag{3.1.2}$$

The parameters ω_i, $i \geq 1$ are chosen so as to prevent the appearance of secular terms.

In a generalization of this method (Lighthill, 1949), one expands not only the dependent variable u but also the independent variable t in powers of ε and in terms of a new independent variable s as

$$\left. \begin{array}{l} u \sim \displaystyle\sum_{m=0}^{N-1} \varepsilon^m u_m(s) + O(\varepsilon^N) \\ t = s + \displaystyle\sum_{m=1}^{N} \varepsilon^m \xi_m(s) + O(\varepsilon^{N+1}) \end{array} \right\}. \tag{3.1.3}$$

The latter expansion can be viewed as a near-identity transformation from t to s. The functions ξ_m are determined such that the expansion for u is uniformly valid.

3.2.   Poincaré-Lindstedt-Lighthill Method of Perturbed Eigenvalues

Suppose we have to find the eigenvalues of the linear operator A perturbed by a small operator, i.e., $(A + \varepsilon B)$, $\varepsilon \ll 1$. We look for solutions of

$$(A + \varepsilon B) x = \lambda x \qquad (3.2.1)$$

and suppose that they can be expressed as power series (Poincaré, 1892, Lindstedt, 1882, Lighthill, 1949) –

$$x \sim \sum_{j=0}^{\infty} \varepsilon^j x_j, \quad \lambda = \sum_{j=0}^{\infty} \varepsilon^j \lambda_j . \qquad (3.2.2)$$

Substituting these expressions into the eigenvalue equation and collecting terms of like powers of ε, we obtain

$$O(1) : (A - \lambda_0 I) x_0 = 0 \qquad (3.2.3)$$

$$O(\varepsilon) : (A - \lambda_0 I) x_1 = \lambda_1 x_0 - B x_0 \qquad (3.2.4)$$

$$\text{etc.}$$

Note that the operator $L \equiv A - \lambda_0 I$, which occurs in every equation, is not invertible, so, according to the Implicit Function Theorem, we cannot solve the equations for $x_1, x_2, \ldots$. However, Fredholm's Alternative Theorem (see Appendix 1) comes to the rescue and implies that solutions can be found if and only if the right-hand sides are orthogonal to the null space of the adjoint operator $L^* = A^* - \overline{\lambda}_0 I$, i.e., if the null space of L_0^* is spanned by y_0, then we have from equation (3.2.4)

$$\lambda_1 \langle x_0, y_0 \rangle - \langle B x_0, y_0 \rangle = 0. \qquad (3.2.5)$$

Since $\langle x_0, y_0 \rangle \neq 0$, equation (3.2.5) uniquely determines λ_1:

$$\lambda_1 = \frac{\langle B x_0, y_0 \rangle}{\langle x_0, y_0 \rangle}. \qquad (3.2.6)$$

For this value of λ_1, the solution x_1 exists and can be made unique by the further requirement $\langle x_1, x_0 \rangle = 0$. Similarly, all of the $x_k's$, $k > 1$, can be solved in succession.

Example 1: Consider

$$
\left.
\begin{aligned}
&\frac{d^2 u}{dt^2} + \omega_0^2 u = \varepsilon u \\
&t = 0 : u = a, \ \frac{du}{dt} = 0
\end{aligned}
\right\}.
\tag{3.2.7}
$$

Construct a regular perturbation expansion of the form

$$
u(t;\varepsilon) \sim u_0(t) + \varepsilon u_1(t) + \cdots.
\tag{3.2.8}
$$

We then obtain from (3.2.7) to various orders of ε:

$$
O(1) : \left.
\begin{aligned}
&\frac{d^2 u_0}{dt^2} + \omega_0^2 u_0 = 0 \\
&t = 0 : u_0 = a, \ \frac{du_0}{dt} = 0
\end{aligned}
\right\}
\tag{3.2.9}
$$

$$
O(\varepsilon) : \left.
\begin{aligned}
&\frac{d^2 u_1}{dt^2} + \omega_0^2 u_1 = u_0 \\
&t = 0 : u_1 = 0, \ \frac{du_1}{dt} = 0
\end{aligned}
\right\}
\tag{3.2.10}
$$

etc.

Thus, we have from (3.2.9),

$$
u_0 = a \cos \omega_0 t.
\tag{3.2.11}
$$

Using (3.2.11), (3.2.10) then becomes

$$
\left.
\begin{aligned}
&\frac{d^2 u_1}{dt^2} + \omega_0^2 u_1 = a \cos \omega_0 t \\
&t = 0 : u_1 = 0, \ \frac{du_1}{dt} = 0
\end{aligned}
\right\}
\tag{3.2.12}
$$

which gives

$$
u_1 = \frac{at}{2\omega_0} \sin \omega_0 t.
\tag{3.2.13}
$$

Using (3.2.11) and (3.2.13), (3.2.8) becomes

$$u(t;\varepsilon) \sim a\cos\omega_0 t + \varepsilon \frac{at}{2\omega_0}\sin\omega_0 t + \cdots. \qquad (3.2.14)$$

Observe the presence of a secular term in (3.2.14); the region of nonuniformity for large t is

$$t = O(1/\varepsilon) \quad \text{as} \quad \varepsilon \to 0. \qquad (3.2.15)$$

Note that the exact solution of (3.2.7) is, however, given by

$$u = a\cos\left(\omega_0\sqrt{1-\varepsilon/\omega_0^2}\ t\right) \qquad (3.2.16)$$

which represents a purely periodic oscillation. Therefore, the unbounded behavior indicated by the regular expansion (3.2.8) is spurious and needs to be rectified. First, note that the origin of the secular term in (3.2.14) can be seen by expanding the exact solution (3.2.16) in powers of t:

$$a\cos\left(\omega_0\sqrt{1-\varepsilon/\omega_0^2}\ t\right) = a\cos\left[\omega_0\left(1-\frac{\varepsilon}{2\omega_0^2}+\cdots\right)t\right] \qquad (3.2.17a)$$

and further expanding the cosine function on the right,

$$a\cos\left(\omega_0\sqrt{1-\varepsilon/\omega_0^2}\ t\right) = a\cos\omega_0 t \cdot \cos\left(\frac{\varepsilon t}{2\omega_0}+\cdots\right) + a\sin\omega_0 t \cdot \sin\left(\frac{\varepsilon t}{2\omega_0}+\cdots\right)$$

$$= a\cos\omega_0 t + \varepsilon \frac{at}{2\omega_0}\sin\omega_0 t + \cdots,$$

$$(3.2.17b)$$

in agreement with (3.2.14). Further, the effect of the perturbation is to modify the frequency from the unperturbed value ω_0 to the perturbed value $\omega_0\sqrt{1-\varepsilon/\omega_0^2}$. Thus, one introduces a new variable

$$s = t\left(1+\varepsilon\omega_1 + \varepsilon^2\omega_2 + \cdots\right) \qquad (3.2.18)$$

and constructs a new expansion

$$u(t;\varepsilon) \sim u_0(s) + \varepsilon u_1(s) + \cdots. \qquad (3.2.19)$$

We then obtain from (3.2.7) to various orders in ε:

$$O(1): \frac{d^2u_0}{ds^2} + \omega_0^2 \, u_0 = 0 \left.\begin{array}{c} \\ \\ \end{array}\right\}$$

$$s = 0 : u_0 = a, \, \frac{du_0}{ds} = 0 \qquad (3.2.20)$$

$$O(\varepsilon): \frac{d^2u_1}{ds^2} + \omega_0^2 \, u_1 = -2\omega_1 \, \frac{d^2u_0}{ds^2} + u_0 \left.\begin{array}{c} \\ \\ \end{array}\right\}$$

$$s = 0 : u_1 = 0, \, \frac{du_1}{ds} = -\omega_1 \, \frac{du_0}{ds} \qquad (3.2.21)$$

etc.

Thus, we have from (3.2.20),

$$u_0 \left(s \right) = a \cos \omega_0 \, s . \qquad (3.2.22)$$

Using (3.2.22), (3.2.21) then becomes

$$\frac{d^2u_1}{ds^2} + \omega_0^2 \, u_1 = \left(1 + 2\omega_1 \, \omega_0^2 \right) a \cos \omega_0 \, s . \qquad (3.2.23)$$

One may now choose ω_1 so as to remove the secular term on the right-hand side, i.e.,

$$1 + 2\omega_1 \, \omega_0^2 = 0 \quad \text{or} \quad \omega_1 = -\frac{1}{2\omega_0^2} . \qquad (3.2.24)$$

Using (3.2.24), we then have, from equation (3.2.23),

$$u_1 \left(s \right) \equiv 0 . \qquad (3.2.25)$$

Using (3.2.22), (3.2.24), and (3.2.25), we obtain from (3.2.18) and (3.2.19),

$$u = a \cos \omega_0 \, s = a \cos \left[\omega_0 \left(1 - \frac{\varepsilon}{2\omega_0^2} + \cdots \right) t \right] \qquad (3.2.26)$$

in agreement with the limiting form (3.2.17a) of the exact solution (3.2.16) for small ε.

Example 2: The equation of motion for a simple pendulum is

$$\ddot{\theta} + \frac{g}{\ell} \sin \theta = 0 . \qquad (3.2.27)$$

For small θ, put

$$\theta = \left(-\sqrt{\varepsilon}/6\right) u\left(\hat{t}\right), \quad \hat{t} = \sqrt{\frac{g}{\ell}}\, t \tag{3.2.28}$$

and drop the hats. We then obtain, from equation (3.2.27), on expanding in powers of ε, Duffing's equation –

$$\frac{d^2 u}{dt^2} + u + \varepsilon u^3 = 0. \tag{3.2.29}$$

Introduce a new independent variable, s, given by the expansion

$$s = t\left(1 + \varepsilon \omega_1 + \varepsilon^2 \omega_2 + \cdots\right) \tag{3.2.30}$$

and consider an expansion

$$u \sim \sum_{n=0}^{\infty} \varepsilon^n u_n(s). \tag{3.2.31}$$

We then obtain from (3.2.29) to various orders in ε:

$$O(1): \frac{d^2 u_0}{ds^2} + u_0 = 0 \tag{3.2.32}$$

$$O(\varepsilon): \frac{d^2 u_1}{ds^2} + u_1 = -u_0^3 - 2\omega_1 \frac{d^2 u_0}{ds^2} \tag{3.2.33}$$

$$O(\varepsilon^2): \frac{d^2 u_2}{ds^2} + u_2 = -3u_0^2\, u_1 - 2\omega_1 \frac{d^2 u_1}{ds^2} - \left(\omega_1^2 + 2\omega_2\right)\frac{d^2 u_0}{ds^2}. \tag{3.2.34}$$

etc.

Solving equation (3.2.32), we obtain

$$u_0 = a\cos\left(s + \varphi\right). \tag{3.2.35}$$

Substituting (3.2.35), equation (3.2.33) becomes

$$\frac{d^2 u_1}{ds^2} + u_1 = -\frac{1}{4}\, a^3 \cos 3\left(s + \varphi\right) - \left(\frac{3}{4}\, a^2 - 2\omega_1\right) a\cos\left(s + \varphi\right). \tag{3.2.36}$$

Removal of secular terms in equation (3.2.36) requires

$$\frac{3}{4}\, a^2 - 2\omega_1 = 0 \quad \text{or} \quad \omega_1 = \frac{3}{8}\, a^2. \tag{3.2.37}$$

We then obtain the following solution for equation (3.2.36) –

$$u_1 = \frac{a^3}{32} \cos 3\left(s + \varphi\right). \tag{3.2.38}$$

Using (3.2.35) and (3.2.38), equation (3.2.34) becomes

$$\frac{d^2 u_2}{ds^2} + u_2 = \left(\frac{21}{128} a^4 + 2\omega_2\right) a\cos(s+\varphi) + \text{nonsecular terms.} \quad (3.2.39)$$

Removal of secular terms in equation (3.2.39) requires

$$\frac{21}{128} a^4 + 2\omega_2 = 0 \quad \text{or} \quad \omega_2 = -\frac{21}{256} a^4. \quad (3.2.40)$$

Using (3.2.35), (3.2.37), (3.2.38), and (3.2.40), we obtain from (3.2.30) and (3.2.31),

$$u \sim a\cos(\omega t + \varphi) + \varepsilon\, \frac{a^3}{32} \cos 3(\omega t + \varphi) + O(\varepsilon^2) \quad (3.2.41)$$

with

$$\omega = 1 + \frac{3}{8} \varepsilon a^2 - \frac{21}{256} \varepsilon^2 a^4 + O(\varepsilon^3). \quad (3.2.42)$$

Observe that the first-order regular perturbation solution gives an error of $O(\varepsilon)$ for $t = O(1)$, but a larger error of $O(1)$ for $t = O(\varepsilon^{-1})$. However, the first-order singular perturbation solution gives an error of only $O(\varepsilon)$ for $t = O(\varepsilon^{-1})$ and does not lead to an error of $O(1)$ until $t = O(\varepsilon^{-2})$.

Example 3: Consider the Mathieu equation

$$\ddot{u} + (\delta + \varepsilon \cos 2t)\, u = 0. \quad (3.2.43)$$

Let us determine transition curves in the (δ, ε) plane, which separate stable and unstable solutions and correspond to periodic solutions (see Appendix 2). Let us consider expansions

$$\delta = n^2 + \varepsilon \delta_1 + \varepsilon^2 \delta_2 + \cdots \quad (3.2.44)$$

$$u \sim u_0 + \varepsilon u_1 + \varepsilon^2 u_2 + \cdots \quad (3.2.45)$$

where n is an integer including zero. We then obtain from equation (3.2.43) to various orders in ε,

$$O(1): \ddot{u}_0 + n^2 u_0 = 0 \quad (3.2.46)$$

$$O(\varepsilon): \ddot{u}_1 + n^2 u_1 = -(\delta_1 + \cos 2t)\, u_0 \quad (3.2.47)$$

$$O(\varepsilon^2): \ddot{u}_2 + n^2 u_2 = -(\delta_1 + \cos 2t)\, u_1 - \delta_2 u_0 \quad (3.2.48)$$

etc.

We now have the following different cases to consider.

Case (i): $n = 0$.

We now obtain from equation (3.2.46),

$$u_0 = 1, \text{ say.} \tag{3.2.49}$$

Using (3.2.49), equation (3.2.47) becomes

$$\ddot{u}_1 = -\delta_1 - \cos 2t. \tag{3.2.50}$$

Uniform validity of the solution of equation (3.2.50) requires

$$\delta_1 = 0. \tag{3.2.51}$$

We then have from equation (3.2.50),

$$u_1 = \frac{1}{4} \cos 2t + A. \tag{3.2.52}$$

Using (3.2.49), (3.2.51), and (3.2.52), equation (3.2.48) becomes

$$\ddot{u}_2 = -\delta_2 - \frac{1}{8} - A\cos 2t - \frac{1}{8} \cos 4t. \tag{3.2.53}$$

Removal of secular terms in equation (3.2.53) requires

$$\delta_2 + \frac{1}{8} = 0 \quad \text{or} \quad \delta_2 = -\frac{1}{8}. \tag{3.2.54}$$

Using (3.2.51) and (3.2.54), (3.2.44) becomes

$$\delta = -\frac{1}{8} \varepsilon^2 + O(\varepsilon^3). \tag{3.2.55}$$

Case (ii): $n = 1$.

We now obtain from equation (3.2.46),

$$u_0 = \cos t. \tag{3.2.56}$$

Using (3.2.56), equation (3.2.47) becomes

$$\ddot{u}_1 + u_1 = -\left(\delta_1 + \frac{1}{2}\right)\cos t - \frac{1}{2} \cos 3t. \tag{3.2.57}$$

Removal of secular terms in equation (3.2.57) requires

$$\delta_1 + \frac{1}{2} = 0 \quad \text{or} \quad \delta_1 = -\frac{1}{2}. \tag{3.2.58}$$

We then have from equation (3.2.57),

$$u_1 = \frac{1}{16} \cos 3t.$$
(3.2.59)

Using (3.2.56), (3.2.58), and (3.2.59), equation (3.2.48) becomes

$$\ddot{u}_2 + u_2 = -\left(\frac{1}{32} + \delta_2\right) \cos t + \frac{1}{32} \cos 3t - \frac{1}{32} \cos 5t.$$
(3.2.60)

Removal of secular terms in equation (3.2.60) requires

$$\delta_2 + \frac{1}{32} = 0 \quad \text{or} \quad \delta_2 = -\frac{1}{32}.$$
(3.2.61)

Using (3.2.58) and (3.2.61), equation (3.2.44) becomes

$$\delta = 1 - \frac{1}{2} \varepsilon - \frac{1}{32} \varepsilon^2 + O(\varepsilon^3).$$
(3.2.62a)

Had we instead taken for the solution of equation (3.2.46),

$$u_0 = \sin t,$$
(3.2.63)

we would have obtained in place of (3.2.62a),

$$\delta = 1 + \frac{1}{2} \varepsilon - \frac{1}{32} \varepsilon^2 + O(\varepsilon^3).$$
(3.2.62b)

Case (iii): $n = 2$.

We now obtain from equation (3.2.46),

$$u_0 = \cos 2t.$$
(3.2.64)

Using (3.2.64), equation (3.2.47) becomes

$$\ddot{u}_1 + 4u_1 = -\frac{1}{2} - \delta_1 \cos 2t - \frac{1}{2} \cos 4t.$$
(3.2.65)

Removal of secular terms in equation (3.2.65) requires

$$\delta_1 = 0.$$
(3.2.66)

We then have from equation (3.2.65),

$$u_1 = -\frac{1}{8} + \frac{1}{24} \cos 4t.$$
(3.2.67)

Using (3.2.64), (3.2.66), and (3.2.67), equation (3.2.48) becomes

$$\ddot{u}_2 + 4u_2 = -\left(\delta_2 - \frac{5}{48}\right) \cos 2t - \frac{1}{48} \cos 6t.$$
(3.2.68)

Removal of secular terms in equation (3.2.68) requires

$$\delta_2 - \frac{5}{48} = 0 \quad \text{or} \quad \delta_2 = \frac{5}{48}. \tag{3.2.69}$$

Using (3.2.66) and (3.2.69), (3.2.44) becomes

$$\delta = 4 + \frac{5}{48}\, \varepsilon^2 + O\!\left(\varepsilon^3\right). \tag{3.2.70}$$

Had we instead taken for the solution of equation (3.2.46),

$$u_0 = \sin 2t \tag{3.2.71}$$

we would have obtained in place of (3.2.70),

$$\delta = 4 - \frac{1}{48}\, \varepsilon^2 + O\!\left(\varepsilon^3\right). \tag{3.2.72}$$

The above results are schematically shown in Figure 3.1.

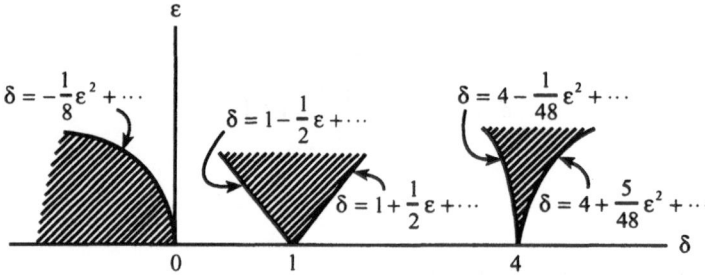

Figure 3.1. Parameter diagram for Mathieu equation:
Shaded regions unstable.

The method of strained parameters gives the transition curves and the periodic solutions along them, as discussed in the foregoing. In order to determine the solutions in the neighborhood of the transition curves, we use Whittaker's method (Whittaker, 1914). For this purpose, we cast the solution first in the normal form

$$u(t) = v(t)\, e^{\sigma t}. \tag{3.2.73}$$

Equation (3.2.43) then becomes

$$\ddot{v} + 2\sigma \dot{v} + \left(\delta + \sigma^2 + \varepsilon \cos 2t\right) v = 0. \tag{3.2.74}$$

Near the transition curves, σ is small, so one may look for a solution of the form –

$$v(t;\varepsilon) \sim v_0(t) + \varepsilon v_1(t) + \varepsilon^2 v_2(t) + \cdots \tag{3.2.75}$$

$$\delta = \delta_0 + \varepsilon \delta_1 + \varepsilon^2 \delta_2 + \cdots \tag{3.2.76}$$

$$\sigma = \varepsilon \sigma_1 + \varepsilon^2 \sigma_2 + \cdots. \tag{3.2.77}$$

Substituting these expansions into equation (3.2.74), we obtain

$$O(1) : \ddot{v}_0 + \delta_0 v_0 = 0 \tag{3.2.78}$$

$$O(\varepsilon) : \ddot{v}_1 + \delta_0 v_1 = -2\sigma_1 \dot{v}_0 - \delta_1 v_0 - v_0 \cos 2t \tag{3.2.79}$$

etc.

We have from equation (3.2.78) –

$$v_0 = a \cos \sqrt{\delta_0} \; t + b \sin \sqrt{\delta_0} \; t. \tag{3.2.80}$$

Now, according to the Floquet theory (see Appendix 2), $v(t)$ is periodic, with period π (2π), if

$$\delta_0 = n^2, \; n \text{ even (odd) integer.} \tag{3.2.81}$$

Using (3.2.81), equation (3.2.79) becomes

$$\ddot{v}_1 + n^2 v_1 = -2\sigma_1 (-an \sin nt + bn \cos nt) - \delta_1 (a \cos nt + b \sin nt)$$
$$- (a \cos nt + b \sin nt) \cos 2t \tag{3.2.82}$$

which may be rewritten, for $n = 1$, as

$$\ddot{v}_1 + v_1 = \left(2\sigma_1 a - \delta_1 b + \frac{b}{2} \right) \sin t$$

$$- \left(2\sigma_1 b + \delta_1 a + \frac{a}{2} \right) \cos t + \text{nonsecular terms.} \tag{3.2.83}$$

Removal of secular terms in equation (3.2.83) requires

$$\left. \begin{aligned} 2\sigma_1 a + \left(\frac{1}{2} - \delta_1 \right) b = 0 \\ \left(\frac{1}{2} + \delta_1 \right) a + 2\sigma_1 b = 0 \end{aligned} \right\}. \tag{3.2.84}$$

The solvability condition for equations (3.2.84) gives

$$4\sigma_1^2 - \left(\frac{1}{4} - \delta_1^2 \right) = 0 \tag{3.2.85}$$

from which,

$$\sigma_1 = \pm\frac{1}{2}\sqrt{\frac{1}{4} - \delta_1^2}. \tag{3.2.86}$$

Using (3.2.86), equations (3.2.84) give

$$b = \mp\sqrt{\frac{\frac{1}{2} + \delta_1}{\frac{1}{2} - \delta_1}}\, a, \quad \delta_1 \neq 1/2. \tag{3.2.87}$$

Using (3.2.73), (3.2.80), (3.2.81), (3.2.86), and (3.2.87), we obtain

$$u \sim a_1\, e^{\frac{1}{2}\varepsilon\sqrt{\frac{1}{4} - \delta_1^2}\, t} \left[\cos t - \sqrt{\frac{\frac{1}{2} + \delta_1}{\frac{1}{2} - \delta_1}}\, \sin t\right]$$

$$+ a_2\, e^{-\frac{1}{2}\varepsilon\sqrt{\frac{1}{4} - \delta_1^2}\, t} \left[\cos t + \sqrt{\frac{\frac{1}{2} + \delta_1}{\frac{1}{2} - \delta_1}}\, \sin t\right] + \cdots, \tag{3.2.88}$$

$$\delta_1 \neq \frac{1}{2}$$

which shows that the solution is bounded when $\delta_1^2 > \frac{1}{4}$ and unbounded when $\delta_1^2 < \frac{1}{4}$. Therefore, the transition from stability to instability is described by $\delta_1^2 = \frac{1}{4}$, and the transition curves emanating from $\delta = 1$ are given, from (3.2.76) and (3.2.81), by

$$\delta = 1 \pm \frac{1}{2}\varepsilon + \cdots \tag{3.2.89}$$

in agreement with the result (3.2.62) given by the method of strained parameters.

3.3. Eigenfunction Expansion Method

Consider a perturbed linear eigenvalue problem

$$u'' + \left[\lambda + \varepsilon f(x)\right] u = 0, \quad f(x) = f(-x), \quad u(0) = u(1) = 0. \quad (3.3.1)$$

Corresponding to $\varepsilon = 0$, we have, from (3.3.1),

$$\left.\begin{array}{c} u_0 = \sqrt{2} \sin n\pi x, \quad \lambda_0 = n^2 \pi^2 \\ n = 1, 2, 3 \end{array}\right\}. \quad (3.3.2)$$

For $\varepsilon \neq 0$, let us put (Schrödinger, 1926)

$$u_n \sim \sqrt{2} \sin n\pi x + \varepsilon u_{n1} + \varepsilon^2 u_{n2} + \cdots \quad (3.3.3)$$

$$\lambda_n = n^2 \pi^2 + \varepsilon \lambda_{n1} + \varepsilon^2 \lambda_{n2} + \cdots. \quad (3.3.4)$$

Using (3.3.3) and (3.3.4), equation (3.3.1) gives to various orders in ε,

$$\left.\begin{array}{c} O(\varepsilon): u''_{n_1} + n^2 \pi^2 u_{n_1} = -f(x) u_{n_0} - \lambda_{n_1} u_{n_0} \\ u_{n_1}(0) = u_{n_1}(1) = 0 \end{array}\right\} \quad (3.3.5)$$

$$\left.\begin{array}{c} O(\varepsilon^2): u''_{n_2} + n^2 \pi^2 u_{n_2} = -f(x) u_{n_1} - \lambda_{n_1} u_{n_1} - \lambda_{n_2} u_{n_0} \\ u_{n_2}(0) = u_{n_2}(1) = 0 \end{array}\right\}. \quad (3.3.6)$$

etc.

Let us expand u_{n_1} in terms of the eigenfunctions (3.3.2) of the $0(1)$ problem, as follows –

$$u_{n_1} = \sum_{m=1}^{\infty} a_{nm} \sqrt{2} \sin m\pi x. \quad (3.3.7)$$

Then equation (3.3.5) gives

$$\sum_{m=1}^{\infty} \sqrt{2} \pi^2 \left(n^2 - m^2\right) a_{nm} \sin m\pi x$$
$$= -\sqrt{2} f(x) \sin n\pi x - \sqrt{2} \lambda_{n_1} \sin n\pi x. \quad (3.3.8)$$

Multiply equation (3.3.8) by $\sqrt{2} \sin k\pi x$, and integrate from 0 to 1; we then obtain

$$\pi^2 \left(n^2 - k^2\right) a_{nk} = -F_{nk} - \lambda_{n_1} \delta_{nk}, \quad (3.3.9)$$

where

$$F_{nk} = 2 \int_0^1 f(x) \sin n\pi x \sin k\pi x \, dx = F_{kn}. \quad (3.3.10)$$

If $k = n$, equation (3.3.9) gives

$$\lambda_{n_1} = -F_{nn} = -2 \int_0^1 f(x) \sin^2 n\pi x \, dx. \tag{3.3.11}$$

If, on the other hand, $k \neq n$, equation (3.3.9) gives

$$a_{nk} = -\frac{F_{nk}}{\pi^2 \left(n^2 - k^2\right)}. \tag{3.3.12}$$

Using (3.3.12), equation (3.3.7) gives

$$u_{n_1} = -\sum_{k \neq n} \frac{F_{nk}}{\pi^2 \left(n^2 - k^2\right)} \sqrt{2} \sin k\pi x + a_{nn} \sqrt{2} \sin n\pi x. \tag{3.3.13}$$

In order to determine a_{nn}, we need to proceed to equation (3.3.6). Let us expand u_{n_2} as follows –

$$u_{n_2} = \sum_{r=1}^{\infty} b_{nr} \sqrt{2} \sin r\pi x. \tag{3.3.14}$$

Equation (3.3.6) then gives

$$\pi^2 \sum_{r=1}^{\infty} \left(n^2 - r^2\right) \sqrt{2} \, b_{nr} \sin r\pi x$$

$$= -\sum_{k=1}^{\infty} a_{nk} \sqrt{2} \, f(x) \sin k\pi x - \sum_{k=1}^{\infty} a_{nk} \lambda_{n_1} \sqrt{2} \sin k\pi x$$

$$- \lambda_{n_2} \sqrt{2} \sin n\pi x. \tag{3.3.15}$$

Multiply equation (3.3.15) by $\sqrt{2} \sin s\pi x$, and integrate from 0 to 1; we then obtain

$$\pi^2 \left(n^2 - s^2\right) b_{ns} = -\sum_{k=1}^{\infty} a_{nk} F_{ks} - a_{ns} \lambda_{n_1} - \lambda_{n_2} \delta_{ns}. \tag{3.3.16}$$

If $s = n$, equation (3.3.16) gives

$$\lambda_{n_2} = -\sum_{k \neq n} a_{nk} F_{kn} = \sum_{k \neq n} \frac{F_{nk}^2}{\pi^2 \left(n^2 - k^2\right)}. \tag{3.3.17}$$

If, on the other hand, $s \neq n$, equation (3.3.16) gives

$$b_{ns} = \sum_{k \neq n} \frac{F_{nk} F_{ks}}{\pi^4 \left(n^2 - k^2\right)\left(n^2 - s^2\right)} - \frac{a_{nn} F_{ns}}{\pi^2 \left(n^2 - s^2\right)} - \frac{F_{nn} F_{ns}}{\pi^4 \left(n^2 - s^2\right)^2}. \tag{3.3.18}$$

Let us normalize u_n as follows –

$$\int_0^1 \left(u_{n_0} + \varepsilon\, u_{n_1} + \varepsilon^2\, u_{n_2} \right)^2 dx = 1, \tag{3.3.19}$$

from which we have

$$\left. \begin{aligned} &\int_0^1 u_{n_0} u_{n_1}\, dx = 0 \\[2mm] &\int_0^1 \left(2u_{n_0} u_{n_2} + u_{n_1}^2 \right) dx = 0 \end{aligned} \right\} \tag{3.3.20}$$

etc.

Using (3.3.7) and (3.3.14), equation (3.3.20) gives

$$\left. \begin{aligned} & a_{nn} = 0 \\[2mm] & b_{nn} = -\frac{1}{2}\sum_{k=1}^{\infty} a_{nk}^2 \end{aligned} \right\}. \tag{3.3.21}$$

Using (3.3.2), (3.3.7), (3.3.17), and (3.3.21), we have from equation (3.3.3),

$$u_n \sim \sqrt{2}\,\sin n\pi x - \varepsilon \sum_{k\neq n} \frac{F_{nk}}{\pi^2 \left(n^2 - k^2\right)} \sqrt{2}\,\sin k\pi x$$

$$+ \varepsilon^2 \sum_{k\neq n} \left\{ \left[\sum_{s\neq n} \frac{F_{ns} F_{ks}}{\pi^4 \left(n^2 - s^2\right)\left(n^2 - k^2\right)} - \frac{F_{nn} F_{nk}}{\pi^4 \left(n^2 - k^2\right)^2} \right] \sqrt{2}\,\sin k\pi x \right.$$

$$\left. - \frac{1}{2} \frac{F_{nk}^2}{\pi^4 \left(n^2 - k^2\right)^2} \sqrt{2}\,\sin n\pi x \right\} + O\!\left(\varepsilon^3\right), \tag{3.3.22a}$$

$$\lambda = n^2 \pi^2 - \varepsilon F_{nn} + \varepsilon^2 \sum_{k\neq n} \frac{F_{nk}^2}{\pi^2 \left(n^2 - k^2\right)} + O\!\left(\varepsilon^3\right). \tag{3.3.22b}$$

3.4. Lighthill's Method of Shifting Singularities

Lighthill's (1949) technique is based on the premise that the source of some non-uniformities may be traced to the fact that the regular perturbative solution may have the right form, but not quite at the right place. Lighthill (1949) then suggested that this non-uniformity may be remedied by straining the independent variable.

Example 4: Consider (Van Dyke, 1975)

$$\left.\begin{array}{c}(x+\varepsilon y)\dfrac{dy}{dx}+y=1\\[2mm]y(1)=2\end{array}\right\}.$$

(3.4.1)

Observe that whereas the singularity for the full equation (3.4.1) is along the line $x=-\varepsilon y$, the limit $\varepsilon \to 0$ shifts it to $x=0$ (see Figure 3.2). The method of strained parameters (Lighthill, 1949) makes it move toward the true position even as $\varepsilon \to 0$, (so long as $\varepsilon \neq 0$).

A straightforward perturbation-expansion gives the following solution –

$$y \sim \frac{1+x}{x} - \varepsilon \frac{(1-x)(1+3x)}{2x^3} + \varepsilon^2 \frac{(1-x^2)(1+3x)}{2x^5} + \cdots$$

(3.4.2)

which breaks down as $x \to 0$. Further, the singularity at $x=0$ is compounded at higher orders.

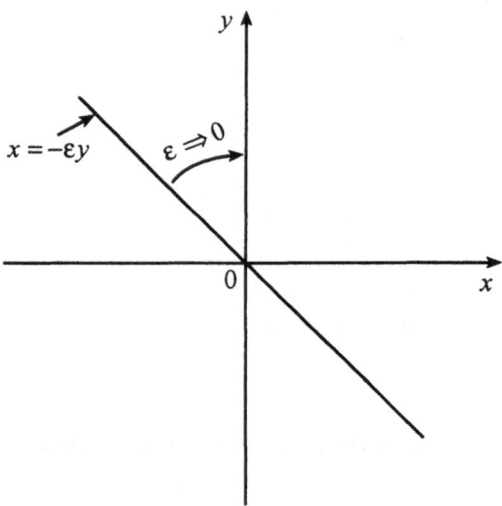

Figure 3.2. Shifting of singularities.

Let us strain the independent variable x as follows –

$$x = x_0 + \varepsilon x_1(x_0) + \cdots.$$

(3.4.3)

Equation (3.4.1) then gives

$$\left[x_0 + \varepsilon\left(x_1 + y\right)\right]\frac{dy}{dx_0} \cdot \left(1 - \varepsilon\,\frac{dx_1}{dx_0}\right) + y = 1, \tag{3.4.4}$$

from which we obtain to various orders in ε –

$$O(1) : x_0\,\frac{dy}{dx_0} + y = 1 \tag{3.4.5}$$

$$O(\varepsilon) : \left(x_1 + y\right)\frac{dy}{dx_0} - x_0\,\frac{dy}{dx_0}\frac{dx_1}{dx_0} = 0 \tag{3.4.6}$$

etc.

Solving equation (3.4.5), we obtain

$$y = \frac{c + x_0}{x_0}, \tag{3.4.7}$$

c being an arbitrary constant.

Using (3.4.7), equation (3.4.6) becomes

$$\frac{dx_1}{dx_0} - \frac{x_1}{x_0} = \frac{c + x_0}{x_0^2}, \tag{3.4.8}$$

from which

$$x_1 = -\frac{c + 2x_0}{2x_0}. \tag{3.4.9}$$

Using (3.4.9), equation (3.4.3) gives

$$x \approx x_0 - \varepsilon\,\frac{c + 2x_0}{2x_0}, \tag{3.4.10}$$

from which

$$x_0 \approx x + \varepsilon\,\frac{c + 2x}{2x}. \tag{3.4.11}$$

Using (3.4.11), we have, from equation (3.4.7),

$$y \approx \frac{c + x + \varepsilon\,\dfrac{c + 2x}{2x}}{x + \varepsilon\,\dfrac{c + 2x}{2x}} \approx \frac{2x^2 + 2\left(c + \varepsilon\right)x + \varepsilon c}{2x^2 + 2\varepsilon x + \varepsilon c}. \tag{3.4.12}$$

Imposing the initial condition

$$y(1) = 2, \tag{3.4.13}$$

equation (3.4.12) gives

$$c = 1 + \frac{3\varepsilon}{2} + O(\varepsilon^2).$$
(3.4.14)

Observe that the singular perturbative solution (3.4.12) no longer shows a singularity on the line $x = 0$!

On expanding in powers of ε, the singular perturbative solution (3.4.12) can be written as

$$y \sim 1 + \frac{1}{x} + \varepsilon \left(\frac{3x^2 - 2x - 1}{2x^3} \right) + O(\varepsilon^2), \quad x \neq 0$$
(3.4.15)

in agreement with the regular perturbation solution (3.4.2)!

Next, in order to find the exact solution for this problem, put

$$z \equiv x + \varepsilon y$$
(3.4.16)

so that equation (3.4.1) becomes

$$z \frac{dz}{dx} = \varepsilon + x.$$
(3.4.17)

The solution of equation (3.4.17) is

$$z = \left[2\varepsilon x + x^2 + c \right]^{1/2},$$
(3.4.18)

c being an arbitrary constant. Thus,

$$y = \left[\frac{2x}{\varepsilon} + \left(\frac{x}{\varepsilon} \right)^2 + c \right]^{1/2} - \frac{x}{\varepsilon}.$$
(3.4.19a)

Imposing again the initial condition (3.4.13), equation (3.4.18) gives

$$c = 4 + \frac{2}{\varepsilon}.$$
(3.4.20)

Using (3.4.20), equation (3.4.19a) becomes

$$y = \left[\left(\frac{x}{\varepsilon} \right)^2 + 2 \left(\frac{1+x}{\varepsilon} \right) + 4 \right]^{1/2} - \frac{x}{\varepsilon}.$$
(3.4.19b)

On expanding this in powers of ε, this can be written as

$$y \sim 1 + \frac{1}{x} + \varepsilon \left(\frac{3x^2 - 2x - 1}{2x^3} \right) + O(\varepsilon^2)$$
(3.4.21)

in agreement with the regular perturbative solution (3.4.2)!

The exact solution (3.4.19b) as well as the regular perturbative solution (3.4.2) are schematically shown in Figure 3.3.

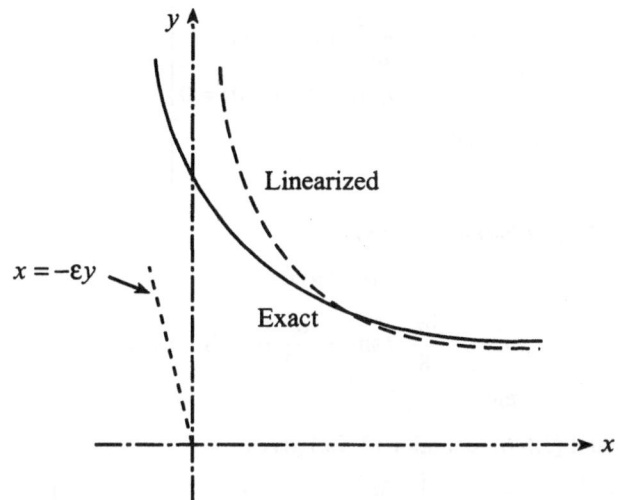

Figure 3.3. Integral curves for equation (3.4.1) (from Van Dyke, 1975).

3.5. Pritulo's Method of Renormalization

Sometimes a perturbation expansion can be rendered uniformly valid by introducing the straining of the independent variable directly in the solution rather than in the equation (Pritulo, 1962).

Example 5: Consider the initial-value problem for Duffing's equation –

$$\frac{d^2u}{dt^2} + u + \varepsilon u^3 = 0, \ u(0) = a, \ \dot{u}(0) = 0. \tag{3.5.1}$$

Look for a straightforward expansion –

$$u(t;\varepsilon) \sim u_0(t) + \varepsilon u_1(t) + \cdots \tag{3.5.2}$$

which leads to

$$O(1): \frac{d^2 u_0}{dt^2} + u_0 = 0 \left.\right\}, \qquad (3.5.3)$$
$$u_0(0) = a, \quad \dot{u}_0(0) = 0$$

$$O(\varepsilon): \frac{d^2 u_1}{dt^2} + u_1 = -u_0^3$$
$$u_1(0) = 0, \quad \dot{u}_1(0) = 0 \left.\right\}. \qquad (3.5.4)$$

etc.

We obtain from (3.5.3) and (3.5.4),

$$u_0 = a \cos t \qquad (3.5.5)$$

$$u_1 = -\frac{3a^3}{8} t \sin t + \frac{a^3}{32} (\cos 3t - \cos t) \qquad (3.5.6)$$

etc.

Using (3.5.5) and (3.5.6), equation (3.5.2) gives

$$u(t;\varepsilon) \sim a \cos t + \varepsilon \left[-\frac{3a^3}{8} t \sin t + \frac{a^3}{32} (\cos 3t - \cos t) \right] + \cdots. \qquad (3.5.7)$$

Pritulo (1962) suggested that one may introduce the straining of the independent variable

$$t = s \left(1 + \varepsilon \omega_1 + \cdots \right) \qquad (3.5.8)$$

directly into the solution (3.5.7). Expanding and collecting the coefficients of equal powers of ε, we obtain

$$u(t;\varepsilon) \sim a \cos s + \varepsilon \left[-a \left(\omega_1 + \frac{3}{8} a^2 \right) s \sin s + \frac{1}{32} a^3 (\cos 3s - \cos s) \right] + \cdots. \qquad (3.5.9)$$

Removal of secular terms in equation (3.5.9) then requires

$$\omega_1 = -\frac{3}{8} a^2. \qquad (3.5.10)$$

Therefore, a uniformly valid expansion is

$$u(t;\varepsilon) \sim a \cos s + \frac{1}{32} \varepsilon a^3 (\cos 3s - \cos s) + \cdots, \qquad (3.5.11)$$

where

$$t = s\left(1 - \frac{3}{8}\varepsilon a^2 + \cdots\right), \tag{3.5.12}$$

in agreement with the result (3.2.41) and (3.2.42) obtained by the method of strained parameters!

Example 6: Consider the nonlinear hyperbolic wave-propagation problem –

$$\left.\begin{array}{l} \dfrac{\partial^2 u}{\partial t^2} - \dfrac{\partial^2 u}{\partial x^2} = \dfrac{\partial u}{\partial x}\dfrac{\partial^2 u}{\partial t^2} \\[2mm] u(0,t) = \varepsilon f(t), \quad f(t) = 0 \quad \text{for } t \leq 0 \\[2mm] u(x,0) = 0 \end{array}\right\}. \tag{3.5.13}$$

Let us expand the solution as follows –

$$u \sim \varepsilon u_1 + \varepsilon^2 u_2 + \cdots. \tag{3.5.14}$$

We then obtain from (3.5.13) to various orders in ε:

$$O(\varepsilon): \left.\begin{array}{l} \dfrac{\partial^2 u_1}{\partial t^2} - \dfrac{\partial^2 u_1}{\partial x^2} = 0 \\[2mm] u_1(0,t) = f(t) \end{array}\right\}, \tag{3.5.15}$$

$$O(\varepsilon^2): \left.\begin{array}{l} \dfrac{\partial^2 u_2}{\partial t^2} - \dfrac{\partial^2 u_2}{\partial x^2} = \dfrac{\partial u_1}{\partial x}\dfrac{\partial^2 u_1}{\partial t^2} \\[2mm] u_2(0,t) = 0 \end{array}\right\} \tag{3.5.16}$$

etc.

Solving (3.5.15), we obtain

$$u_1(x,t) = f(t-x). \tag{3.5.17}$$

Let us introduce the following characteristics of the linearized problem (3.5.15) –

$$\xi = t - x, \quad \eta = t + x. \tag{3.5.18}$$

Equation (3.5.16) then becomes

$$\frac{\partial^2 u_2}{\partial \xi \partial \eta} = -\frac{1}{4}f'(\xi)f''(\xi), \tag{3.5.19}$$

from which we have

$$u_2(x,t) = -\frac{1}{8}\left[f'(\xi)\right]^2 \eta + G(\xi). \tag{3.5.20}$$

Using the condition $u_2(0,t) = 0$, we obtain

$$G(t) = \frac{1}{8}\left[f'(t)\right]^2 t. \tag{3.5.21}$$

Using (3.5.21), equation (3.5.20) becomes

$$u_2(x,t) = \frac{1}{8}\left[f'(\xi)\right]^2 (\xi - \eta) = -\frac{1}{4}\left[f'(\xi)\right]^2 x. \tag{3.5.22}$$

Using (3.5.17) and (3.5.22), equation (3.5.14) gives

$$u(x,t) \sim \varepsilon f(\xi) - \varepsilon^2 \left\{\frac{1}{4}\left[f'(\xi)\right]^2 x\right\} + \cdots. \tag{3.5.23}$$

Observe the nonuniformity in (3.5.23). Physically, the latter is due to the erroneous assumption of constant speed of propagation in the far field. The cumulative effects of variation in the speed of propagation produce errors in the far field. In order to remove this nonuniformity, let us strain one of the linearized characteristics as follows (Lin, 1954 and Fox, 1955) –

$$\xi = \xi_0 + \varepsilon \xi_1 + \cdots. \tag{3.5.24}$$

Equation (3.5.23) then becomes

$$u(x,t) \sim \varepsilon f(\xi_0) + \varepsilon^2 \left\{\xi_1 f'(\xi_0) - \frac{1}{4}\left[f'(\xi_0)\right]^2 x\right\} + \cdots. \tag{3.5.25}$$

The nonuniform term in (3.5.25) can be eliminated by choosing

$$\xi_1 = \frac{x}{4} f'(\xi_0). \tag{3.5.26}$$

Using (3.5.26), equation (3.5.25) then becomes

$$u(x,t) \sim \varepsilon f\left([t-x] - \frac{\varepsilon x}{4} f'[t-x]\right) + \cdots. \tag{3.5.27}$$

Thus, the uniformly valid first-order solution for the nonlinear problem is simply the solution for the corresponding linear problem with the respective characteristic replaced by the characteristic calculated by including the first-order nonlinearities in the problem (Whitham, 1952). The variation in the speed of propagation is thus described by "bending" the linearized characteristics.

3.6. Wave Propagation in an Inhomogeneous Medium

The method of strained parameters is used in the following to treat the general problem of wave propagation in an inhomogeneous medium through the model example (Shivamoggi, 1978c) –

$$u_{tt} - \left\{1 + \varepsilon a(x,t)\right\}^2 u_{xx} = 0, \quad \varepsilon \ll 1 \left.\right\}$$
$$u(x,0) = f(x), \quad u_t(x,0) = g(x) \qquad \qquad (3.6.1)$$

where the inhomogeneities in the medium are modeled by the parameter $a(x,t)$. The main idea is to obtain a uniformly valid solution by straining the characteristic corresponding to the wave propagation in a homogeneous medium – in analogy to that in Lin's method (1954) for a nonlinear hyperbolic system (Section 3.5).

Thus, introduce the characteristic parameters

$$\xi, \eta = x \mp kt \qquad \qquad (3.6.2)$$

and seek solutions of the form

$$u(x,t;\varepsilon) \sim \sum_{n=0}^{\infty} \varepsilon^n u_n(\xi, \eta) \left.\right\}$$
$$k(x,t;\varepsilon) = 1 + \varepsilon \kappa(\xi, \eta) + O(\varepsilon^2) \qquad (3.6.3)$$

where the functions $\kappa(\xi, \eta)$ are determined by imposing the condition that (3.6.3) be uniformly valid for large distances.

Putting $a(x,t) = \hat{a}(\xi, \eta)$, one obtains from equation (3.6.1),

$$O(1) : \frac{\partial^2 u_0}{\partial \xi \partial \eta} = 0, \qquad \qquad (3.6.4)$$

$$O(\varepsilon) : \frac{\partial^2 u_1}{\partial \xi \partial \eta} = -\frac{\hat{a}(\xi, \eta)}{2} \left[u_{0\xi\xi} + 2u_{0\xi\eta} + u_{0\eta\eta} \right]$$
$$+ \frac{\kappa(\xi, \eta)}{2} \left[u_{0\xi\xi} - 2u_{0\xi\eta} + u_{0\eta\eta} \right]$$
$$+ \frac{1}{2} \frac{\partial \kappa(\xi, \eta)}{\partial \xi} u_{0\xi} + \frac{1}{2} \frac{\partial \kappa(\xi, \eta)}{\partial \eta} u_{0\eta}, \qquad (3.6.5)$$

etc.

From equation (3.6.4),

$$u_0(\xi,\eta) = Ap(\xi) + Bq(\eta), \tag{3.6.6}$$

where A and B are arbitrary constants determinable by the given initial conditions. For the sake of determining $\kappa(\xi,\eta)$, one may consider only the right-running waves so that $u_0 = u_0(\xi)$; then the removal of the secular terms in equation (3.6.5) requires that

$$\frac{\partial \kappa(\xi,\eta)}{\partial \xi} + \kappa(\xi,\eta)\frac{u_{0\xi\xi}}{u_{0\xi}} - \hat{a}(\xi,\eta)\frac{u_{0\xi\xi}}{u_{0\xi}} = 0, \tag{3.6.7}$$

from which

$$\kappa(\varepsilon,\eta) = \exp\left(-\int^{\xi}\frac{u_{0\xi\xi}}{u_{0\xi}}\,d\xi\right)\left[\int^{\xi}\exp\left(\int^{\xi}\frac{u_{0\xi\xi}}{u_{0\xi}}\,d\xi\right)\hat{a}(\xi,\eta)\frac{u_{0\xi\xi}}{u_{0\xi}}\,d\xi\right]$$

or

$$\kappa(\xi,\eta) = \frac{1}{u_{0\xi}}\int^{\xi}\hat{a}(\xi,\eta)\,u_{0\xi\xi}\,d\xi. \tag{3.6.8}$$

Thus, the uniformly valid first-order solution for wave propagation in an inhomogeneous medium is simply the solution corresponding to the wave propagation in a homogeneous medium with the respective characteristic replaced by the characteristic calculated by including the first-order inhomogeneities in the medium. In other words, the solution corresponding to a homogeneous medium, when the first-order inhomogeneities are included, may still have the right form, but not quite at the right place.

For the case $\hat{a}(\xi,\eta) \equiv 1$, equation (3.6.8) gives

$$\kappa = 1, \tag{3.6.9}$$

and if $g(x) = -f'(x)$, equation (3.6.6) becomes

$$u_0(\xi,\eta) = \frac{1}{2}\frac{2+\varepsilon}{1+\varepsilon}\,f\left[x - (1+\varepsilon)\,t\right]$$
$$+ \frac{1}{2}\frac{\varepsilon}{1+\varepsilon}\,f\left[x + (1+\varepsilon)\,t\right] \tag{3.6.10}$$

which is the exact solution to (3.6.1) with $a(x,t) = 1$!

3.7. Applications to Solid Mechanics: Nonlinear Buckling of Elastic Columns

The buckling problem involves the investigation of a potentially unstable equilibrium between the external loading and the internal resistance of the column. A sudden breakdown of the internal resistance constitutes a characteristic feature of the buckling phenomenon, regardless of whether at the instant of failure the elastic limit is exceeded or not.

Under a compressive load P, a possible state for the column is that of pure compression; but experience shows that for sufficiently large values of P, transverse deflections can occur. The classical linear theory for infinitesimal deflections due to Euler (see Timoshenko, 1959) establishes the conditions under which a column under the action of a concentrated axial load can take up equilibrium configurations other than the one corresponding to a uniform compression. This phenomenon is discussed here as a bifurcation scenario.

A characteristic feature of a bifurcation scenario is that the load is a monotonically increasing function of the deformation.[1] It is to be noted that although the critical load of bifurcation problems is a characteristic value corresponding to the linear problem, the study of the *supercritical* behavior in a bifurcation scenario requires the consideration of the nonlinear problem. Besides, the linear theory leads to lateral deflections whose magnitude remains indeterminate. The bifurcation problem associated with the buckling of elastic columns is treated here using the method of strained parameters.

Consider a long, slender, inextensible column simply supported at both ends, and subjected to a centrally applied compressive load P at its upper end (see Figure 3.4). The column is assumed to be perfectly straight and of uniform cross-section, while the material is homogeneous and it behaves elastically. The shape of the deflected central line of the column is described by $y(x)$, which satisfies the following differential equation and the boundary conditions:

---

[1] The buckling of a column can also occur via a snap-through scenario which is characterized by the occurrence of a jump from one stable state to another stable state, by passing unstable stables between the two.

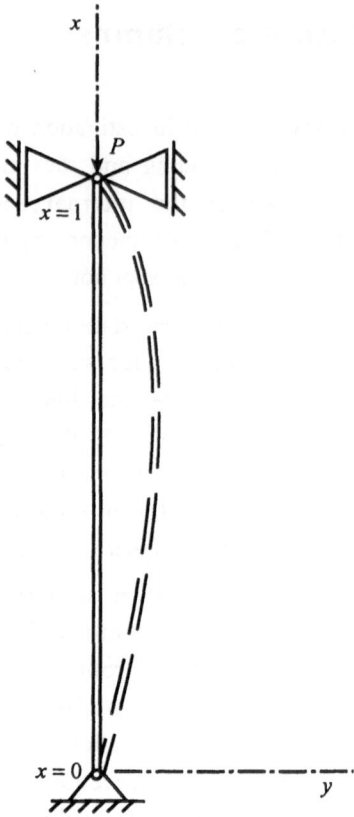

Figure 3.4. Buckled column.

$$\frac{d^2y}{dx^2} + \frac{PL^2y}{EI}\left[1+\left(\frac{dy}{dx}\right)^2\right]^{3/2} = 0 \Bigg\}$$

$$y(0) = y(1) = 0 \tag{3.7.1}$$

where all distances have been nondimensionalized using L, E being the modulus of elasticity, and I the moment of inertia of the cross section.

If $(dy/dx) \ll 1$, (3.7.1) may be written as

$$\frac{d^2y}{dx^2} + \lambda_0^2\, y = -\frac{3\lambda_0^2}{2}\, y\left(\frac{dy}{dx}\right)^2 \Bigg\}$$

$$y(0) = y(1) = 0 \tag{3.7.2}$$

where

$$\lambda_0^2 \equiv \frac{PL^2}{EI}.$$

Look for solutions of the form

$$\left. \begin{aligned} y(x,\varepsilon) &\sim \sum_{n=1}^{\infty} \varepsilon^n y_n(s), \quad \varepsilon \ll 1 \\ x &= s\left[1 + \varepsilon \lambda_1 + \varepsilon^2 \lambda_2 + O(\varepsilon^3)\right] \end{aligned} \right\}. \tag{3.7.3}$$

One then obtains from (3.7.2),

$$\left. \begin{aligned} O(\varepsilon): \quad &\frac{d^2 y_1}{ds^2} + \lambda_0^2 \, y_1 = 0 \\ &y_1(0) = y_1(1) = 0 \end{aligned} \right\}, \tag{3.7.4}$$

$$\left. \begin{aligned} O(\varepsilon^2): \quad &\frac{d^2 y_2}{ds^2} + \lambda_0^2 \, y_2 = -2\lambda_0^2 \, \lambda_1 y_1 \\ &y_2(0) = y_2(1) = 0 \end{aligned} \right\} \tag{3.7.5}$$

$$\left. \begin{aligned} O(\varepsilon^3): \quad &\frac{d^2 y_3}{ds^2} + \lambda_0^2 \, y_3 = -2\lambda_0^2 \, \lambda_1 y_2 - \lambda_0^2 \left(\lambda_1^2 + 2\lambda_2\right) y_1 - \frac{3\lambda_0^2}{2} \, y_1 \left(\frac{dy_1}{ds}\right)^2 \\ &y_3(0) = y_3(1) = 0 \\ &\text{etc.} \end{aligned} \right\}. \tag{3.7.6}$$

From (3.7.4), one obtains the characteristic values and the characteristic solutions corresponding to the linear problem

$$\left. \begin{aligned} y_1(s) &= A_1 \sin n\pi s \\ \lambda_0^2 &= n^2 \pi^2; \quad n = 1, 2, \ldots \end{aligned} \right\}, \tag{3.7.7}$$

The plot $|y_1|$ vs. λ_0^2 is sketched in Figure 3.5 – a picture, which on physical grounds is not acceptable. Therefore, in order to have a reasonable description of the buckling phenomenon, one proceeds to include the nonlinear effects.

Using (3.7.7), it is obvious that the removal of the secular terms in (3.7.5) requires

$$\lambda_1 = 0. \tag{3.7.8}$$

Equation (3.7.5) then yields

$$y_2(s) \equiv 0. \tag{3.7.9}$$

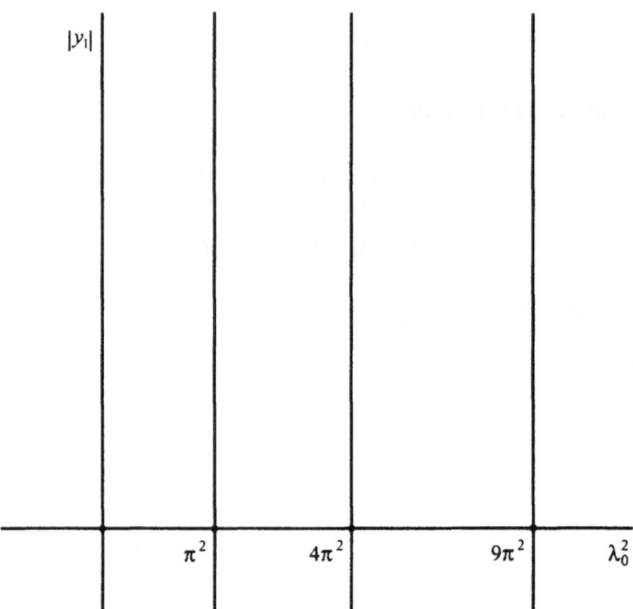

Figure 3.5. Buckling amplitude vs. load: Linear theory.

Using (3.7.7)–(3.7.9), equation (3.7.6) becomes

$$\left.\begin{aligned}\frac{d^2 y_3}{ds^2} + \lambda_0^2\, y_3 &= \left(-2\lambda_0^2\, \lambda_2\, A_1 - \frac{3\lambda_0^4\, A_1^3}{8}\right)\sin n\pi s - \frac{3\lambda_0^4\, A_1^3}{8}\sin 3n\pi s \\ y_3(0) &= y_3(1) = 0\end{aligned}\right\}, \qquad (3.7.10)$$

so that the removal of the secular terms in (3.7.10) requires

$$\lambda_2 = -\frac{3}{16}\,\lambda_0^2\, A_1^2 = -\frac{3}{16}\, n^2\pi^2 A_1^2. \qquad (3.7.11)$$

Equation (3.7.10) then yields

$$y_3(s) = -\frac{3n^2\pi^2 A_1^3}{64}\sin 3n\pi s. \qquad (3.7.12)$$

Thus, on using (3.7.7)-(3.7.9) and (3.7.11), (3.7.3) becomes

$$y(x;\varepsilon) \sim \varepsilon\, A_1 \sin \lambda x + O(\varepsilon^2), \qquad (3.7.13)$$

where

$$\lambda = \frac{n\pi}{1 - \varepsilon^2 (3/16) \, n^2 \pi^2 \, A_1^2 + O(\varepsilon^3)}$$

or

$$\lambda^2 = n^2 \pi^2 \left[1 + \varepsilon^2 (3/8) \, n^2 \pi^2 A_1^2 + O(\varepsilon^3) \right], \qquad (3.7.14)$$

so that the column buckles into a shape given by (3.7.13) with an amplitude A_1. Note the bifurcation phenomenon, (see Appendix 3 for a review of bifurcation theory) which corresponds to the branching of a nonlinear solution from the solution corresponding to the undeflected configuration at $\lambda^2 = \lambda_0^2 = n^2 \pi^2$, as shown graphically in Figure 3.6.

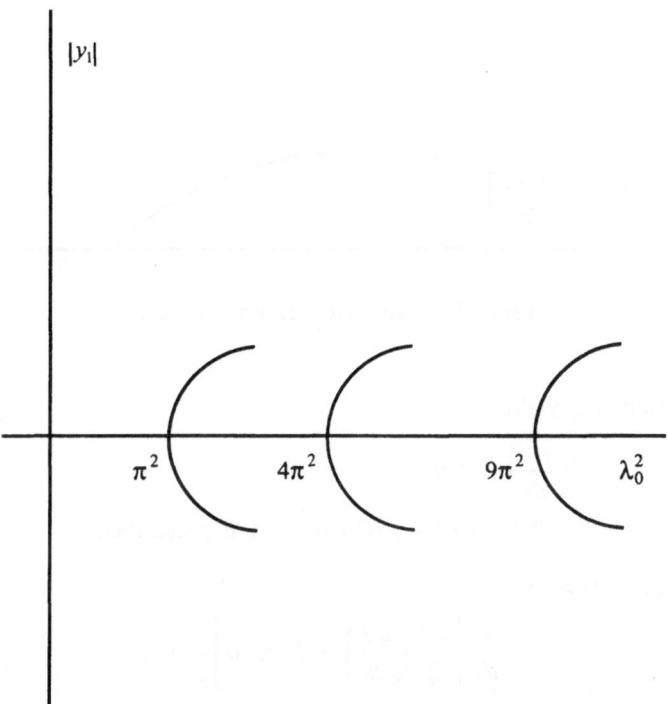

Figure 3.6. Buckling amplitude vs. load: Nonlinear theory.

This simple example also affords a somewhat exact formulation which may be used to check the accuracy of the results obtained in the foregoing using the method of strained parameters. The coordinates s and θ are introduced to describe the deformed configuration (see Figure 3.7). The potential energy of the deformed column is given by

$$U = \frac{EI}{2} \int_0^L \left(\frac{d\theta}{ds}\right)^2 ds + P \left\{\int_0^L \cos\theta\, ds - L\right\}. \tag{3.7.15}$$

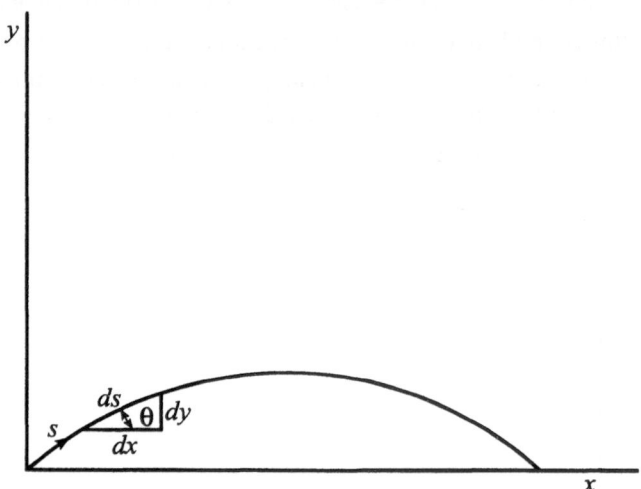

Figure 3.7. Buckled column geometry.

Extremization of U yields

$$\left.\begin{array}{l} EI \dfrac{d^2\theta}{ds^2} + P\sin\theta = 0 \\[2mm] x = 0, L:\ EI\,(d\theta/ds) = 0,\ \text{and}\ \theta\ \text{prescribed} \end{array}\right\}. \tag{3.7.16a}$$

Writing (3.7.16a) as

$$\frac{d}{ds}\left[\frac{EI}{2}\left(\frac{d\theta}{ds}\right)^2 - P\cos\theta\right] = 0,$$

one obtains

$$\frac{EI}{2}\left(\frac{d\theta}{ds}\right)^2 = P\cos\theta + c \tag{3.7.16b}$$

where c is an arbitrary constant. Let the boundary conditions be

$$x = 0, L:\ \frac{d\theta}{ds} = 0,\quad \theta = \pm\alpha, \tag{3.7.17}$$

so that equation (3.7.16b) becomes

$$\frac{d\theta}{ds} = \left[\frac{2P}{EI}\left(\cos\theta - \cos\alpha\right)\right]^{1/2}. \tag{3.7.18}$$

Also noting that

$$\sin\theta = dy/ds, \tag{3.7.19}$$

one obtains from equation (3.7.16a),

$$y = -\frac{EI}{P}\frac{d\theta}{ds},$$

and using equation (3.7.18), this becomes

$$y = -\sqrt{2EI/P}\,\left(\cos\theta - \cos\alpha\right)^{1/2}. \tag{3.7.20}$$

We have from equation (3.7.20),

$$\frac{y_{max}}{L} = \frac{2}{\pi}\frac{\sin\left(\alpha/2\right)}{\sqrt{P/\left(\pi^2 EI/L^2\right)}}, \tag{3.7.21}$$

a trend which agrees with that given by (3.7.14) using the method of strained parameters.

3.8. Applications to Fluid Dynamics

3.8.1 Nonlinear Hyperbolic Waves in a Gas

Uniformly valid simple-wave solutions are developed here for nonlinear hyperbolic equations using the method of strained parameters (Shivamoggi, 1977a and 1978a).

The potential function ϕ (the fluid velocity $v = \nabla\phi$) of the flow due to the propagation of nonlinear acoustic plane waves in an inviscid, irrotational, and initially quiescent medium is governed by

$$\frac{\partial^2\phi}{\partial t^2} - c^2\frac{\partial^2\phi}{\partial x^2} = -\frac{\partial}{\partial t}\left(\frac{\partial\phi}{\partial x}\right)^2 + (1-\gamma)\frac{\partial\phi}{\partial t}\frac{\partial^2\phi}{\partial x^2} - \frac{1}{2}(1+\gamma)\left(\frac{\partial\phi}{\partial x}\right)^2\frac{\partial^2\phi}{\partial x^2}, \tag{3.8.1}$$

where c is the speed of sound in the undisturbed medium and γ the ratio of specific heats of the gas. We consider only simple right-running waves subjected to the boundary condition

$$\phi(0,t) = \varepsilon\,f(t), \tag{3.8.2}$$

where ε is a small nondimensional parameter.

Since we are looking for travelling waves, introduce characteristic coordinates –

$$\xi, \eta = t \mp \frac{kx}{c},$$
(3.8.3)

and seek solutions of the boundary–value problems (3.8.1) and (3.8.2) of the form

$$\phi(x,t;\varepsilon) \sim \sum_{n=1}^{\infty} \varepsilon^n \phi_n(\xi,\eta),$$

$$k(\xi;\varepsilon) = 1 + \varepsilon \kappa(\xi) + O(\varepsilon)^2,$$
(3.8.4)

so that one obtains from equation (3.8.1),

$$O(\varepsilon) : \frac{\partial^2 \phi_1}{\partial s_1 \partial s_2} = 0$$
(3.8.5)

$$O(\varepsilon^2) : 4 \frac{\partial^2 \phi_2}{\partial s_1 \partial s_2} = \left[2 \frac{\partial \kappa}{\partial s_1} + 2\kappa \frac{d^2 \phi_1/ds_1^2}{d\phi_1/ds_1} - \frac{\gamma+1}{c^2} \frac{d^2 \phi_1}{ds_1^2} \right] \frac{d\phi_1}{ds_1},$$
(3.8.6)

etc.,

where

$$s_{1,2} = t \mp \frac{x}{c}.$$

Using (3.8.2), one obtains from equation (3.8.5),

$$\phi_1(s_1) = f(s_1).$$
(3.8.7)

The removal of the secular terms in equation (3.8.6) requires

$$2 \frac{\partial \kappa}{\partial s_1} + 2\kappa \frac{d^2 \phi_1/ds_1^2}{d\phi_1/ds_1} - \frac{\gamma+1}{c^2} \frac{d^2 \phi_1}{ds_1^2} = 0$$
(3.8.8)

from which,

$$\kappa = \exp\left(-\int \frac{d^2\phi_1/ds_1^2}{d\phi_1/ds_1} ds_1\right) \left[\int \exp\left(2\int \frac{d^2\phi_1/ds_1^2}{d\phi_1/ds_1} ds_1\right) \frac{\gamma+1}{2c^2} \frac{d^2\phi_1}{ds_1^2} ds_1\right]$$

or

$$\kappa = \frac{1}{d\phi_1/ds_1} \int^{s_1} \left(\frac{d\phi_1}{ds_1}\right) \frac{\gamma+1}{2c^2} \frac{d^2\phi_1}{ds_1^2} ds_1$$

or

$$\kappa = \frac{\gamma+1}{4c^2}\left(\frac{d\phi_1}{ds_1}\right). \tag{3.8.9}$$

Using (3.8.4) and (3.8.9), we have from (3.8.3),

$$s_1 = t - \frac{x}{c} = \xi + \varepsilon \frac{\gamma+1}{4c^2} x \left(\frac{d\phi_1}{ds_1}\right). \tag{3.8.10}$$

Writing $\phi_1(s_1)$ as a Taylor series about $s_1 = \xi$, and using equation (3.8.7), one readily obtains

$$\phi_1 = f(\xi) + \varepsilon \frac{\gamma+1}{4c^3} x \left[\frac{df(\xi)}{d\xi}\right]^2. \tag{3.8.11}$$

The flow velocity is then given by

$$\phi_x \sim -\frac{\varepsilon}{c} f'(\xi)\left[1 + \varepsilon \frac{\gamma+1}{4c^3} xf''(\xi)\right] + O(\varepsilon^3), \tag{3.8.12}$$

which contains a secular term. In order to eliminate it and render the expression uniformly valid, we use Pritulo's method, and put

$$\xi = t - \frac{\zeta}{c} = t - \frac{\delta}{c}\left[1 - \varepsilon\sigma(\delta) + O(\varepsilon^3)\right], \tag{3.8.13}$$

so that the removal of the secular term in (3.8.12) requires

$$\sigma(\delta) = -\frac{\gamma+1}{4c^2} f'\left(t - \frac{\delta}{c}\right) + O(\varepsilon). \tag{3.8.14}$$

Using equation (3.8.14), we have from (3.8.10) and (3.8.13),

$$\left.\begin{array}{l} x = \delta\left[1 - \varepsilon \frac{\gamma+1}{4c^2} f'\left(t - \frac{\delta}{c}\right)\right] \\[3mm] \zeta = \delta\left[1 + \varepsilon \frac{\gamma+1}{4c^2} f'\left(t - \frac{\delta}{c}\right)\right] \end{array}\right\}, \tag{3.8.15}$$

from which

$$\delta = \frac{1}{2}(\zeta + x). \tag{3.8.16}$$

Using (3.8.15), we have from equation (3.8.15),

$$\zeta = \frac{1}{2}(\zeta + x)\left[1 + \varepsilon \frac{\gamma+1}{4c^2} f'\left(t - \frac{\delta}{c}\right)\right] + O(\varepsilon^3)$$

or

$$\zeta = x\left[1 + \varepsilon \frac{\gamma+1}{2c^2} f'\left(t - \frac{x}{c}\right)\right] + O(\varepsilon^3). \tag{3.8.17}$$

Using (3.8.13) and (3.8.17), we have from equation (3.8.12)

$$\phi_x \sim -\frac{\varepsilon}{c} f'\left\{t - \frac{x}{c}\left[1 + \varepsilon \frac{\gamma+1}{2c^2} f'\left(t - \frac{x}{c}\right)\right]\right\} + O(\varepsilon^3) \tag{3.8.18}$$

which is uniformly valid.

Equation (3.8.18) implies that the uniformly valid first-order solution for the nonlinear problem is simply the solution for the corresponding linear problem, with the respective characteristic replaced by the characteristic calculated by including the first-order nonlinearities in the problem, as we saw previously in Example 6.

For compressive waves, the solution (3.8.11) contains regions of multivalues, which have to be resolved by the introduction of shocks. Since the potential function is continuous across shocks, their locations are given by

$$\left. \begin{aligned} \phi(\xi_a, x) &= \phi(\xi_b, x) \\ t(\xi_a, x) &= t(\xi_b, x) \end{aligned} \right\}, \tag{3.8.19}$$

where ξ_a and ξ_b denote the characteristics ahead and behind the shocks. One then has

$$\left. \begin{aligned} f(\xi_a) - f(\xi_b) &= \varepsilon\left(\frac{\gamma+1}{4c^3}\right) x \left[f'^2(\xi_a) - f'^2(\xi_b)\right] \\ \xi_b - \xi_a &= \varepsilon\left(\frac{\gamma+1}{2c^3}\right) x \left[f'(\xi_a) - f'(\xi_b)\right] \end{aligned} \right\}. \tag{3.8.20a}$$

Eliminating εx, one obtains

$$f(\xi_b) - f(\xi_a) = \frac{1}{2}(\xi_b - \xi_a)\left[f'(\xi_b) + f'(\xi_a)\right], \tag{3.8.20b}$$

which is essentially the equal area rule (Landau, 1945; Whitham, 1952).

Incidentally, equation (3.8.20b) shows that the slope of the line $\phi_1 = $ const is equal to the arithmetic mean of the slope of the characteristics in the unperturbed flow and the slope of the characteristic at the point in question. For the case of a shock running into the unperturbed flow, this is one form of the bisector rule as pointed out earlier by Van Dyke (1952).

3.8.2 Rayleigh-Taylor Instability of Superposed Fluids

The instability of the interface between two fluids having different densities and accelerated towards each other is called the Rayleigh-Taylor instability (Chandrasekhar, 1961 and Shivamoggi, 1998). A static state, in which an incompressible fluid of variable density subjected to a vertical acceleration is arranged in horizontal strata and the pressure and the density ρ are functions of the vertical coordinate z only, is obviously a kinematically realizable one. However, whether this state is also dynamically realizable is related to the issue of stability with respect to small disturbances.

If viscosity is neglected, the flow can be assumed to be irrotational and the velocity potential of each fluid satisfies the Laplace equation. The difficulty in solving this type of problem, however, arises from the nonlinear boundary conditions at an unknown interface. On the other hand, if one assumes the disturbance-amplitude to be infinitesimal, the boundary conditions can be linearized and a solution to the linear problem can be readily obtained. For the linear problem of two different inviscid and incompressible fluids in contact and subjected to an acceleration directed from the heavier fluid towards the lighter one, it turns out, for the case of zero surface tension, that any slight disturbance in the plane interface grows exponentially with time. The effect of the surface tension in the linear problem is to produce a critical wave number k_c, so that the interface is unstable or stable according as the wave number is less or greater than k_c.

(a) The Linear Problem

The two fluids are taken to be inviscid and incompressible, and if the motion of the whole system is supposed to start from rest, it may be assumed to be irrotational. The applied acceleration g' is directed from heavier to lighter fluid (see Figure 3.8). Initially, the interface is taken to be disturbed according to a simple sinusoidal standing wave with an amplitude a' and a wavelength λ'.

If $y = \eta$ denotes the disturbed shape of the interface, then, one has with the velocity potentials, Φ_j, such that the fluid velocity v is given by $v = -\nabla\Phi$,

$$y \lessgtr \eta : \nabla^2\Phi_j = 0; \quad j = 1, 2. \tag{3.8.21}$$

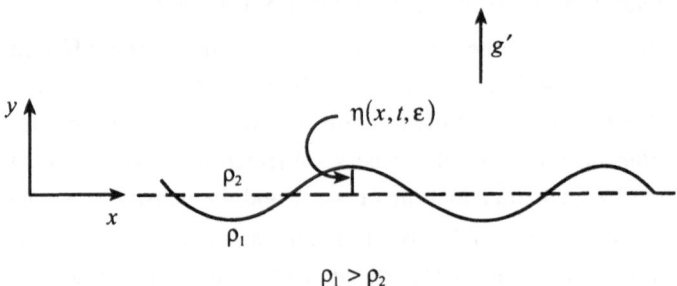

Figure 3.8. The interface accelerated normally and subjected to a perturbation.

Further, one has the following boundary conditions at the interface:

(i) Kinematic Condition: This expresses the fact that a fluid particle initially on the interface must stay on the interface during the course of perturbation. This requires

$$y = \eta : \eta_t - \eta_x \Phi_{jx} + \Phi_{jy} = 0; \quad j = 1, 2. \tag{3.8.22}$$

(ii) Dynamic Condition: This refers to the force balance across the interface. In order to derive the dynamic condition, note that in the presence of a surface tension T', at the interface, for equilibrium, the normal stresses acting on the two sides of an interfacial element dS' must differ by an amount given by

$$\left[\left(-p'\delta_{ik} \right)_1 - \left(-p'\delta_{ik} \right)_2 \right] \hat{n}_k = -T' \left(\frac{1}{R_1'} \right)$$

where p' is the normal pressure exerted by the fluid; $\hat{n}_k$ is the outward normal to dS', and R_1' is the principal radius of curvature of dS'. The convention regarding signs is that R_1' is to be considered positive if the corresponding center of curvature lies on side 1 of dS'. For the case under consideration, note that

$$\frac{1}{R_1'} = \eta'_{x'x'} \left(1 + \eta'^2_{x'} \right)^{-3/2}.$$

The dynamic condition at the interface becomes

$$y = \eta : \Phi_{1t} - \frac{1}{2} \left(\nabla \Phi_1 \right)^2 + \eta - s \left[\Phi_{2t} - \frac{1}{2} \left(\nabla \Phi_2 \right)^2 + \eta \right] = -k^2 \eta_{xx} \left(1 + \eta_x^2 \right)^{-3/2}, \tag{3.8.23}$$

where

$$k^2 \equiv \left(\frac{2\pi}{\lambda'}\right)\frac{T'}{\rho_1' g'}, \quad s = \frac{\rho_2'}{\rho_1'}$$

and the initial conditions are

$$t = 0 : \eta = \varepsilon \cos x, \quad \eta_t = 0, \tag{3.8.24}$$

where

$$\varepsilon \equiv a'\left(2\pi/\lambda'\right).$$

The conditions at infinity are

$$y \to (-1)^j \infty : \Phi_{jx} \to 0. \tag{3.8.25}$$

All quantities in equations (3.8.21)–(3.8.25) have been nondimensionalized with respect to a reference length $\lambda'/2\pi$, a time $(\lambda'/2\pi g')^{1/2}$, where primes denote dimensional quantities.

Make a change of variable

$$\tau = \sigma t, \tag{3.8.26}$$

so that the boundary-value problem (3.8.21)–(3.8.23) becomes

$$y \lessgtr \eta : \nabla^2 \Phi_j = 0; \quad j = 1, 2 \tag{3.8.27}$$

$$y = \eta : \sigma \eta_\tau - \eta_x \Phi_{jx} + \Phi_{jy} = 0 \tag{3.8.28}$$

$$y = \eta : \sigma \Phi_{1\tau} - \frac{1}{2}\left(\nabla \Phi_1\right)^2 + \eta - s\left[\sigma \Phi_{2\tau} - \frac{1}{2}\left(\nabla \Phi_2\right)^2 + \eta\right]$$

$$= -k^2 \eta_{xx}\left(1 + \eta_x^2\right)^{-3/2}. \tag{3.8.29}$$

Upon linearizing the system (3.8.27)–(3.8.29), (3.8.24) and (3.8.25) we obtain

$$y \lessgtr 0 : \nabla^2 \phi_j = 0; \quad j = 1, 2 \tag{3.8.30}$$

$$y = 0 : \sigma \eta_\tau + \phi_{jy} = 0 \tag{3.8.31}$$

$$y = 0 : \sigma \phi_{1\tau} + \eta - s\left(\sigma \phi_{2\tau} + \eta\right) + k^2 \eta_{xx} = 0 \tag{3.8.32}$$

$$y \to (-1)^j \infty : \phi_{jy} \to 0 \tag{3.8.33}$$

$$\tau = 0 : \eta = \cos x, \quad \eta_\tau = 0. \tag{3.8.34}$$

From (3.8.30), (3.8.31), (3.8.33), and (3.8.34), one obtains

$$\eta(x, \tau) = \cos x \cdot \cos \tau \tag{3.8.35}$$

$$\phi_j(x,y,\tau) = -(-1)^j \cos x \cdot \sin \tau \cdot e^{-(-1)^j y}. \tag{3.8.36}$$

Using (3.8.35) and (3.8.36), equation (3.8.32) gives the dispersion relation

$$\sigma^2 = \frac{k^2 + s - 1}{s + 1}. \tag{3.8.37}$$

Note from (3.8.37) that –

(i) in the absence of the surface tension, the interface is stable or unstable according as $\rho_1' \lessgtr \rho_2'$;

(ii) in the presence of the surface tension, the interface is stable or unstable according as $k' \gtrless k_c'$ where

$$k_c' = \sqrt{\frac{g'(\rho_1' - \rho_2')}{T'}}. \tag{3.8.38}$$

(b) The Nonlinear Problem

When the interface is unstable according to the linear theory, the evolution of the interface quickly gets out of the linear regime because the predicted growth is exponential. It turns out that the nonlinear problem of instability of a vertically accelerated horizontal two-dimensional interface between two different fluids is one of singular perturbation type. We therefore use the method of strained parameters to develop a uniformly valid solution for the above problem for wave numbers near the linear cut-off value (Nayfeh, 1969 and Shivamoggi, 1979).

A few remarks about the issue of transferring the boundary conditions to known boundaries in a perturbation scheme are in order. Often a boundary condition is imposed at a boundary, which is unknown without the solution and varies slightly with the perturbation parameter ε. A knowledge of the way in which the solution varies in the vicinity of the basic configuration of the boundary corresponding to $\varepsilon = 0$ is necessary in order to effect the transfer of a boundary condition. If the solution is analytic in spatial coordinates, this transfer is simply accomplished by an expansion in a Taylor's series about the values at the basic configuration. If this is not possible, ε will appear both implicitly and explicitly in the perturbation expansion so that the resulting operations will not be consistent.

Seek solutions to the boundary-value problem (3.8.27)–(3.8.29), (3.8.24) and (3.8.25), of the form, for wavenumbers near the linear cut-off value k_c,

$$\Phi_j(x, y, \tau; \varepsilon) \sim \sum_{n=1}^{\infty} \varepsilon^n \, \phi_j^{(n)}(x, y, \tau) \tag{3.8.39}$$

$$\eta(x, \tau; \varepsilon) = \sum_{n=1}^{\infty} \varepsilon^n \, \eta_n(x, \tau) \tag{3.8.40}$$

$$\sigma(k; \varepsilon) = \sum_{n=1}^{\infty} \varepsilon^{n-1} \, \sigma_n(k) \tag{3.8.41}$$

$$k^2(\varepsilon) = k_c^2 + \varepsilon^2 K + O(\varepsilon^3). \tag{3.8.42}$$

In (3.8.42), one could include an $O(\varepsilon)$ term on the right-hand side, but it turns out to be zero anyway.

One obtains upon substitution of (3.8.39)–(3.8.42) into (3.8.27)–(3.8.29), (3.8.24) and (3.8.25):

$$O(\varepsilon): y \leqq 0 : \phi_{jxx}^{(1)} + \phi_{jyy}^{(1)} = 0 \tag{3.8.43}$$

$$y = 0 : \sigma_1 \, \eta_{1\tau} + \phi_{jy}^{(1)} = 0 \tag{3.8.44}$$

$$y = 0 : \sigma_1 \, \phi_{1\tau}^{(1)} + \eta_1 - s\left(\sigma_1 \, \phi_{2\tau}^{(1)} + \eta_1\right) + k_c^2 \, \eta_{1xx} = 0 \tag{3.8.45}$$

$$y \to (-1)^j \infty : \phi_{jy}^{(1)} \to 0 \tag{3.8.46}$$

$$\tau = 0 : \eta_1 = \cos x, \quad \eta_{1\tau} = 0 \tag{3.8.47}$$

$$O(\varepsilon^2): y \leqq 0 : \phi_{jxx}^{(2)} + \phi_{jyy}^{(2)} = 0 \tag{3.8.48}$$

$$y = 0 : \sigma_1 \, \eta_{2\tau} + \phi_{jy}^{(2)} = \eta_{1x} \, \phi_{jx}^{(1)} - \phi_{jyy}^{(1)} \, \eta_1 - \sigma_2 \, \eta_{1\tau} \tag{3.8.49}$$

$$y = 0 : \sigma_1 \, \phi_{1\tau}^{(2)} + \eta_2 - s\left(\sigma_1 \, \phi_{2\tau}^{(2)} + \eta_2\right) + k_c^2 \, \eta_{2xx} = \frac{1}{2}\left[\left(\phi_{1x}^{(1)}\right)^2 + \left(\phi_{1y}^{(1)}\right)^2\right]$$
$$- s\frac{1}{2}\left[\left(\phi_{2x}^{(1)}\right)^2 + \left(\phi_{2y}^{(1)}\right)^2\right] - \sigma_1 \, \phi_{1\tau y}^{(1)} \, \eta_1 - \sigma_2 \, \phi_{1\tau}^{(1)}$$
$$+ s\left(\sigma_1 \, \phi_{2\tau y}^{(1)} \, \eta_1 + \sigma_2 \, \phi_{2\tau}^{(1)}\right) \tag{3.8.50}$$

$$y \to (-1)^j \infty : \phi_{jy}^{(2)} \to 0 \tag{3.8.51}$$

$$\tau = 0 : \eta_2 = 0, \quad \eta_{2\tau} = 0 \tag{3.8.52}$$

$$O(\varepsilon^3): y \leqq 0 : \phi_{jxx}^{(3)} + \phi_{jyy}^{(3)} = 0 \tag{3.8.53}$$

$$y = 0 : \sigma_1 \, \eta_{3\tau} + \phi_{jy}^{(3)} = -\phi_{jyy}^{(1)} \, \eta_2 - \phi_{jyy}^{(2)} \, \eta_1 - \frac{1}{2} \, \phi_{jyyy}^{(1)} \, \eta_1^2 + \phi_{jx}^{(2)} \, \eta_{1x}$$
$$+ \phi_{jxy}^{(1)} \, \eta_1 \, \eta_{1x} + \phi_{jx}^{(1)} \, \eta_{2x} - \sigma_2 \, \eta_{2\tau} - \sigma_3 \, \eta_{1\tau} \tag{3.8.54}$$

$$y = 0 : \sigma_1 \phi_{1\tau}^{(3)} + \eta_3 - s\left(\sigma_1 \phi_{2\tau}^{(3)} + \eta_3\right) + k_c^2 \eta_{3xx} = -\sigma_2\left(\phi_{1\tau}^{(2)} + \phi_{1\tau y}^{(1)} \eta_1\right)$$

$$-\sigma_3 \phi_{1\tau}^{(1)} - \sigma_1\left(\phi_{1\tau y}^{(2)} \eta_1 + \phi_{1\tau y}^{(1)} \eta_2 + \frac{1}{2} \phi_{1\tau yy}^{(1)} \eta_1^2\right) + s\left[\sigma_2\left(\phi_{2\tau}^{(2)} + \phi_{2\tau y}^{(1)} \eta_1\right)\right.$$

$$-\sigma_3 \phi_{2\tau}^{(1)} - \sigma_1\left(\phi_{2\tau y}^{(2)} \eta_1 + \phi_{2\tau y}^{(1)} \eta_2 + \frac{1}{2} \phi_{2\tau yy}^{(1)} \eta_1^2\right)\right] + \left(\phi_{1x}^{(1)} \phi_{1x}^{(2)}\right.$$

$$\left. + \phi_{1x}^{(1)} \phi_{1xy}^{(1)} \eta_1 + \phi_{1y}^{(1)} \phi_{1y}^{(2)} + \phi_{1y}^{(1)} \phi_{1yy}^{(1)} \eta_1\right) - s\left(\phi_{2x}^{(2)} \phi_{2x}^{(2)} + \phi_{2x}^{(1)} \phi_{2xy}^{(1)} \eta_1\right.$$

$$\left. + \phi_{2y}^{(1)} \phi_{2y}^{(2)} + \phi_{2y}^{(1)} \phi_{2yy}^{(1)} \eta_1\right) + \frac{3}{2} k_c^2 \eta_{1xx} \eta_{1x}^2 - K\eta_{1xx} \tag{3.8.55}$$

$$y \to (-1)^j \infty : \phi_{jy}^{(3)} \to 0 \tag{3.8.56}$$

$$\tau = 0 : \eta_3 = 0, \qquad \eta_{3\tau} = 0. \tag{3.8.57}$$

etc.

One obtains for the $O(\varepsilon)$ problem, the linearized solution –

$$\eta_1(x, \tau) = \cos x \cdot \cos \tau \tag{3.8.58}$$

$$\phi_j^{(1)}(x, y, \tau) = -(-1)^j \cos x \cdot \sin \tau \cdot e^{-(-1)^j y} \tag{3.8.59}$$

$$\sigma_1^2 = \frac{k_c^2 + s - 1}{s + 1}, \tag{3.8.60}$$

the linear cut-off wave number k_c being given by

$$\sigma_1^2 = 0 \quad \text{or} \quad k_c^2 + s - 1 = 0. \tag{3.8.38}$$

One may expect to construct uniformly valid solutions only for wavenumbers larger than k_c. But it turns out, upon a consideration of the nonlinear problem in the following, that this is possible only for wavenumbers greater than k_c by a definite amount, namely, $\varepsilon^2 K$.

For wavenumbers near k_c, using (3.8.58) and (3.8.59), the system (3.8.48)–(3.8.52) becomes

$$y \lessgtr 0 : \phi_{jxx}^{(2)} + \phi_{jyy}^{(2)} = 0 \tag{3.8.61}$$

$$y = 0 : \phi_{jy}^{(2)} = \sigma^2 \cos x \cdot \sin \tau \tag{3.8.62}$$

$$y = 0 : (1 - s) \eta_2 + k_c^2 \eta_{2xx} = 0 \tag{3.8.63}$$

$$y \to (-1)^j \infty : \phi_{jy}^{(2)} \to 0 \tag{3.8.64}$$

$$\tau = 0 : \eta_2 = 0, \qquad \eta_{2\tau} = 0 \tag{3.8.65}$$

from which

$$\eta_2(x,\tau) = \cos x \cdot (1 - \cos \tau) \tag{3.8.66}$$

$$\phi_j^{(2)}(x,y,\tau) = -(-1)^j \sigma_2 \cos x \cdot \sin \tau \cdot e^{-(-1)^j y}, \tag{3.8.67}$$

where σ_2 is now non-zero, but is arbitrary in the $O(\varepsilon^2)$ problem.

Next, using (3.8.58), (3.8.59), (3.8.66), and (3.8.67), the system (3.8.53)–(3.8.57) becomes

$$y \lessgtr 0 : \phi_{jxx}^{(3)} + \phi_{jyy}^{(3)} = 0 \tag{3.8.68}$$

$$y = 0 : \phi_{jy}^{(3)} = -\frac{\sigma_2}{2} \cos 2x \cdot \sin 2\tau - (\sigma_2 - \sigma_3) \cos x \cdot \sin \tau \tag{3.8.69}$$

$$y = 0 : (1-s) \eta_3 + k_c^2 \eta_{3xx} = \left[-(1+s)\sigma_2^2 + K - \frac{3}{8}(1-s) \right] \cos x \cdot \cos \tau \tag{3.8.70}$$

$$y \to (-1)^j \infty : \phi_{jy}^{(3)} \to 0 \tag{3.8.71}$$

$$\tau = 0 : \eta_3 = 0, \quad \eta_{3\tau} = 0. \tag{3.8.72}$$

The removal of the secular terms in (3.8.70) requires

$$(1+s)\sigma_2^2 - K + \frac{3}{8}(1-s) = 0$$

or

$$\sigma_2 = \pm \left[\frac{K - \frac{3}{8}(1-s)}{1+s} \right]^{1/2} \tag{3.8.73}$$

so that one has

$$\left. \begin{array}{l} K < \dfrac{3}{8}(1-s) : \text{instability} \\[2mm] K = \dfrac{3}{8}(1-s) : \text{neutral stability} \\[2mm] K > \dfrac{3}{8}(1-s) : \text{stability} \end{array} \right\}$$

and corresponding to neutral stability, one has

$$k^2 = k_c^2 + \frac{3}{8}(1-s)\varepsilon^2 + O(\varepsilon^3) \tag{3.8.74}$$

which is graphically represented in Figure 3.9.

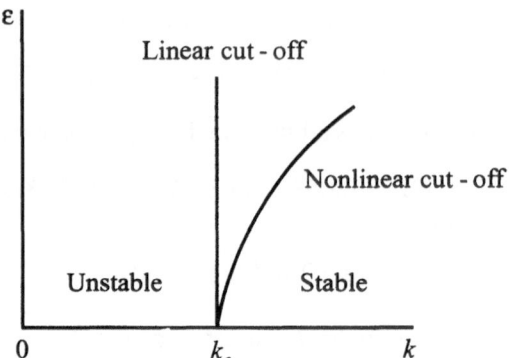

Figure 3.9. Linear- and nonlinear-cutoffs for Rayleigh-Taylor instability.

It is to be noted that the interfacial waves grow even at $k = k_c$ despite the cutoff predicted by the linear theory. On the other hand, the onset of instability even below the linear stability threshold when the disturbance has finite amplitude implies that this instability is a *subcritical* instability.

3.9. Applications to Plasma Physics

3.9.1 Nonlinear Waves in an Electron-Plasma

Consider a one-dimensional wave motion in a warm isothermal electron plasma, with the ions forming an immobile neutralizing background (Infeld and Rowlands,1979; Shivamoggi 1988a). The equations governing such a wave motion are:

(i)   conservation of mass:

$$\frac{\partial n}{\partial t} + \frac{\partial}{\partial x}(nv) = 0 \qquad (3.9.1)$$

(ii) conservation of momentum:

$$\frac{\partial v}{\partial t} + v\frac{\partial v}{\partial x} = -\frac{1}{n}\frac{\partial n}{\partial x} + \frac{\partial \phi}{\partial x} \qquad (3.9.2)$$

(iii) Gauss' law:

$$\frac{\partial^2 \phi}{\partial x^2} = n - 1 \qquad (3.9.3)$$

where n is the electron number density normalized by the mean number density n_0, v is the electron fluid velocity non-dimensionalized by the electron thermal speed $V_{T_e} = \sqrt{KT_e/m_e}$, ϕ is the electric potential normalized by T_e/e, the time t is normalized by ω_p^{-1} and x by the Debye length $\lambda_D = V_{T_e}/\omega_p$. T_e is the temperature of the electron plasma and K is the Boltzmann constant.

If we look for a stationary wave traveling in the x-direction, the various physical quantities will depend on x and t only through the combination $\xi = kx - \omega t$. Then on putting

$$n = 1 + N \tag{3.9.4}$$

we may derive from equations (3.9.1)-(3.9.3):

$$\left[\frac{\omega^2}{(1+N)^3} - \frac{k^2}{1+N}\right]\frac{\partial^2 N}{\partial \xi^2} + \left[\frac{k^2}{(1+N)^2} - \frac{3\omega^2}{(1+N)^4}\right]\left(\frac{\partial N}{\partial \xi}\right) + N = 0. \tag{3.9.5}$$

Let us now introduce a small parameter $\varepsilon \ll 1$, which may characterize a typical wave amplitude, and seek solutions to equation (3.9.5) of the form

$$\left.\begin{aligned}
N(\xi;\varepsilon) &\sim \sum_{n=1}^{\infty} \varepsilon^n N_n(\xi) \\
\omega(k;\varepsilon) &= \sum_{n=0}^{\infty} \varepsilon^n \omega_n(k)
\end{aligned}\right\}, \tag{3.9.6}$$

Using (3.9.6), equation (3.9.5) gives

$$O(\varepsilon): \left(\omega_0^2 - k^2\right)\frac{\partial^2 N_1}{\partial \xi^2} + N_1 = 0 \tag{3.9.7}$$

$$O(\varepsilon^2): \left(\omega_0^2 - k^2\right)\frac{\partial^2 N_2}{\partial \xi^2} + N_2 = \left(3\omega_0^2 - K^2\right)\left[N_1\frac{\partial^2 N_1}{\partial \xi^2} + \left(\frac{\partial N_1}{\partial \xi}\right)^2\right]$$

$$-2\omega_0\omega_1\frac{\partial^2 N_1}{\partial \xi^2} \tag{3.9.8}$$

$$O(\varepsilon^3): \left(\omega_0^2 - k^2\right)\frac{\partial^2 N_3}{\partial \xi^2} + N_3 = \left(3\omega_0^2 - k^2\right)\left[N_2\frac{\partial^2 N_1}{\partial \xi^2} + N_1\frac{\partial^2 N_2}{\partial \xi^2} + 2\frac{\partial N_1}{\partial \xi}\frac{\partial N_2}{\partial \xi}\right]$$

$$-\left(6\omega_0^2 - k^2\right)\left[N_2\frac{\partial^2 N_1}{\partial \xi^2} + 2N_1\left(\frac{\partial N_1}{\partial \xi}\right)^2\right]$$

$$-\left(2\omega_0\omega_2 + \omega_1^2\right)\frac{\partial^2 N_1}{\partial \xi^2} - 2\omega_0\omega_1\frac{\partial^2 N_2}{\partial \xi^2}. \tag{3.9.9}$$

etc.

We obtain from equation (3.9.7), the linear result:

$$\left.\begin{array}{l} N_1 = a_0 \cos\xi \\ \omega_0^2 = 1 + k^2 \end{array}\right\}. \tag{3.9.10}$$

Using (3.9.10), the removal of secular terms on the right-hand side of equation (3.9.8) requires

$$\omega_1 \equiv 0 \tag{3.9.11}$$

and then the solution to equation (3.9.8) is given by

$$N_2 = \frac{a_0^2}{3}\left(3 + 2k^2\right)\cos 2\xi. \tag{3.9.12}$$

Using (3.9.10)-(3.9.12), the condition for the removal of secular terms on the right-hand side of the equation (3.9.9) gives

$$\omega_2 = \frac{k^2\left(8k^2 + 9\right)}{24\sqrt{1+k^2}}\,a_0^2. \tag{3.9.13}$$

Thus, the nonlinear dispersion relation is given by

$$\omega^2 = \left(1 + k^2\right) + \varepsilon^2\,\frac{k^2}{12}\left(8k^2 + 9\right)a_0^2 + 0\left(\varepsilon^3\right). \tag{3.9.14}$$

3.9.2 Rayleigh-Taylor Instability of a Plasma in a Magnetic Field

Consider wave motions at the surface of an incompressible, infinitely-conducting plasma of infinite depth supported by a magnetic field H', (Figure 3.10).

If the linear problem can be used as a guide it (see Chandrasekhar, 1961 and Shivamoggi, 1986) shows that –

(i)   the magnetic field affects the development of Rayleigh-Taylor instability only if the disturbances propagate along the magnetic field;

(ii)   the plasma motion is irrotational.

Though the validity of these results for the nonlinear problem is not established, we shall, in the following, for the sake of simplicity, consider only wave motions parallel to the magnetic field and take the plasma motion to be irrotational (Shivamoggi, 1982a).

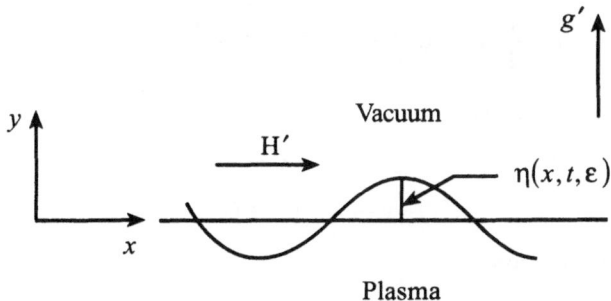

**Figure 3.10. The plasma surface accelerated normally
and subjected to a perturbation.**

The applied acceleration g' is directed away from the plasma. Initially, the surface of the plasma is taken to be disturbed according to a simple sinusoidal standing wave with an amplitude a' and a wavelength $\lambda'/2\pi$. Nondimensionalize the various physical quantities with respect to a reference length $\lambda'/2\pi$, a time $(\lambda'/2\pi g')^{1/2}$ and a magnetic field H'. Here the primes denote the dimensional quantities. Introduce the velocity potential ϕ and the magnetic potential ψ for the perturbations according to

$$\mathbf{v} = -\nabla\phi, \quad \mathbf{h} = -\nabla\psi. \tag{3.9.15}$$

If $y = \eta(x,t;\varepsilon)$, where $\varepsilon = a'(2\pi/\lambda')$, denotes the disturbed shape of the plasma surface, one has the following boundary-value problem.

(i) Laplace equation for the velocity potential –

$$y < \eta : \phi_{zz} + \phi_{yy} = 0. \tag{3.9.16}$$

(ii) Laplace equation for the magnetic-field potential –

$$y > \eta : \psi_{xx} + \psi_{yy} = 0. \tag{3.9.17}$$

(iii) Kinematic condition at the plasma surface –

$$y = \eta : \eta_t - \eta_x\phi_x + \phi_y = 0. \tag{3.9.18}$$

(iv) Frozen-in field constraint at the plasma surface –

$$y = \eta : -\eta_x + \eta_x\psi_x - \psi_y = 0. \tag{3.9.19}$$

(v) Dynamic condition at the plasma surface –

$$\phi_t - \frac{1}{2}\left(\phi_x^2 + \phi_y^2\right) + \eta = k^2\left(-\psi_x + \frac{1}{2}\psi_x^2 + \frac{1}{2}\psi_y^2\right). \qquad (3.9.20)$$

(vi) Infinity conditions –

$$y \to -\infty : \phi \to 0 \qquad (3.9.21)$$

$$y \to \infty : \psi \to 0. \qquad (3.9.22)$$

(vii) Initial condition –

$$t = 0 : \eta = \varepsilon \cos x, \quad \eta_t = 0. \qquad (3.9.23)$$

Here

$$k^2 = \frac{H'^2}{\rho' g'}\left(\frac{2\pi}{\lambda'}\right)^2,$$

ρ' being the mass density of the plasma. Note that (3.9.19) expresses the fact that the plasma surface remains a magnetic-field line even in a perturbed state.[2]

Make a change of variable

$$\tau = \sigma t \qquad (3.9.24)$$

so that the boundary-value problem (3.9.16)-(3.9.23) becomes

$$y < \eta : \phi_{xx} + \phi_{yy} = 0 \qquad (3.9.25)$$

$$y > \eta : \psi_{xx} + \psi_{yy} = 0 \qquad (3.9.26)$$

$$y = \eta : \sigma\eta_\tau - \eta_x\phi_x + \phi_y = 0 \qquad (3.9.27)$$

$$-\eta_x + \eta_x\psi_x - \psi_y = 0 \qquad (3.9.28)$$

$$\sigma\phi_\tau - \frac{1}{2}\left(\phi_x^2 + \phi_y^2\right) + \eta = k^2\left(-\psi_x + \frac{1}{2}\psi_x^2 + \frac{1}{2}\psi_y^2\right) \quad (3.9.29)$$

$$y \to -\infty : \phi \to 0 \qquad (3.9.30)$$

$$y \to \infty : \psi \to 0 \qquad (3.9.31)$$

$$\tau = 0 : \eta = \varepsilon \cos x, \quad \eta_\tau = 0. \qquad (3.9.32)$$

For wavenumbers k near the linear cut-off value k_c, seek solutions to (3.9.25)-(3.9.32) of the form (Shivamoggi, 1982a)

---

[2] The variational basis of the governing equations (3.9.16) and (3.9.17) and the boundary conditions (3.9.18)-(3.9.22) for this problem was given by Shivamoggi (1983).

$$\phi(x,y,\tau;\varepsilon) \sim \sum_{n=1}^{\infty} \varepsilon^n \phi_n(x,y,\tau) \tag{3.9.33}$$

$$\psi(x,y,\tau;\varepsilon) \sim \sum_{n=1}^{\infty} \varepsilon^n \psi_n(x,y,\tau) \tag{3.9.34}$$

$$\eta(x,t;\varepsilon) \sim \sum_{n=1}^{\infty} \varepsilon^n \eta_n(x,\tau) \tag{3.9.35}$$

$$\sigma(k;\varepsilon) = \sum_{n=1}^{\infty} \varepsilon^{n-1} \sigma_n(k) \tag{3.9.36}$$

$$k^2(\varepsilon) = k_e^2 + \varepsilon^2 \kappa + O(\varepsilon^3). \tag{3.9.37}$$

In (3.9.37) one could include an $O(\varepsilon)$ term on the right-hand side, but it turns out to be zero anyway.

One obtains upon substitution of (3.9.33)-(3.9.37) into (3.9.25)-(3.9.32), and equating coefficients of equal powers of ε, a hierarchy of problems of various orders in ε:

$$O(\varepsilon): y<0: \phi_{1xx} + \phi_{1yy} = 0 \tag{3.9.38}$$

$$y>0: \psi_{1xx} + \psi_{1yy} = 0 \tag{3.9.39}$$

$$y=0: \sigma_1 \eta_{1\tau} + \phi_{1y} = 0 \tag{3.9.40}$$

$$\psi_{1y} + \eta_{1x} = 0 \tag{3.9.41}$$

$$\sigma_1 \phi_{1\tau} + \eta_1 + k_c^2 \psi_{1x} = 0 \tag{3.9.42}$$

$$y \to -\infty: \phi_1 \to 0 \tag{3.9.43}$$

$$y \to +\infty: \psi_1 \to 0 \tag{3.9.44}$$

$$\tau=0: \eta_1 = \cos x, \quad \eta_{1\tau} = 0. \tag{3.9.45}$$

$$O(\varepsilon^2): y<0: \phi_{2xx} + \phi_{2yy} = 0 \tag{3.9.46}$$

$$y>0: \psi_{2xx} + \psi_{2yy} = 0 \tag{3.9.47}$$

$$y=0: \sigma_1 \eta_{2\tau} + \phi_{2y} = \eta_{1x}\phi_{1x} - \phi_{1yy}\eta_1 - \sigma_2 \eta_{1\tau} \tag{3.9.48}$$

$$\psi_{2y} + \eta_{2x} = \eta_{1x}\psi_{1x} - \psi_{1yy}\eta_1 \tag{3.9.49}$$

$$\sigma_1\phi_{2\tau} + \eta_2 + k_c^2\,\psi_{2x} = \frac{1}{2}\left(\phi_{1x}^2 + \phi_{1y}^2\right) - \sigma_2\phi_{1\tau} - \sigma_1\phi_{1y}\eta_1$$

$$+ k_c^2\left[-\psi_{1xy}\eta_1 + \frac{1}{2}\left(\psi_{1x}^2 + \psi_{1y}^2\right)\right] \qquad (3.9.50)$$

$$y \to -\infty : \phi_2 \to 0 \qquad (3.9.51)$$

$$y \to +\infty : \psi_2 \to 0 \qquad (3.9.52)$$

$$\tau = 0 : \eta_2 = 0, \quad \eta_{2\tau} = 0. \qquad (3.9.53)$$

$$O(\varepsilon^3) : y < 0 : \phi_{3xx} + \phi_{3yy} = 0 \qquad (3.9.54)$$

$$y > 0 : \psi_{3xx} + \psi_{3yy} = 0 \qquad (3.9.55)$$

$$y = 0 : \sigma_1\eta_{3\tau} + \phi_{3y} = -\phi_{1yy}\eta_2 - \phi_{2yy}\eta_1 - \frac{1}{2}\phi_{1yyy}\eta_1^2 + \phi_{2x}\eta_{1x}$$

$$+ \phi_{1x}\eta_{2x} + \phi_{1xy}\eta_1\eta_{1x} - \sigma_2\eta_{2\tau} - \sigma_3\eta_{1\tau} \qquad (3.9.56)$$

$$\psi_{3y} + \eta_{3x} = \eta_{2x}\psi_{1x} + \eta_{1x}\psi_{2x} + \psi_{1xy}\eta_1\eta_{1x} - \psi_{2yy}\eta_1 - \psi_{1yy}\eta_2 - \frac{1}{2}\psi_{1yyy}\eta_1^2$$

$$\qquad (3.9.57)$$

$$\sigma_1\phi_{3\tau} + \eta_3 + k_c^2\,\psi_{3x} = -\sigma_2\left(\phi_{2\tau} + \phi_{1y}\eta_1\right) - \sigma_3\phi_{1\tau} - \sigma_1\left(\phi_{2y}\eta_1 + \phi_{1y}\eta_2\right.$$

$$\left. + \frac{1}{2}\phi_{1yy}\eta_1^2\right) + \left(\phi_{1x}\phi_{2x} + \phi_{1x}\phi_{1xy}\eta_1 + \phi_{1y}\phi_{2y}\right.$$

$$\left. + \phi_{1y}\phi_{1yy}\eta_1\right) + k_c^2\left[-\psi_{1xy}\eta_2 - \psi_{2xy}\eta_1 - \frac{1}{2}\psi_{1xyy}\eta_1^2 + \psi_{1x}\psi_{2x}\right.$$

$$\left. + \psi_{1x}\psi_{1xy}\eta_1 + \psi_{1y}\psi_{2y} + \psi_{1y}\psi_{1yy}\eta_1\right] - \kappa\psi_{1x} \qquad (3.9.58)$$

$$y \to -\infty : \phi_3 \to 0 \qquad (3.9.59)$$

$$y \to +\infty : \psi_3 \to 0 \qquad (3.9.60)$$

$$\tau = 0 : \eta_3 = 0, \quad \eta_{3\tau} = 0 \qquad (3.9.61)$$

etc.

For $k \approx k_c$ where $\sigma_1 = 0$, the system (3.9.38)-(3.9.45) becomes

$$y < 0 : \phi_{1xx} + \phi_{1yy} = 0 \qquad (3.9.62)$$

$$y > 0 : \psi_{1xx} + \psi_{1yy} = 0 \qquad (3.9.63)$$

$$y = 0 : \phi_{1y} = 0 \qquad (3.9.64)$$

$$\psi_{1y} + \eta_{1x} = 0 \qquad (3.9.65)$$

$$\eta_1 + k_c^2 \, \psi_{1x} = 0 \tag{3.9.66}$$

$$y \to -\infty : \phi_1 \to 0 \tag{3.9.67}$$

$$y \to +\infty : \psi_1 \to 0 \tag{3.9.68}$$

$$\tau = 0 : \eta_1 = \cos x, \quad \eta_{1\tau} = 0, \tag{3.9.69}$$

from which one obtains

$$\eta_1 = \cos x \cdot \cos \tau \tag{3.9.70}$$

$$\psi_1 = -e^{-y} \sin x \cdot \cos \tau \tag{3.9.71}$$

$$\phi_1 \equiv 0 \tag{3.9.72}$$

$$k_c^2 = 1. \tag{3.9.73}$$

One may expect to construct uniformly valid solutions only for wavenumbers larger than k_c.

Using (3.9.70)-(3.9.73), the system (3.9.46)-(3.9.53) becomes

$$y < 0 : \phi_{2xx} + \phi_{2yy} = 0 \tag{3.9.74}$$

$$y > 0 : \psi_{2xx} + \psi_{2yy} = 0 \tag{3.9.75}$$

$$y = 0 : \phi_{2y} = \sigma_2 \cos x \cdot \sin \tau \tag{3.9.76}$$

$$\psi_{2y} + \eta_{2x} = \frac{1}{2} \sin 2x \, (1 + \cos 2\tau) \tag{3.9.77}$$

$$\eta_2 + \psi_{2x} = -\frac{1}{4} \cos 2x \cdot (1 + \cos 2\tau) \tag{3.9.78}$$

$$y \to -\infty : \phi_2 \to 0 \tag{3.9.79}$$

$$y \to +\infty : \psi_2 \to 0 \tag{3.9.80}$$

$$\tau = 0 : \eta_2 = 0, \quad \eta_{2\tau} = 0, \tag{3.9.81}$$

from which one obtains

$$\eta_2 = -\frac{1}{4} \cos 2x \, (1 + \cos 2\tau) \tag{3.9.82}$$

$$\psi_2 \equiv 0 \tag{3.9.83}$$

$$\phi_2 = +\sigma_2 \, e^y \cos x \cdot \sin \tau \tag{3.9.84}$$

where σ_2 is nonzero but is arbitrary in the $O(\varepsilon^2)$ problem.

Next, using (3.9.70)-(3.9.73) and (3.9.82)-(3.9.84), the system (3.9.54)-(3.9.61) becomes

$$y < 0 : \phi_{3xx} + \phi_{3yy} = 0 \tag{3.9.85}$$

$$y > 0 : \psi_{3xx} + \psi_{3yy} = 0 \tag{3.9.86}$$

$$y = 0 : \phi_{3y} = \sigma_3 \cos x \cdot \sin \tau - \sigma_2 \cos 2x \sin 2\tau \tag{3.9.87}$$

$$\psi_{3y} + \eta_{3x} = -\frac{3}{2} \sin x \cdot \cos^2 x \cdot \cos^3 \tau +$$

$$- \frac{1}{4} \sin x \cdot \cos \tau \cdot \cos 2x \cdot (1 + \cos 2\tau) +$$

$$+ \frac{1}{2} \cos x \cdot \cos \tau \cdot \sin 2x \cdot (1 + \cos 2\tau) \tag{3.9.88}$$

$$\eta_3 + \psi_{3x} = \left(-\sigma_2^2 + \kappa - \frac{3}{8} \right) \cos x \cdot \cos \tau + \text{Higher Harmonics} \tag{3.9.89}$$

$$y \to -\infty : \phi_3 \to 0 \tag{3.9.90}$$

$$y \to +\infty : \psi_3 \to 0 \tag{3.9.91}$$

$$\tau = 0 : \eta_3 = 0, \quad \eta_{3\tau} = 0. \tag{3.9.92}$$

The removal of the secular terms in (3.9.89) requires

$$\sigma_2^2 - \kappa + \frac{3}{8} = 0$$

or

$$\sigma_2 = \pm \sqrt{\kappa - \frac{3}{8}} \tag{3.9.93}$$

so that one has

$$\kappa < \frac{3}{8} : \text{instability}$$

$$\kappa = \frac{3}{8} : \text{neutral stability}$$

$$\kappa > \frac{3}{8} : \text{stability}$$

and corresponding to neutral stability, one has

$$k^2 = 1 + \frac{3}{8} \varepsilon^2 + O(\varepsilon^3) \tag{3.9.94}$$

which is graphically represented in Figure 3.11.

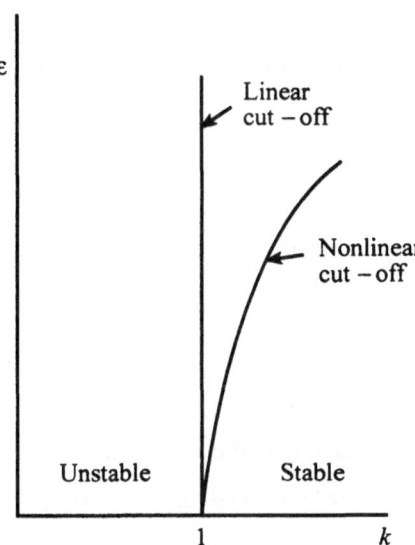

Figure 3.11. Linear and nonlinear cut-offs for
Rayleigh-Taylor instability of a plasma.

It thus appears that the waves on the plasma surface grow even at $k = k_c = 1$, despite the cut-off predicted by the linear theory.

In the present case, we have a fluid, which is subject to a magnetic field, and there is no surface tension. The opposite case wherein the fluid is subject to a surface tension and there is no magnetic field was considered in section 3.8.2, where we found for the cut-off wavenumber the same expression as (3.9.94)! This seems to indicate that the magnetic field and the surface tension have dynamically equivalent effects on the present problem even in the nonlinear case.

3.10. Limitations of the Method of Strained Parameters

Generally, the method of strained parameters succeeds when the singularity predicted by the regular perturbative solution actually exists, though at a slightly different location (Van Dyke, 1975). However, this method, which is effective in determining periodic solutions, is incapable of determining transient responses of dissipative systems because any expansion procedure that is based on approximating $e^{-\varepsilon t}$ by a finite number of terms in a series expansion is doomed to fail for large t!

Example 7: Consider the initial-value problem

$$\left.\begin{array}{l} \dfrac{d^2u}{dt^2} + u = -\varepsilon\dfrac{du}{dt} \\[4mm] t = 0 : u = 1, \quad \dfrac{du}{dt} = 0 \end{array}\right\}. \tag{3.10.1}$$

Let us construct a solution to the initial-value problem (3.10.1) of the form

$$u(t;\varepsilon) \sim u_0(t) + \varepsilon u_1(t) + O(\varepsilon^2). \tag{3.10.2}$$

Then, we obtain from (3.10.1), to various orders in ε:

$$O(1) : \left.\begin{array}{l} \dfrac{d^2u_0}{dt^2} + u_0 = 0 \\[4mm] t = 0 : u_0 = 1, \quad \dfrac{du_0}{dt} = 0 \end{array}\right\} \tag{3.10.3}$$

$$O(\varepsilon) : \left.\begin{array}{l} \dfrac{d^2u_1}{dt^2} + u_1 = -\dfrac{du_0}{dt} \\[4mm] t = 0 : u_1 = 0, \quad \dfrac{du_1}{dt} = 0 \end{array}\right\} \tag{3.10.4}$$

etc.

Solving (3.10.3) and (3.10.4), we obtain

$$u_0 = \cos t \tag{3.10.5}$$

$$u_1 = \frac{1}{2}(-t\cos t + \sin t) \tag{3.10.6}$$

etc.

Using (3.10.5) and (3.10.6), (3.10.2) becomes

$$u \sim \cos t + \varepsilon\frac{1}{2}(-t\cos t + \sin t) + O(\varepsilon^2). \tag{3.10.7}$$

In order to remove the secular term in (3.10.7), we use Pritulo's method and renormalize according to

$$t = s + \varepsilon f_1(s) + O(\varepsilon^2) \tag{3.10.8}$$

so that (3.10.7) becomes

$$u \sim \cos s - \varepsilon f_1 \sin s - \frac{1}{2} \varepsilon s \cos s + \frac{1}{2} \varepsilon \sin s + \cdots. \qquad (3.10.9)$$

In order to remove the secular terms in (3.10.7), one may elect to choose

$$f_1 = -\frac{1}{2} s \cot s \qquad (3.10.10)$$

which is, however, unacceptable because f_1 becomes unbounded for $s = \pi, 2\pi, 3\pi, \ldots$. Thus, the method of strained parameters cannot handle dissipative systems.

3.11. Exercises

1. The equation of motion for a pendulum rotating about its vertical axis with angular velocity Γ is

$$\frac{d^2 \Theta}{dt^2} + \left(\frac{g}{l} - \Gamma^2 \cos \theta \right) \sin \Theta = 0.$$

Use the method of strained parameters to calculate the amplitude-dependent frequency.

2. A point mass moves freely along a parabola $x^2 = 2pz$ $(p > 0)$ rotating about its axis with angular velocity ω. The equation of motion of the mass is

$$\left(1 + \frac{x^2}{p^2} \right) \ddot{x} + \frac{x \dot{x}^2}{p^2} + \left(\frac{g}{p} - \omega^2 \right) x = 0.$$

Solve it using the method of strained parameters.

3. For the Mathieu equation

$$\ddot{u} + \left(\delta + \varepsilon \cos 2t \right) u = 0,$$

use Whittaker's method to determine the solutions in the neighborhood of the transition curve given by

$$\delta = -\frac{1}{8} \varepsilon^2 + O(\varepsilon^3).$$

4. Solve using the eigenfunction expansion method

$$\frac{d^2\varphi}{dx^2} + \lambda\varphi = -\varepsilon\varphi^3,$$

$$\varphi(0) = \varphi(1) = 0.$$

5. Obtain the two-term straightforward expansion of the solution of (Nayfeh, 1973)

$$(x + \varepsilon y)\frac{dy}{dx} + y = 0,$$

$$y(1) = 1.$$

Use Lighthill's method to remove the singularity at $x = 0$.

6. Obtain the two-term regular perturbation expansion of the solution of

$$(x + \varepsilon y)\frac{dy}{dx} + 2y = 2,$$

$$y(1) = 2.$$

Use the method of strained parameters to obtain the uniformly valid one-term expansion

$$y \sim 1 + \frac{C}{s^2} + O(\varepsilon)$$

where,

$$x = s - \varepsilon\left(1 + \frac{C}{3s^2}\right) + O(\varepsilon^2),$$

$$C = 1 + \frac{8}{3}\varepsilon + O(\varepsilon^2).$$

Show that it does not have the problem that the regular perturbation expansion has, as $x \Rightarrow 0$.

7. Solve using the method of strained parameters

$$(1 + \varepsilon u)\, u_x + u_y = 0,$$

$$t = 0 : u = \varepsilon\phi(x).$$

8. Solve using the method of strained parameters

$$u_{tt} - u_{xx} = \varepsilon u_x u_{xx},$$

$$t = 0 : u = f(x) + g(x), \quad u_t = g'(x) - f'(x).$$

9. Consider the weakly nonlinear wave equation

$$u_{tt} - u_{xx} + u + \varepsilon u^3 = 0.$$

For $\varepsilon = 0$, this equation has the solution

$$u_0 = a_0 \sin(\theta + \phi_0), \quad \theta = kx - \sqrt{1 + k^2}\, t$$

for arbitrary constants a_0, ϕ_0 and k. Construct a periodic solution, for $0 < \varepsilon \ll 1$, of the form

$$u(x, t; \varepsilon) \sim a_0 \sin(\Theta + \phi_0) + \varepsilon u_1(\Theta) + \cdots$$

where

$$\Theta = kx - \sqrt{1 + k^2}\ \omega(\varepsilon)t,$$

$$\omega(\varepsilon) = 1 + \varepsilon\omega_1 + \varepsilon^2\omega_2 + \cdots,$$

$$\omega_1 = \frac{3a_0^2}{8(1 + k^2)}, \text{ etc.,}$$

$$u_1 = -\frac{a_0^3}{32} \sin 3(\Theta + \phi_0), \text{ etc.}$$

3.12. Appendix 1

Fredholm's Alternative Theorem

Let L be a bounded linear operator, and suppose we wish to solve the equation –

$$L f = g, \quad x \in [a, b]. \tag{A1.1}$$

We want to know if there is a solution, and if so, how many solutions are possible.

Theorem (Uniqueness). *The solution of equation (A1.1) (if it exists) is unique if and only if the only solution of $L f_0 = 0$ is $f_0 \equiv 0$.*

Proof: Suppose $L f_0 = 0$ for some $f_0 \neq 0$. If $L f = g$, then $f_1 = f + \alpha f_0$ also solves $L f_1 = g$ for any α so that the solution is not unique. Conversely, if solutions of $L f = g$ are not unique, then there are solutions f_1 and f_2 with $f = f_1 - f_2 \neq 0$ satisfying $L f_1 - L f_2 = 0$.

Theorem (Existence): *Equation (A1.1) has a solution if and only if*

$$\langle g, v \rangle = \int_a^b g(x)\, v(x)\, dx = 0 \tag{A1.2}$$

for every $v(x)$ *satisfying* $L^* v = 0$, *where* L^* *is the adjoint of L.*

Proof: Suppose R_L denotes the range of L and N_L denotes the null space of L while $R_L^\perp$ denotes the orthogonal complement of R_L. The proof is done if we show that $R_L^\perp = N_{L^*}$. Suppose $v \in R_L^\perp$. This implies that

$$\langle v, L f \rangle = 0, \quad \forall f \in H,$$

or

$$\langle L^* v, f \rangle = 0,$$

from which

$$L^* v = 0.$$

Therefore,

$$v \in N_{L^*}.$$

Conversely, suppose $v \in N_{L^*}$. Then,

$$\langle L^* v, h \rangle = 0, \quad \forall h \in H$$

or

$$\langle v, L h \rangle = 0,$$

so

$$v \in R_L^\perp.$$

Example A1: Consider

$$A = \begin{pmatrix} 1 & 2 \\ 3 & 6 \end{pmatrix}. \tag{A1.3}$$

$N(A)$ is spanned by the vector –

$$x_0 = \begin{pmatrix} 2 \\ -1 \end{pmatrix} \tag{A1.4}$$

so that solutions of $Ax = b$, if they exist, are not unique. Since N_{A} is spanned by the vector –

$$v = \begin{pmatrix} 3 \\ -1 \end{pmatrix},$$
(A1.5)

solutions of $Ax = b$ exist only if b is of the form

$$b = \alpha \begin{pmatrix} 1 \\ 3 \end{pmatrix}.$$
(A1.6)

This is no surprise since the second row of A is three times the first row.

3.13. Appendix 2

Floquet Theory

Consider non-autonomous linear differential equations with periodic coefficients, i.e.,

$$\left.\begin{array}{c} \dot{x} = a(t)x, \quad a(t+T) = a(t), \\ x(0) = x_0 \end{array}\right\},$$
(A2.1)

where $x \in \mathbb{R}$.

Though $a(t)$ is periodic, $x(t)$ may not be periodic.

We have, from equation (A2.1),

$$x(t) = x_0 \, e^{\int_0^t a(s)\, ds} = \varphi(t)\, x_0,$$
(A2.2)

where

$$\varphi(t) \equiv e^{\int_0^t a(s)\, ds}$$
(A2.3)

is a fundamental solution of the problem.

Theorem: *Let ρ be a Floquet multiplier for the system –*

$$x' = a(t)x, \quad a(t+T) = a(t), \quad \forall t$$
(A2.1)

and let μ be the corresponding Floquet exponent so that $\rho = e^{\mu T}$. Then, there exists a solution $x(t)$ such that

$$x(t+T) = \rho x(t), \ \forall t. \tag{A2.4}$$

Further,

$$x(t) = e^{\mu t} v(t), \ \forall t \tag{A2.5}$$

where $v(t+T) = v(t)$.

Proof: Noting that

$$\int_0^{t+T} a(s) \, ds = \int_0^T a(s) \, ds + \int_T^{t+T} a(s) \, ds$$

$$= \int_0^T a(s) \, ds + \int_T^{t+T} a(s-T) \, ds$$

$$= \int_0^T a(s) \, ds + \int_0^t a(s) \, ds,$$

we have

$$\varphi(t+T) = e^{\int_0^{t+T} a(s) \, ds}$$

$$= e^{\int_0^T a(s) \, ds} \cdot e^{\int_0^t a(s) \, ds} \tag{A2.6}$$

$$= \varphi(T) \, \varphi(t).$$

Thus,

$$x(t+T) = \rho x(t) \tag{A2.7}$$

where ρ is the Floquet multiplier

$$\rho \equiv \varphi(T). \tag{A2.8}$$

If $\rho = e^{\mu T}$, μ being the Floquet exponent, we have

$$\mu = \frac{1}{T} \log \varphi(T) + \frac{2k\pi i}{T} = \frac{1}{T} \int_0^T a(s) \, ds + \frac{2k\pi i}{T} \tag{A2.9}$$

so that μ is determined only up to multiples of $2\pi i/T$.

Further, if we put

$$v(t) = x(t) e^{-\mu t} \tag{A2.5}$$

we have, on using (A2.7) and (A2.8),

$$\begin{aligned}
v(t+T) &= x(t+T) e^{-\mu t} \cdot e^{-\mu T} \\
&= \rho x(t) e^{-\mu t} \cdot e^{-\mu T} \\
&= x(t) e^{-\mu t} = v(t) \tag{A2.10}
\end{aligned}$$

so that $v(t)$ is periodic and bounded.

The Floquet exponents enable one to determine the stability of the origin (which is a stationary point). If $\operatorname{Re}(\mu) < 0$, solutions tend to zero, while if $\operatorname{Re}(\mu) > 0$, solutions become unbounded as $t \Rightarrow \infty$.

Example A2: Consider

$$\dot{x} = (\delta + \cos t) \, x.$$

Then, we have

$$a(t) = \delta + \cos t, \quad a(t+2\pi) = a(t).$$

Noting

$$\int_0^{2\pi} a(s) \, ds = 2\pi\delta,$$

we have that the Floquet exponent is

$$\mu = \frac{1}{2\pi} \int_0^{2\pi} a(s) \, ds = \delta.$$

Therefore, the origin is stable if $\delta < 0$ and unstable if $\delta > 0$.

Let us now generalize the above discussion to consider the case $x \in \mathbb{R}^n$,

$$\dot{x} = A(t) \cdot x, \quad A(t+T) = A(t). \tag{A2.11}$$

Definition: Let $x^1(t), \cdots, x^n(t)$ be n linearly independent solutions of

$$\dot{x} = A(T)\, x \tag{A2.11}$$

on $t \in \mathcal{G}$, and put

$$\Phi(t) = \left[x^1(t), \cdots, x^n(t) \right] \tag{A2.12}$$

where $\Phi(t)$ is an $n \times n$ matrix solution of

$$\dot{\Phi} = A\Phi. \tag{A2.13}$$

Then, Φ is called a fundamental matrix, and if $\Phi(t_0) = I$, I being the unit matrix, then $\Phi(t)$ is the principal fundamental matrix. We have

$$x(t) = \Phi(t) \cdot x_0. \tag{A2.14}$$

Further, if we introduce

$$W(t) = det\Phi(t), \tag{A2.15}$$

$W(t)$ is called the Wronskian.

Lemma: *The Wronskian $W(t)$ is given by*

$$W(t) = W(t_0)\, e^{\int_{t_0}^{t} tr A(s)\, ds} \tag{A2.16}$$

Proof: Let us write

$$\Phi(t) = \Phi(t_0) + (t - t_0)\, \Phi'(t_0) + o(t - t_0)$$

or

$$\Phi(t) = \Phi(t_0) + (t - t_0)\, A(t_0)\, \Phi(t_0) + o(t - t_0). \tag{A2.17}$$

Using (A2.15), we then have

$$W(t) = W(t_0)\, det\left[I + (t - t_0)\, A(t_0) + o(t - t_0) \right]$$

or

$$W(t) = W(t_0)\left[1 + (t - t_0)\, tr A(t_0) \right] + o(t - t_0). \tag{A2.18)a}$$

Writing

$$W(t) = W(t_0) + (t - t_0)\, W'(t_0) + o(t - t_0), \tag{A2.18)b}$$

we have, from (A2.18)a,

$$\dot{W}(t_0) = W(t_0)\, tr A(t_0). \tag{A2.19}$$

So
$$\dot{W} = W \, tr \, A, \tag{A2.20}$$

from which
$$W(t) = W(t_0) \, e^{\int_{t_0}^{t} tr \, A(s) \, ds} \tag{A2.16}$$

Theorem: *Let* $\Phi(t)$ *be a fundamental matrix for*
$$\dot{x} = A(t)x, \quad A(t+T) = A(t), \quad \forall t. \tag{A2.11}$$

Then, $\Phi(t+T)$ *is also a fundamental matrix, and there exists a non-singular constant matrix B such that*
$$\Phi(t+T) = \Phi(t) \, B, \quad \forall t. \tag{A2.21}$$

Also,
$$detB = e^{\int_{0}^{T} tr \, A(s) \, ds} \tag{A2.22}$$

Proof: Since $\Phi(t)$ is a fundamental matrix, we have
$$\dot{\Phi} = A(t) \, \Phi. \tag{A2.13}$$

Let
$$\Psi(t) \equiv \Phi(t+T). \tag{A2.23}$$

Then
$$\dot{\Psi}(t) = \dot{\Phi}(t+T) \tag{A2.24}$$

and using equation (A2.13), we have
$$\dot{\Psi}(t) = A(t+T) \, \Phi(t+T) = A(t)\Psi(t). \tag{A2.25}$$

Equation (A2.25) implies that $\Psi(t)$ is also a fundamental matrix, and hence,
$$\Psi(t) \equiv \Phi(t+T) = \Phi(t) \, B \tag{A2.26}$$

for some constant non-singular matrix B.

We have, from (A2.15),

$$W(t+T) = W(0)\left(e^{\int_0^t tr\,A(s)\,ds} \int_t^{t+T} tr\,A(s)\,ds \right). \tag{A2.27}$$

If we write

$$W(t+T) = W(t)\det B, \tag{A2.28}$$

we then have from (A2.27),

$$\det B = e^{\int_t^{t+T} tr\,A(s)\,ds} = e^{\int_0^T tr\,A(s)\,ds}. \tag{A2.22}$$

Equation (A2.21) implies that

$$B = \Phi^{-1}(0)\Phi(T). \tag{A2.29}$$

If $\Phi(t)$ is the principal fundamental matrix, so $\Phi(0) = I$, then we have

$$B = \Phi(T). \tag{A2.30}$$

Definition: Let the eigenvalues of B be $\rho_1, \cdots, \rho_n$, which are the Floquet multipliers. If $\rho_i = e^{\mu_i T}$; $i = 1, \cdots, n$, then μ_i are the Floquet exponents.

Note that

$$\mu_i = \frac{1}{T}\log \rho_i + \frac{2k\pi i}{T}; \quad k = 1, 2, \cdots. \tag{A2.31}$$

So, μ_i are only determined up to multiples of $2\pi i/T$.

Let $e_1, \cdots, e_n$ be the eigenvectors of B. Writing

$$x_0 = \sum_i a_i e_i, \tag{A2.32}$$

we have

$$x(t+T) = \Phi(t+T) \cdot x_0$$

$$= \Phi(t) \cdot \Phi(T) \cdot \left(\sum_i a_i e_i \right) \tag{A2.33}$$

$$= \Phi(t) \cdot \left(\sum_i e^{\mu_i T} a_i e_i \right).$$

Thus, if

$$x_k(t) = \Phi(t) \cdot a_k e_k, \quad 1 \le k \le n, \tag{A2.34}$$

we obtain

$$x_k(t+T) = \Phi(t) \cdot a_k \, e^{\mu_k T} \, e_k. \tag{A2.35}$$

If we introduce

$$v_k(t) = e^{-\mu_k t} x_k(t), \tag{A2.36}$$

we have from (A2.35),

$$
\begin{aligned}
v_k(t+T) &= e^{-\mu_k(t+T)} x_k(t+T) \\
&= e^{-\mu_k(t+T)} \Phi(t) \cdot a_k \, e^{\mu_k T} \, e_k \\
&= e^{-\mu_k t} x_k(t) = v_k(t)
\end{aligned} \tag{A2.37}
$$

so that $v_k(t)$, $1 \le k \le n$, is periodic and bounded.

Furthermore, noting that

$$x(t) = \sum_k x_k(t) = \sum_k e^{\mu_k t} v_k(t), \tag{A2.38}$$

we have that, if $\mathrm{Re}(\mu_k) \le 0$, $\forall k$, the origin is asymptotically stable, while if $\mathrm{Re}(\mu_k) \ge 0$ for any k, $1 \le k \le n$, the origin is unstable.

Example A3: Consider Hill's equation

$$\ddot{u} + a(t)u = O,$$

where

$$a(t+T) = a(t), \quad \forall t.$$

Putting

$$x_1 = u, \quad x_2 = \dot{u},$$

we have

$$
\left.
\begin{aligned}
\dot{x}_1 &= x_2 \\
\dot{x}_2 &= -a(t)x_1
\end{aligned}
\right\}
$$

which may be rewritten as,

$$\dot{x} = A(t)x, \quad A(t) = \begin{bmatrix} 0 & 1 \\ -a(t) & 0 \end{bmatrix}.$$

The fundamental matrix $\Phi(t)$ is given by

$$\Phi(t) = \begin{bmatrix} u^1 & u^2 \\ \dot{u}^1 & \dot{u}^2 \end{bmatrix}$$

where u^1 and u^2 are linearly independent solutions such that

$$\left. \begin{array}{ll} u^1(0) = 1, & u^2(0) = 0 \\ \dot{u}^1(0) = 0, & \dot{u}^2(0) = 1 \end{array} \right\}.$$

So,

$$\Phi(0) = I.$$

Note,

$$\Phi(t + T) = \Phi(t)B$$

where, from (A2.22) and (A2.29),

$$B = \Phi^{-1}(0)\,\Phi(T) = \begin{bmatrix} u^1(T) & u^2(T) \\ \dot{u}^1(T) & \dot{u}^2(T) \end{bmatrix},$$

and

$$\det B = e^{\int_0^T tr\, A(s)\,ds} = 1.$$

The Floquet multipliers ρ, which are the eigenvalues of B, are then given by

$$\rho^2 - 2\phi\rho + 1 = O,$$

where

$$\phi = \text{trace } B = \frac{1}{2}\left[u^1(T) + \dot{u}^2(T)\right].$$

Thus,

$$\rho_{1,2} = \phi \pm \sqrt{\phi^2 - 1}$$

with

$$\rho_1\rho_2 = 1, \quad \rho_1 + \rho_2 = 2\phi.$$

The Floquet exponents $\mu_{1,2}$, where $\rho_{1,2} = e^{\mu_{1,2}T}$, are therefore given by

$$\mu_1 + \mu_2 = 0, \quad \cosh \mu_1 T = \phi.$$

There are now several cases to consider.

(i) $\phi > 1$: Here, $\rho_{1,2}$ are both real and positive, and $\rho_1 > 1 > \rho_2 > 0$. Thus, μ_1 is real and positive while μ_2 is real and negative. The general solution is

$$u = c_1 e^{\mu_1 t} v_1(t) + c_2 e^{\mu_2 t} v_2(t)$$

where

$$v_{1,2}(t+T) = v_{1,2}(t).$$

There are no periodic solutions, and in general, $|u| \Rightarrow \infty$, as $t \Rightarrow \infty$.

(ii) $\phi < -1$: Here, $\rho_{1,2}$ are both real and negative, and $\rho_2 < -1 < \rho_1 < 0$. Thus,

$$\left.\begin{array}{c} \mu_1 = \dfrac{i\pi}{T} - \gamma, \;\; \mu_2 = -\dfrac{i\pi}{T} + \gamma \\ \text{and} \;\; \cosh \gamma T = -\phi \end{array}\right\}.$$

The general solution is

$$u = c_1 e^{-\gamma t} K_1(t) + c_2 e^{\gamma t} K_2(t)$$

where

$$\left.\begin{array}{c} K_{1,2}(t) \equiv e^{\pm \frac{i\pi t}{T}} v_{1,2}(t) \\ \text{and} \\ K_{1,2}(t+2T) = K_{1,2}(t), \;\; \forall t \end{array}\right\}.$$

Again, there are no periodic solutions, and in general, $|u| \Rightarrow \infty$, as $t \Rightarrow \infty$.

(iii) $-1 < \phi < 1$: Here, $\rho_{1,2}$ are both complex, with $|\rho_{1,2}| = 1$. We have

$$\rho_{1,2} = e^{\pm i\sigma T}, \;\; 0 < \sigma T < \pi$$

and

$$\mu_{1,2} = \pm i\sigma, \;\; \cos \sigma T = \phi.$$

The general solution is

$$u = c_1 \operatorname{Re}\left[e^{i\sigma t} v(t)\right] + c_2 \operatorname{Im}\left[e^{i\sigma t} v(t)\right]$$

where

$$v(t+T) = v(t), \;\; \forall t.$$

The solutions are bounded and oscillatory.

(iv) $\phi = 1$: This case is the boundary between cases (i) and (iii). Here, we have

$$\rho_1 = \rho_2 = 1,$$

and

$$\mu_1 = \mu_2 = 0.$$

The general solution is

$$u = c_1 v_1(t) + c_2 v_2(t)$$

where

$$v_{1,2}(t+T) = v_{1,2}(t), \quad \forall t.$$

The solutions are purely oscillatory.

(v) $\phi = -1$: This case is the boundary between cases (ii) and (iii). Here, we have

$$\rho_1 = \rho_2 = -1,$$

and

$$\mu_{1,2} = \pm \frac{i\pi}{T}.$$

The general solution is

$$u = c_1 K_1(t) + c_2 K_2(t)$$

where

$$\left. \begin{array}{l} K_{1,2}(t) \equiv e^{\pm \frac{i\pi t}{T}} v_{1,2}(t), \\ \text{and} \\ K_{1,2}(t+2T) = K_{1,2}(t) \end{array} \right\}.$$

The solutions are again purely oscillatory.

Thus, the boundary between bounded and unbounded solutions is characterized by periodic solutions with period T or $2T$.

As a special case of Hill's equation, consider a parametric resonance system

$$\ddot{u} + \left\{ \omega^2 + \varepsilon b(t) \right\} u = 0$$

where

$$b(t+T) = b(t).$$

When $\varepsilon = 0$, this equation describes a simple harmonic oscillation of period $T_o \equiv 2\pi/\omega$.

We have,

$$\phi = \phi\left(\omega^2, \varepsilon\right)$$

so, the stability boundaries are now curves in the ω^2, ε-plane.

Further, we have, for the case $\varepsilon = 0$,

$$u^1 = \cos \omega t, \quad u^2 = \frac{1}{\omega} \sin \omega t.$$

So,

$$\phi = \cos \omega T,$$

and hence, $-1 \le \phi \le 1$, for $\varepsilon = 0$. The stability boundaries are given by

$$\phi = 1 \Rightarrow \omega T = 2k\pi \text{ or } T = kT_o,$$
$$\phi = -1 \Rightarrow \omega T = (2k+1)\pi \text{ or } 2T = (2k+1)T_0,$$

where $k = 0, 1, 2, \ldots$. The analyticity of the function $\phi\left(\omega^2, \varepsilon\right)$ then implies that the above stability boundaries can be extended into the $\varepsilon \ne 0$, but small ε-regime by expanding the solutions as a power series in ε. The unbounded behavior of the solutions, when T or $2T$ is an integer multiple of T_o, is called parametric resonance.

3.14. Appendix 3

Bifurcation Theory

Bifurcation or branching occurs in a system when the state of the system depends on some parameter μ, and as that parameter varies the state branches to another state at some critical value μ_c of the parameter μ with usually a concomitant change of stability. Stability considerations determine the solution sought out by the system when there is a multiplicity of solutions. The bifurcation theory is concerned with how the multiplicity of solutions varies with the parameter and the stability properties of the bifurcating solutions.

Bifurcations are classified according to how the stability of an equilibrium solution changes. There are two ways in which this can occur. An eigenvalue of the system linearized about this solution can pass through zero, or a pair of non-zero eigenvalues may cross the imaginary axis. The first case corresponds to a saddle-node or tangent bifurcation and describes the birth or collapse of two equilibria (like a stable node coalescing with a saddle and annihilating it). This corresponds to a manifold associated with a given fixed point intersecting itself. On the other hand, when manifolds associated with different fixed points intersect, an exchange of stability occurs – this corresponds to a transcritical bifurcation or a pitchfork bifurcation. The second case corresponds to the so-called Hopf bifurcation and describes the birth of a family of periodic orbits (limit cycle) following the change in stability of a focus.

Consider the first-order autonomous differential equation describing flow in a one-dimensional phase space R,

$$\frac{du}{dt} = f(u;\mu), \quad t > 0 \tag{A3.1}$$

where μ is a real parameter, and f is a given analytic function of u and μ with continuous partial derivatives of all order with respect to u and μ.

The equilibrium solution $u = u_0$ of equation (A3.1) is found from

$$f(u;\mu) = 0. \tag{A3.2}$$

If $\frac{\partial f}{\partial u} = 0$ in an open neighborhood of $\mu = \mu_c$, then corresponding to one value of μ, several equilibrium solutions u_o may exist. This is guaranteed by the Implicit Function Theorem[3] which allows the solutions of equation (A3.2) to become non-unique whenever $\partial f/\partial u = 0$.

The graph of equation (A3.2) is called the branching or bifurcation diagram. The intersecting branches are the bifurcating solutions and the points of intersection, which correspond to change of stability, are called bifurcation points.

The stability of equilibrium solutions changes as μ varies. Often an equilibrium solution $u_o(\mu)$ will be stable for $\mu < \mu_c$ and unstable for $\mu \geq \mu_c$. Thus, as μ is increased slowly, the equilibrium solution $u_o(\mu)$ becomes unstable at μ_c, the system may therefore branch to another stable solution. The bifurcation theory seeks to explore how the stability of various equilibria

---

[3] THEOREM: Suppose that $f(u;\mu): R \times R \to R$ is a C^1 function satisfying

$$f(u_o;\mu_c) = 0 \quad \text{and} \quad \frac{\partial f}{\partial u}(u_o;\mu_c) \neq 0.$$

Then, there exists a unique solution of the implicit equation $f(u;\mu) = 0$ given by $u = g(\mu)$ in some open subset W of μ_c.

changes as μ is varied near μ_c. This issue, of course, depends in an essential way on the nonlinear nature of the problem.

In order to determine stability of the equilibrium solution $u_o(\mu)$, consider a small perturbation $\hat{u}(t)$ about $u_o(\mu)$, so that

$$u(t) = u_o(\mu) + \hat{u}(t). \tag{A3.3}$$

Equation (A3.1) then becomes

$$\frac{d\hat{u}}{dt} = f(u_o + \hat{u}; \mu). \tag{A3.4}$$

Linearizing f in $\hat{u}$, equation (A3.4) becomes

$$\frac{d\hat{u}}{dt} \approx f_u(u_o; \mu)\,\hat{u}, \tag{A3.5}$$

from which we have

$$\hat{u}(t) = ce^{f_u(u_o;\mu)t}. \tag{A3.6}$$

Thus, if $f_u(u_o; \mu) < 0$, u_o is stable and vice versa. The bifurcation point μ_c is here defined by $f_u(u_o; \mu) = 0$ along with $f(u_o; \mu) = 0$. Now, by the Implicit Function Theorem, $f(u_o; \mu) = 0$ implies $\mu = \mu(u_o)$ whenever $f_\mu(u_o; \mu) \neq 0$. We have, on differentiating $f(u_o; \mu) = 0$ with respect to u_o,

$$f_{u_o} + f_\mu \frac{d\mu}{du_o} = 0. \tag{A3.7}$$

Equation (A3.7) shows that $\dfrac{d\mu}{du_o} = 0$ at a bifurcation point (where $f_{u_o} = 0$), if $f_\mu \neq 0$ there.

In higher dimensions, one has, in place of equation (A3.1),

$$\frac{du}{dt} = f(u; \mu). \tag{A3.8}$$

Let $u_o(\mu)$ be an equilibrium point of equation (A3.8), so that

$$f(u_o(\mu); \mu) = 0. \tag{A3.9}$$

The stability of $u_o(\mu)$ is then determined by the eigenvalues $\lambda_1(\mu), \lambda_2(\mu), \cdots, \lambda_n(\mu)$ of the Jacobian matrix

$$A(\mu) = \frac{\partial f(u_o(\mu); \mu)}{\partial u}. \tag{A3.10}$$

If all the eigenvalues have a negative real part, then $u_o(\mu)$ is stable. On the other hand, if one or more eigenvalues have a positive real part, then $u_o(\mu)$ is unstable. Further, if the eigenvalues depend on the parameter μ, this stability may change as the parameter μ varies. In fact, the value of $\mu = \mu_c$, say, for which

$$\left.\begin{array}{l} \operatorname{Re}\lambda_i(\mu_c) = 0 \quad \text{for some } i \\ \operatorname{Re}\lambda_j(\mu_c) < 0 \quad \text{for all } j \neq i \end{array}\right\}$$

define the bifurcation points. It is apparent that a bifurcation point can arise in two ways:

(i) $\lambda_1(\mu)$ is real-valued, $\lambda_1(\mu_c) = 0$, and $\operatorname{Re}\lambda_i(\mu_c) < 0$ for $i = 2,...,n$;

(ii) $\lambda_1(\mu)$ and $\lambda_2(\mu)$ form a complex conjugate pair, so that $\lambda_1(\mu) = \overline{\lambda_2(\mu)} = \alpha(\mu) + i\beta(\mu)$, $\alpha(\mu_c) = 0$, $\beta(\mu_c) \neq 0$, and $\operatorname{Re}\lambda_i(\mu_c) < 0$ for $i = 3,...,n$.

Note that for a one-dimensional system, only Case (i) can occur. Case (i) is called the saddle-node bifurcation. Case (ii) is called the Hopf bifurcation. Figure 3.12 shows the schematic diagram of the variation of the eigenvalues as μ varies through μ_c in cases (i) and (ii) for a two-dimensional system.

Case (i) corresponds to the transition of the critical point $u_o(\mu)$ from a stable node into a saddle point. Case (ii) corresponds to the transition of the critical point $u_o(\mu)$ from a stable focus into an unstable focus and appearance of a periodic solution (see Shivamoggi, 1997, for futher details).

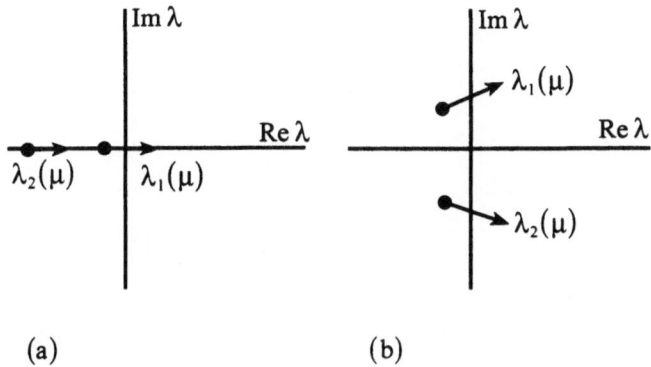

(a)   (b)

Figure 3.12. Schematic diagrams for (a) saddle-node bifurcation and (b) Hopf bifurcation.

Chapter 4

Method of Averaging

4.1. Introduction

Consider a weakly nonlinear oscillation problem

$$\ddot{x} + \varepsilon h(x, \dot{x}) + x = 0, \quad \varepsilon \ll 1. \tag{4.1.1}$$

The effect of the perturbation is to cause the parameters of the unperturbed system to vary with time. If the perturbation is small, the variation of the parameters within one period of the unperturbed motion will also be small. Thus, the perturbation theory is based on the premise that during small intervals of time the perturbed system moves along a route, which has the same functional form as the unperturbed system but with parameters changing with time. The latter variation may be periodic or secular. In the latter case, the change in a parameter is cumulative so that no matter how small this change may be over a small interval of time, eventually the perturbed route will differ greatly from the unperturbed route. The periodic variations in the parameters, on the other hand, do not materialize over long intervals of time because they can be eliminated by averaging the perturbation effects over the unperturbed period!

Assume that this system has at least one periodic solution, so that its phase diagram consists of either a limit cycle or a center. Such an equation is in a sense "close" to the linear equation

$$\ddot{x} + x = 0 \tag{4.1.2}$$

and the paths will be close to being circular. One may use this fact in constructing approximate solutions. In order to determine periodic solutions, in particular, we assume the existence of a nearly circular *closed path* in the phase diagram. This method calculates the average value of the amplitude of a slowly-varying oscillation over a cycle of the path. It can calculate approximate periods of limit cycles and the shape of the spiral paths around limit cycles, or it can calculate amplitude-frequency relations in the case of a center.

4.2. Krylov-Bogoliubov Method of Averaging

Introduce the phase-plane coordinates $(x, y = \dot{x})$, then equation (4.1.1) can be written as the following system of first-order differential equations:

$$\left.\begin{aligned} \dot{x} &= y \\ \dot{y} &= -\varepsilon h - x \end{aligned}\right\}. \tag{4.2.1}$$

Let $\big(a(t),\ \theta(t)\big)$ be the polar coordinates of a representative point on a phase curve of the system. Thus,

$$\left.\begin{aligned} x &= a\cos\theta \\ y &= a\sin\theta \end{aligned}\right\}, \tag{4.2.2}$$

from which

$$a^2 = x^2 + y^2 \tag{4.2.3}$$

and

$$\theta = \tan^{-1}(y/x). \tag{4.2.4}$$

We then have from (4.2.3) the following result –

$$a\dot{a} = x\dot{x} + y\dot{y} = -\varepsilon\, y\, h(x, y)$$

or

$$\dot{a} = -\varepsilon h(a\cos\theta,\ a\sin\theta)\sin\theta. \tag{4.2.5}$$

Further, we have from (4.2.4) the following result –

$$\dot{\theta} = -1 - \frac{\varepsilon}{a}\, h(a\cos\theta,\ a\sin\theta)\cos\theta. \tag{4.2.6}$$

Putting

$$\beta(t) \equiv \theta(t) + t, \tag{4.2.7}$$

we obtain from (4.2.6),

$$\dot{\beta} = -\frac{\varepsilon}{a}\, h(a\cos\theta,\ a\sin\theta)\cos\theta. \tag{4.2.8}$$

Thus, both $\beta(t)$ and $a(t)$ are slowly-varying with t. Figure 4.1 schematically shows these variables for the Duffing equation, for which $h(x,\dot{x}) = x^3$.

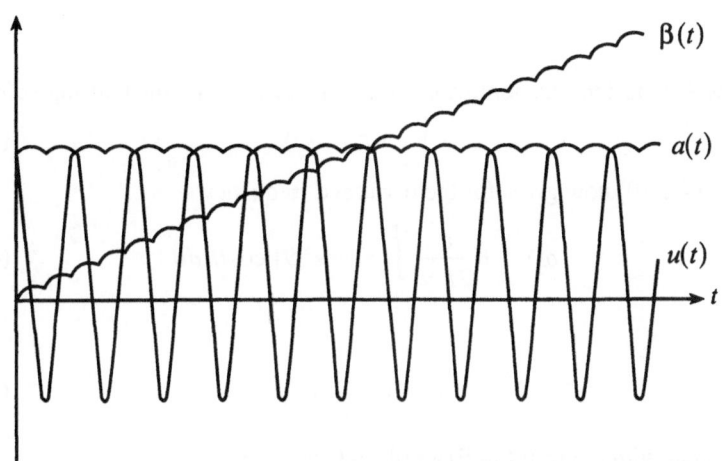

Figure 4.1. The variation of a, β, and u with t for $a(0) = 1.5$, $\beta(0) = 0.0$, and $\epsilon = 0.05$, for the Duffing equation with $h(x,\dot{x}) = x^3$ (From Nayfeh, 1981).

The perturbed period is given by (Krylov and Bogoliubov, 1947)

$$\tau = -\int_0^{2\pi} \frac{d\theta}{\theta} \approx 2\pi - \frac{\varepsilon}{a} \int_0^{2\pi} h(a\cos\theta,\ a\sin\theta)\cos\theta\ d\theta. \qquad (4.2.9)$$

The perturbed frequency is then given by

$$\omega = \frac{2\pi}{\tau} \approx 1 + \frac{\varepsilon}{2\pi a} \int_0^{2\pi} h(a\cos\theta,\ a\sin\theta)\cos\theta\ d\theta. \qquad (4.2.10)$$

We have on the other hand, using (4.2.5) and (4.2.6),

$$\frac{da}{d\theta} = \frac{\dot{a}}{\theta} \approx \varepsilon h(a\cos\theta,\ a\sin\theta)\sin\theta. \qquad (4.2.11)$$

If the solution is periodic with period nearly equal to 2π, we obtain

$$\int_0^{2\pi} \frac{da}{d\theta}\ d\theta = 0$$

or

$$\int_0^{2\pi} h(a\cos\theta,\ a\sin\theta)\sin\theta\ d\theta = 0 \qquad (4.2.12)$$

which gives $a = \bar{a}$, say.

Example 1: Consider the nonlinear oscillator described by the Duffing equation

$$\ddot{x} + x + \varepsilon x^3 \approx 0. \qquad (4.2.13)$$

Equation (4.2.10) now gives for the perturbed frequency

$$\omega \approx 1 + \frac{\varepsilon}{2\pi a}\int_0^{2\pi}\left(a^3\cos^3\theta\right)\cos\theta\ d\theta \qquad (4.2.14)$$

from which

$$\omega \approx 1 + \frac{3\varepsilon}{8}a^2, \qquad (4.2.15)$$

in agreement with the result in Example 2, Chapter 3.

Let us expand the nonlinear term in a Fourier series as follows –

$$h\left[a(\theta)\cos\theta,\ a(\theta)\sin\theta\right]\sin\theta$$

$$= \frac{p_0[a(\theta)]}{2} + \sum_{n=1}^{\infty}\left\{p_n[a(\theta)]\cos n\theta + q_n[a(\theta)]\sin n\theta\right\},\ \ 0 < \theta < 2\pi \qquad (4.2.16)$$

where

$$\left.\begin{array}{l} p_o[a(\theta)] = \dfrac{1}{\pi}\displaystyle\int_0^{2\pi} h[a(\theta)\cos u,\ a(\theta)\sin u]\sin u\ du \\[3mm] q_n[a(\theta)] = \dfrac{1}{\pi}\displaystyle\int_0^{2\pi} h[a(\theta)\cos u,\ a(\theta)\sin u]\sin u\ \sin(nu)\ du \\[3mm] p_n[a(\theta)] = \dfrac{1}{\pi}\displaystyle\int_0^{2\pi} h[a(\theta)\cos u,\ a(\theta)\sin u]\sin u\ \cos(nu)\ du \end{array}\right\}. \quad (4.2.17)$$

Using (4.2.16) in (4.2.5) and integrating over a cycle, we obtain

$$\langle \dot{a}\rangle = -\frac{1}{2}\varepsilon p_0(a). \qquad (4.2.18)$$

Similarly, we obtain

$$\langle \dot{\theta} \rangle = -1 - \frac{\varepsilon}{2\pi a} \int_{0}^{2\pi} h(a\cos u,\, a\sin u) \cos u \; du. \qquad (4.2.19)$$

Example 2: Consider the van der Pol oscillator described by

$$\ddot{x} + \varepsilon \left(x^2 - 1 \right) \dot{x} + x = 0, \quad \varepsilon > 0. \qquad (4.2.20)$$

This system is unstable for small-amplitude perturbations, with a linear growth rate $\varepsilon/2$. The nonlinear dissipative term $\varepsilon x^2 \dot{x}$ eventually limits this growth, and the amplitude of the oscillation saturates (as shown by the numerical solution sketched in Figure 4.2).

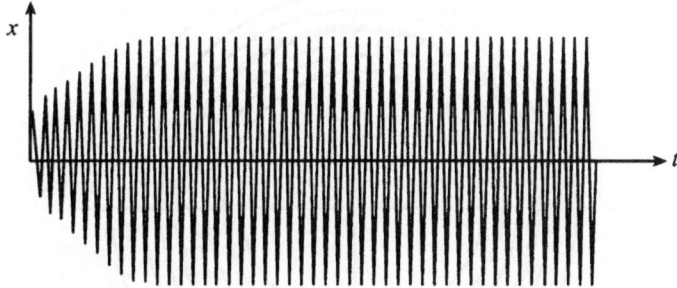

Figure 4.2. Exact solution of equation (4.22) for $x(0) = 0.5$, $\dot{x}(0) = 0$ and $\epsilon = 0.1$ (From Nayfeh, 1981).

We have from (4.2.17) and (4.2.18), for this case,

$$\dot{a} \approx \langle \dot{a} \rangle = -\frac{\varepsilon}{2\pi} a \int_{0}^{2\pi} \left(a^2 \cos^2 u - 1 \right) \sin^2 u \; du = -\frac{1}{2} \varepsilon a \left(\frac{a^2}{4} - 1 \right). \quad (4.2.21)$$

Separating the variables and integrating, we obtain from equation (4.2.21)

$$\int \frac{da}{a(a^2 - 4)} = -\frac{1}{8} \varepsilon (t + c)$$

where c is a constant of integration. This leads to

$$-\frac{1}{4} \log a + \frac{1}{8} \log \left| a^2 - 4 \right| = -\frac{1}{8} \varepsilon (t + c),$$

from which

$$a(t) = \frac{2}{\left[1 - \left(1 - \frac{4}{a_0^2}\right) e^{-\varepsilon t}\right]^{1/2}}, \qquad (4.2.22)$$

where $a_0 \equiv a(0)$.

Equation (4.2.22) indicates trajectories spiraling inward or outward (according to $a_0 > 2$ or $a_0 < 2$) into a limit cycle, which is a circle given by $a = 2$, in the phase plane (see Figure 4.3). This result shows the relaxation nature of the oscillations of the van der Pol oscillator.

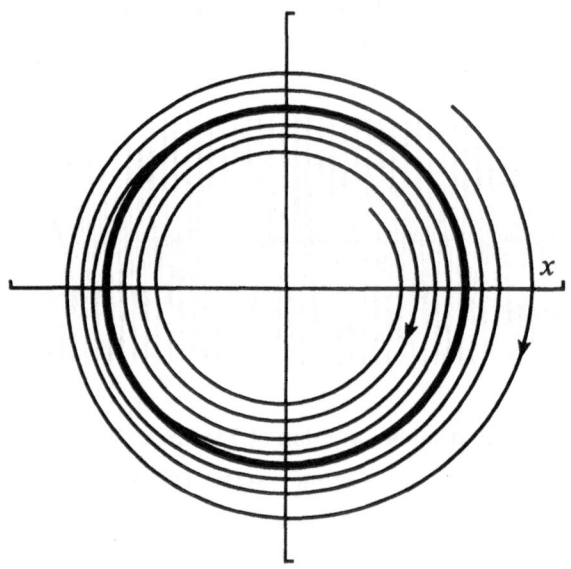

Figure 4.3. Orbits approaching the limit cycle for the van der Pol equation.

4.3. Krylov-Bogoliubov-Mitropolski Generalized Method of Averaging

In this method, one assumes an asymptotic expansion of the form

$$u = a \cos\psi + \sum_{n=1}^{N} \varepsilon^n u_n(a, \psi) + O(\varepsilon^{N+1}) \qquad (4.3.1)$$

where each u_n is a periodic function of ψ with a period 2π, and a and ψ are taken to vary with time according to (Bogoliubov and Mitropolski, 1961)

$$\left. \begin{aligned} \frac{da}{dt} &= \sum_{n=1}^{N} \varepsilon^n A_n(a) + O(\varepsilon^{N+1}) \\ \frac{d\psi}{dt} &= \omega_0 + \sum_{n=1}^{N} \varepsilon^n \psi_n(a) + O(\varepsilon^{N+1}) \end{aligned} \right\} \tag{4.3.2}$$

while the derivatives are transformed according to

$$\left. \begin{aligned} \frac{d}{dt} &= \frac{da}{dt}\frac{\partial}{\partial a} + \frac{d\psi}{dt}\frac{\partial}{\partial \psi} + \frac{\partial}{\partial t} \\ \frac{d^2}{dt^2} &= \left(\frac{da}{dt}\right)^2 \frac{\partial^2}{\partial a^2} + \frac{d^2 a}{dt^2}\frac{\partial}{\partial a} + 2\frac{da}{dt}\frac{d\psi}{dt}\frac{\partial^2}{\partial a \partial \psi} \\ &\quad + \left(\frac{d\psi}{dt}\right)^2 \frac{\partial^2}{\partial \psi^2} + \frac{d^2\psi}{dt^2}\frac{\partial}{\partial \psi} + \frac{\partial^2}{\partial t^2} \end{aligned} \right\}. \tag{4.3.3}$$

Example 3: Consider the Duffing oscillator described by

$$\ddot{u} + \omega_0^2 u = -\varepsilon u^3. \tag{4.3.4}$$

Using (4.25)-(4.27), we obtain from equation (4.3.4),

$$\begin{aligned} &\left[\left(\varepsilon A_1 + \varepsilon^2 A_2 + \cdots\right)^2 \frac{\partial^2}{\partial a^2} + \left(\varepsilon A_1 + \varepsilon^2 A_2 + \cdots\right)\left(\varepsilon A_1' + \varepsilon^2 A_2' + \cdots\right)\frac{\partial}{\partial a} \right. \\ &+ 2\left(\varepsilon A_1 + \varepsilon^2 A_2 + \cdots\right)\left(\omega_0 + \varepsilon \psi_1 + \varepsilon^2 \psi_2 + \cdots\right)\frac{\partial^2}{\partial a \partial \psi} \\ &+ \left(\omega_0 + \varepsilon \psi_1 + \varepsilon^2 \psi_2 + \cdots\right)^2 \frac{\partial^2}{\partial \psi^2} + \left.\left(\varepsilon A_1 + \varepsilon^2 A_2 + \cdots\right)\left(\varepsilon \psi_1' + \varepsilon^2 \psi_2' + \cdots\right)\frac{\partial}{\partial \psi}\right] \\ &\times \left(a\cos\psi + \varepsilon u_1 + \varepsilon^2 u_2 + \cdots\right) + \omega_0^2\left(a\cos\psi + \varepsilon u_1 + \varepsilon^2 u_2 + \cdots\right) \\ &= -\varepsilon\left(a\cos\psi + \varepsilon u_1 + \varepsilon^2 u_2 + \cdots\right)^3 \end{aligned}$$

$$\tag{4.3.5}$$

from which, we obtain to various orders in ε,

$$O(\varepsilon): \omega_0^2\left(\frac{\partial^2 u_1}{\partial \psi^2} + u_1\right) = 2\omega_0 \psi_1 a\cos\psi + 2\omega_0 A_1 \sin\psi - a^3\cos^3\psi \tag{4.3.6}$$

$$O(\varepsilon^2): \omega_0^2 \left(\frac{\partial^2 u_2}{\partial \psi^2} + u_2 \right) = \left[\left(2\omega_0 \, \psi_2 + \psi_1^2 \right) a - A_1 \frac{dA_1}{da} \right] \cos \psi$$

$$+ \left[2 \left(\omega_0 \, A_2 + A_1 \, \psi_1 \right) + a A_1 \frac{d\psi_1}{da} \right] \sin \psi - 3u_1 \, a^2 \cos^2 \psi$$

$$- 2\omega_0 \, \psi_1 \frac{\partial^2 u_1}{\partial \psi^2} - 2\omega_0 \, A_1 \frac{\partial^2 u_1}{\partial a \, \partial \psi} \qquad (4.3.7)$$

etc.

Removal of secular terms in equation (4.3.6) requires

$$A_1 = 0, \quad \psi_1 = \frac{3a^2}{8\omega_0}. \qquad (4.3.8)$$

Equation (4.3.6) then yields

$$u_1 = \frac{a^3}{32 \, \omega_0^2} \cos 3\psi. \qquad (4.3.9)$$

Using (4.3.8) and (4.3.9), equation (4.3.7) becomes

$$\omega_0^2 \left(\frac{\partial^2 u_2}{\partial \psi^2} + u_2 \right) = \left(2\omega_0 \psi_2 + \frac{15a^4}{128 \, \omega_0^2} \right) a \cos \psi + 2\omega_0 \, A_2 \sin \psi$$

$$+ \frac{a^5}{128 \, \omega_0^2} \left(21 \cos 3\psi - 3 \cos 5\psi \right). \qquad (4.3.10)$$

Removal of secular terms in equation (4.3.10) requires

$$A_2 = 0, \quad \psi_2 = -\frac{15a^4}{256 \, \omega_0^3}. \qquad (4.3.11)$$

Equation (4.3.10) then yields,

$$u_2 = -\frac{a^5}{1024 \, \omega_0^4} \left(21 \cos 3\psi - \cos 5\psi \right). \qquad (4.3.12)$$

Using (4.3.8), (4.3.9), (4.3.11), and (4.3.12), we have from (4.3.1) and (4.3.2)

$$\left. \begin{array}{l} u \sim a \cos \psi + \dfrac{\varepsilon a^3}{32 \, \omega_0^2} \cos 3\psi - \dfrac{\varepsilon^2 a^5}{1024 \, \omega_0^4} \left(21 \cos 3\psi - \cos 5\psi \right) + O(\varepsilon^3) \\[4mm] \dfrac{da}{dt} = 0 \quad \text{or} \quad a = a_0 = \text{constant} \\[4mm] \dfrac{d\psi}{dt} = \omega_0 + \dfrac{3\varepsilon a^2}{8\omega_0} - \dfrac{15\varepsilon^2 \, a^4}{256 \, \omega_0^3} + O(\varepsilon^3) \end{array} \right\}. \quad (4.3.13)$$

Example 4: Consider the van der Pol oscillator described by

$$\ddot{u} + u = \varepsilon\left(1 - u^2\right)\dot{u}. \tag{4.3.14}$$

Using (4.3.1)-(4.3.3), we obtain from equation (4.3.14),

$$\left[\left(\varepsilon A_1 + \varepsilon^2 A_2 + \cdots\right)^2 \frac{\partial^2}{\partial a^2} + \left(\varepsilon A_1 + \varepsilon^2 A_2 + \cdots\right)\left(\varepsilon A_1' + \varepsilon^2 A_2' + \cdots\right)\frac{\partial}{\partial a}\right.$$

$$+ 2\left(\varepsilon A_1 + \varepsilon^2 A_2 + \cdots\right)\left(1 + \varepsilon \psi_1 + \varepsilon^2 \psi_2 + \cdots\right)\frac{\partial^2}{\partial a\, \partial \psi} + \left(1 + \varepsilon \psi_1 + \varepsilon^2 \psi_2 + \cdots\right)^2$$

$$\times \frac{\partial^2}{\partial \psi^2} + \left(\varepsilon A_1 + \varepsilon^2 A_2 + \cdots\right)\left(\varepsilon \psi_1' + \varepsilon^2 \psi_2' + \cdots\right)\frac{\partial}{\partial \psi}\left.\right]\left(a\cos\psi + \varepsilon u_1 + \varepsilon^2 u_2 + \cdots\right)$$

$$+ \left(a\cos\psi + \varepsilon u_1 + \varepsilon^2 u_2 + \cdots\right) = \varepsilon\left[1 - \left(a\cos\psi + \varepsilon u_1 + \varepsilon^2 u_2 + \cdots\right)^2\right]$$

$$\times \left[\left(\varepsilon A_1 + \varepsilon^2 A_2 + \cdots\right)\frac{\partial}{\partial a} + \left(1 + \varepsilon \psi_1 + \varepsilon^2 \psi_2 + \cdots\right)\frac{\partial}{\partial \psi}\right]$$

$$\times \left(a\cos\psi + \varepsilon u_1 + \varepsilon^2 u_2 + \cdots\right) \tag{4.3.15}$$

from which, we obtain to various orders in ε,

$$O(\varepsilon): \frac{\partial^2 u_1}{\partial \psi_1^2} + u_1 = 2\psi_1\, a\cos\psi + 2 A_1 \sin\psi - a\left(1 - \frac{1}{4}a^2\right)\sin\psi + \frac{1}{4}a^3 \sin 3\psi \tag{4.3.16}$$

$$O(\varepsilon^2): \frac{\partial^2 u_2}{\partial \psi^2} + u_2 = \left[\left(2\psi_2 + \psi_1^2\right)a - A_1\frac{dA_1}{da}\right]\cos\psi$$

$$+ \left[2\left(A_2 + A_1 \psi_1\right) + a A_1\frac{d\psi_1}{da}\right]\sin\psi$$

$$- 2\psi_1\frac{\partial^2 u_1}{\partial \psi^2} - 2 A_1\frac{\partial^2 u_1}{\partial a\, \partial \psi}$$

$$+ \left(1 - a^2 \cos^2\psi\right)\left(A_1 \cos\psi - a\psi_1 \sin\psi + \frac{\partial u_1}{\partial \psi}\right)$$

$$+ a^2 u_1 \sin 2\psi \tag{4.3.17}$$

etc.

Removal of secular terms in equation (4.3.16) requires

$$\psi_1 = 0, \qquad A_1 = \frac{1}{2}a\left(1 - \frac{1}{4}a^2\right). \tag{4.3.18}$$

Equation (4.3.16) then yields

$$u_1 = -\frac{a^3}{32} \sin 3\psi .$$ (4.3.19)

Using (4.3.18) and (4.3.19), equation (4.3.17) becomes

$$\frac{\partial^2 u_2}{\partial \psi^2} + u_2 = \left[2a\psi_2 - A_1 \frac{dA_1}{da} + \left(1 - \frac{3}{4} a^2\right) A_1 + \frac{a^5}{128} \right] \cos \psi$$

$$+ 2 A_2 \sin \psi + \frac{a^3 \left(a^2 + 8\right)}{128} \cos 3\psi + \frac{5a^6}{128} \cos 5\psi .$$ (4.3.20)

Removal of secular terms in equation (4.3.42) requires

$$A_2 = 0, \quad \psi_2 = \frac{A_1}{2a} \left(\frac{dA_1}{da} - 1 + \frac{3}{4} a^2 \right) - \frac{a^4}{256} .$$ (4.3.21)

Equation (4.3.20) then yields

$$u_2 = -\frac{5a^5}{3072} \cos 5\psi - \frac{a^3 \left(a^2 + 8\right)}{1024} \cos 3\psi .$$ (4.3.22)

Using (4.3.18), (4.3.19), (4.3.21), and (4.3.22), we have from (4.3.1) and (4.3.2),

$$u \sim a \cos \psi - \frac{\varepsilon a^3}{32} \sin 3\psi - \frac{\varepsilon^2 a^3}{1024} \left[\frac{5}{3} a^2 \cos 5\psi \right.$$

$$\left. \left(a^2 + 8\right) \cos 3\psi \right] + O\left(\varepsilon^3\right)$$

$$\frac{da}{dt} = \varepsilon \frac{a}{2} \left(1 - \frac{1}{4} a^2\right)$$ (4.3.23a,b,c)

$$\frac{d\psi}{dt} = 1 + \varepsilon^2 \left[\frac{A_1}{2a} \left(\frac{dA_1}{da} - 1 + \frac{3}{4} a^2 \right) - \frac{a^4}{256} \right].$$

Observe that equation (4.3.23b) is identical to equation (4.2.21) that was obtained by using the method of averaging.

4.4. Whitham's Average Lagrangian Method

The averaged Lagrangian method (Whitham, 1965, 1967) is based on variational principles. The main appeal of the variational principle is its power of synthesis. The whole dynamics of the problem is expressed in terms of a single function through which the governing equations as well as all the necessary boundary conditions follow by taking the functional derivatives.

In order to illustrate this method, consider the following nonlinear dispersive wave propagation problem (Bretherton, 1964),

$$\phi_{tt} + \phi_{xxxx} + \phi_{xx} + \phi = \varepsilon\phi^3, \quad \varepsilon \ll 1. \tag{4.4.1}$$

The linear problem associated with equation (4.4.1) has the following solution

$$\left.\begin{array}{l} \varphi = a\cos\theta, \quad \theta = kx - \omega t \\ \omega^2 = k^4 - k^2 + 1 \end{array}\right\}. \tag{4.4.2}$$

It may be readily verified that equation (4.4.1) has a variational character set forth by the following Lagrangian –

$$L = \frac{1}{2}\,\phi_t^2 - \frac{1}{2}\,\phi_{xx}^2 + \frac{1}{2}\,\phi_x^2 - \frac{1}{2}\,\phi^2 + \frac{1}{4}\,\varepsilon\phi^4 \tag{4.4.3}$$

because extremization of L, namely,

$$\delta \iint L\,dx\,dt = 0 \tag{4.4.4}$$

leads to (see Appendix 1 for a review of calculus of variations),

$$\frac{\partial}{\partial t}\left(\frac{\partial L}{\partial \phi_t}\right) + \frac{\partial}{\partial x}\left(\frac{\partial L}{\partial \phi_x}\right) - \frac{\partial^2}{\partial x^2}\left(\frac{\partial L}{\partial \phi_{xx}}\right) - \frac{\partial L}{\partial \phi} = 0 \tag{4.4.5}$$

which, on using (4.4.3), in turn gives equation (4.4.1).

In the presence of a weak nonlinearity, one may argue that the form of the solution (4.4.2) is maintained, but the amplitude a will not be a constant, and θ will not be linear in x and t. The wave number k and frequency ω may now be generalized by defining them as

$$k = \frac{\partial\theta}{\partial x}, \quad \omega = -\frac{\partial\theta}{\partial t}. \tag{4.4.6}$$

The parameters a, k, and ω are now slowly-varying functions of x and t. We have from (4.4.6) the following compatibility condition –

$$\frac{\partial k}{\partial t} + \frac{\partial\omega}{\partial x} = 0. \tag{4.4.7}$$

The variations of the parameters a, k, and ω are governed by the variational equations for the averaged Lagrangian. Thus, putting $\phi = a\cos\theta$, using (4.4.6), and averaging the Lagrangian (4.4.3) over a period, we obtain

$$\mathcal{L} = \frac{1}{2\pi} \int_0^{2\pi} L \, d\theta$$

$$= \frac{1}{4} \left(\omega^2 - k^4 + k^2 - 1 \right) a^2 + \frac{3\varepsilon}{32} a^4 + O(\varepsilon^2). \tag{4.4.8}$$

The variational equations for $\mathcal{L}$ are obtained by extremizing $\mathcal{L}$, first with respect to a,

$$\frac{\partial \mathcal{L}}{\partial a} = 0 \tag{4.4.9}$$

which gives on using (4.4.8)

$$\omega^2 = k^4 - k^2 + 1 + \frac{3}{4} \varepsilon a^2 + O(\varepsilon^2). \tag{4.4.10}$$

Next, extremizing $\mathcal{L}$ with respect to θ, we have

$$\frac{\partial}{\partial t} \left(\frac{\partial \mathcal{L}^2}{\partial \theta_t} \right) + \frac{\partial}{\partial x} \left(\frac{\partial \mathcal{L}}{\partial \theta_x} \right) - \frac{\partial \mathcal{L}}{\partial \theta} = 0 \tag{4.4.11}$$

which, on using (4.4.6), becomes

$$-\frac{\partial}{\partial t} \left(\frac{\partial \mathcal{L}}{\partial \omega} \right) + \frac{\partial}{\partial x} \left(\frac{\partial \mathcal{L}}{\partial k} \right) - \frac{\partial \mathcal{L}}{\partial \theta} = 0. \tag{4.4.12}$$

Using (4.4.8), equation (4.4.12) gives

$$\frac{\partial}{\partial t} \left(\omega \, a^2 \right) + \frac{\partial}{\partial x} \left[\left(2k^3 - k \right) a^2 \right] = 0. \tag{4.4.13}$$

We have from (4.4.10)

$$\omega \frac{\partial \omega}{\partial k} = 2k^3 - k + O(\varepsilon^2). \tag{4.4.14}$$

Using (4.4.14), equation (4.4.13) becomes

$$\frac{\partial}{\partial t} \left(\omega a^2 \right) + \frac{\partial}{\partial x} \left(\omega \frac{\partial \omega}{\partial k} a^2 \right) = 0. \tag{4.4.15}$$

from which

$$\omega \left[\frac{\partial a^2}{\partial t} + \frac{\partial}{\partial x} \left(\frac{\partial \omega}{\partial k} a^2 \right) \right] + a^2 \left(\frac{\partial \omega}{\partial t} + \frac{\partial \omega}{\partial k} \frac{\partial \omega}{\partial x} \right) = 0. \tag{4.4.16}$$

We have from equation (4.4.7)

$$\frac{\partial \omega}{\partial t} + \frac{\partial \omega}{\partial k} \frac{\partial \omega}{\partial x} = 0. \qquad (4.4.17)$$

Using equation (4.4.17), equation (4.4.16) becomes

$$\frac{\partial}{\partial t}\left(a^2\right) + \frac{\partial}{\partial x}\left(\frac{\partial \omega}{\partial k}\, a^2\right) = 0. \qquad (4.4.18)$$

Equation (4.4.18) implies that the quantity a^2 (which may be energy in some sense) propagates with the group velocity $\dfrac{\partial \omega}{\partial k}$.

4.5. Hamiltonian Perturbation Method

4.5.1 Systems with Constant Parameters

This method applies to systems with near-integrable Hamiltonians of the form

$$H(I,\theta) = H_0(I) + \varepsilon\, H_1(I,\theta) + \varepsilon^2\, H_2(I,\theta) + \cdots \qquad (4.5.1a)$$

where I is the action and θ is the angle (see Appendix 2 for the action-angle formulation). The system associated with $H_0(I)$ is soluble (or integrable), and its solution is considered known. Canonical perturbation theory is based on the assumption that there is a region of phase space in which the phase curves of H and H_0 may be continuously deformed into each other. (In other words, during small intervals of time, the perturbed system moves in an orbit of the same functional form as the unperturbed system but with parameters changing with time.) The basic idea of the present method[1] is to find a new set of action-angle variables (J,φ) for the perturbed system $H(I,\theta)$ such that the new Hamiltonian $K(J)$ found by means of a canonical transformation is a function only of J.

For the unperturbed system, Hamilton's equations are

$$\dot{I} = -\frac{\partial H_0}{\partial \theta} = 0 \qquad (4.5.2)$$

$$\dot{\theta} = \frac{\partial H_0}{\partial I} = \omega_0(I) \qquad (4.5.3)$$

---

[1] While the present procedure uses canonical variables, it may be mentioned that other perturbation theories, such as the Lie transform theory (Cary, 1981) are also available for this purpose.

and for the perturbed system, Hamilton's equations are

$$\dot{I} = -\frac{\partial H}{\partial \theta} = -\varepsilon \frac{\partial H_1}{\partial \theta} - \varepsilon^2 \frac{\partial H_2}{\partial \theta} + O(\varepsilon^3) \qquad (4.5.4)$$

$$\dot{\theta} = \frac{\partial H}{\partial I} = \omega_0(I) + \varepsilon \frac{\partial H_1}{\partial I} + \varepsilon^2 \frac{\partial H_2}{\partial I} + O(\varepsilon^3). \qquad (4.5.5)$$

The object is to find the generating function $S(\theta, J)$ which produces the transformation $H(I, \theta) \to K(J)$ through the relations

$$I = \frac{\partial S}{\partial \theta}, \quad \varphi = \frac{\partial S}{\partial J}, \qquad (4.5.6)$$

where φ is the new angle variable.

Let us expand S in a power series in ε as follows:

$$S = S_0 + \varepsilon S_1 + \varepsilon^2 S_2 + \cdots \qquad (4.5.7)$$

where S_0 is the identity transformation, $S_0 = J\theta$ so that, from equations (4.5.6), (4.5.7) gives a near-identity transformation –

$$I = \frac{\partial S}{\partial \theta} = J + \varepsilon \frac{\partial S_1}{\partial \theta} + \varepsilon^2 \frac{\partial S_2}{\partial \theta} + \cdots. \qquad (4.5.8)$$

Let us further expand $K(J)$ in another power series in ε,

$$K(J) = K_0(J) + \varepsilon K_1(J) + \varepsilon^2 K_2(J) + \cdots. \qquad (4.5.1b)$$

Using the expansions (4.5.7), (4.5.8), and (4.5.1b), the Hamilton-Jacobi equation

$$H\left(\frac{\partial S}{\partial \theta}, \theta\right) = K(J), \qquad (4.5.9)$$

on expanding each term in these expansions in Taylor series, becomes,

$$H_0(J) + \varepsilon \left[\frac{\partial H_0}{\partial J} \frac{\partial S_1}{\partial \theta} + H_1\right] + \varepsilon^2 \left[\frac{1}{2} \frac{\partial^2 H_0}{\partial J^2} \left(\frac{\partial S_1}{\partial \theta}\right)^2 \right.$$

$$\left. + \frac{\partial H_0}{\partial J} \frac{\partial S_2}{\partial \theta} + \frac{\partial H_1}{\partial J} \frac{\partial S_1}{\partial \theta} + H_2\right] + O(\varepsilon^3) = K_0(J) + \varepsilon K_1(J) + \varepsilon^2 K_2(J) + O(\varepsilon^3).$$

$$(4.5.10)$$

Equating coefficients in equation (4.5.10), of equal powers of ε, we obtain

$$O(1): H_0(J) = K_0(J) \qquad (4.5.11)$$

$$O(\varepsilon) : \frac{\partial H_0}{\partial J} \frac{\partial S_1}{\partial \theta} + H_1(J,\theta) = K_1(J) \tag{4.5.12}$$

$$O(\varepsilon^2) : \frac{1}{2} \frac{\partial^2 H_0}{\partial J^2} \left(\frac{\partial S_1}{\partial \theta}\right)^2 + \frac{\partial H_0}{\partial J} \frac{\partial S_2}{\partial \theta} + \frac{\partial H_1}{\partial J} \frac{\partial S_1}{\partial \theta} + H_2(J,\theta) = K_2(J) \tag{4.5.13}$$

etc.

Recalling that the motion is periodic in θ, we average equation (4.5.12) over θ; we then obtain

$$K_1(J) = \overline{H_1}(J,\theta) \tag{4.5.14}$$

where

$$\overline{H_1}(J,\theta) = \frac{1}{2\pi} \int_0^{2\pi} H_1(J,\theta) \, d\theta.$$

Using (4.5.14), equation (4.5.12) gives

$$\frac{\partial}{\partial \theta} S_1(\theta,J) = -\frac{1}{\omega_0(J)} \tilde{H}_1(J,\theta) \tag{4.5.15}$$

where

$$\left. \begin{aligned} \tilde{H}_1(J,\theta) &= H_1(J,\theta) - \overline{H_1}(J,\theta) \\ \omega_0(J) &\equiv \frac{dH_0}{dJ} \end{aligned} \right\}.$$

Writing $S_1(\theta,J)$ and $\tilde{H}_1(J,\theta)$ as Fourier series in θ,

$$\left. \begin{aligned} \tilde{H}_1(J,\theta) &= \sum_{k=1}^{\infty} A_k(J) e^{ik\theta} \\ S_1(J,\theta) &= \sum_{k=1}^{\infty} B_k(J) e^{ik\theta} \end{aligned} \right\}, \tag{4.5.16}$$

we obtain from equation (4.5.15)

$$B_k(J) = \frac{i}{k\omega_0(J)} A_k(J). \tag{4.5.17}$$

Using (4.5.17), (4.5.16) gives

$$S_1(J,\theta) = \sum_{k=1}^{\infty} \frac{i A_k(J)}{k\omega_0(J)} e^{ik\theta}. \tag{4.5.18}$$

Using (4.5.18) in (4.5.6), the new angle is given by

$$\varphi = \frac{\partial S}{\partial J} = \theta + \varepsilon \frac{\partial S_1}{\partial J} + \cdots \qquad (4.5.19)$$

while, from (4.5.8), the new action is given by

$$J = I - \varepsilon \frac{\partial S_1}{\partial \theta} + \cdots \qquad (4.5.20)$$

The perturbed frequency from (4.5.16) is given by

$$\omega(J) = \omega_0(J) + \varepsilon \frac{\partial K_1}{\partial J} + \cdots. \qquad (4.5.21)$$

Note that if ω_0 is small, the effect of the perturbation can be quite large. Since the nth-order correction has terms proportional to $1/\omega_0^2$, this difficulty gets compounded at the higher-order corrections. Usually, ω_0 is small near a separatrix dividing phase-space into regions containing different types of motion. This problem becomes much more serious, as we will see later, for systems of n degrees of freedom $(n \geq 2)$ because, there are n fundamental frequencies. These frequencies and all their integer linear combinations occur in the denominator of the expressions for the various orders of corrections, so that even if all the frequencies are large, a particular linear combination of them may be small. Thus, the convergence of perturbation expansions for systems of more than one degree of freedom is a very tricky problem.

Example 5: Consider the motion of a simple pendulum executing oscillations about the downward vertical. For this problem, the unperturbed Hamiltonian is

$$H_0(p,q) = \frac{1}{2}\left(p^2 + \omega_0^2 \, q^2\right) \qquad (4.5.22)$$

while the perturbation is

$$H_1(p,q) = -\omega_0^2\left(\cos q - 1 + \frac{1}{2}q^2\right). \qquad (4.5.23)$$

The Hamilton-Jacobi equation for the linear problem given by $H_0(p,q)$ is

$$\frac{1}{2}\left(\frac{\partial S_0}{\partial q}\right)^2 + \frac{1}{2}\,\omega_0^2\,q^2 = E \qquad (4.5.24)$$

where $S(q,I)$ is the generating function such that $\dfrac{\partial S_0}{\partial q} = p$.

The action I is given by

$$I = \frac{1}{2\pi} \oint_C \sqrt{2\left(E - \frac{1}{2}\omega_0^2 q^2\right)} \, dq = \frac{E}{\omega_0} \qquad (4.5.25)$$

where C is the closed path connecting the turning points $\pm\frac{\sqrt{2E}}{\omega_0}$. Thus, the generating function $S_0(q,I)$ with I, the action, as the new "momentum" is given from equation (4.5.24) by

$$S_0(q,I) = \int_{q_0}^{q} \sqrt{2\left(I\omega_0 - \frac{1}{2}\omega_0^2 q^2\right)} \, dq. \qquad (4.5.26)$$

The new coordinate, which is merely the angle θ, is then given by

$$\theta = \frac{\partial S_0}{\partial I} = \sin^{-1}\left(q\sqrt{\frac{\omega_0}{2I}}\right) \qquad (4.5.27)$$

from which

$$q = \sqrt{\frac{2I}{\omega_0}} \sin\theta. \qquad (4.5.28)$$

Next, we have, from (4.5.26),

$$p = \frac{\partial S_0}{\partial q} = 2\left(I\omega_0 - \frac{1}{2}\omega_0^2 q^2\right) = \sqrt{2I\omega_0} \cos\theta. \qquad (4.5.29)$$

Using (4.5.28) and (4.5.29), we have from (4.5.22) and (4.5.23),

$$\left.\begin{array}{l} H_0(I) = I\omega_0 \\[1mm] H_1(I,\theta) = -\frac{1}{6} I^2 \sin^4\theta \end{array}\right\}. \qquad (4.5.30)$$

On averaging over θ, we have from (4.5.14),

$$K_1(J) = \overline{H}_1(J,\theta) = -\frac{1}{16} I^2 \approx -\frac{1}{16} J^2. \qquad (4.5.31)$$

Using (4.5.10), (4.5.28), and (4.5.31), (4.5.1b) gives the new Hamiltonian to $0(\varepsilon)$,

$$K(J) = \omega_0 J - \varepsilon \frac{J^2}{16} + 0(\varepsilon^2), \qquad (4.5.32)$$

and hence the new frequency,

$$\omega(J) = \frac{\partial K}{\partial J} = \omega_0 \sim \varepsilon \frac{J}{8} + 0(\varepsilon^2). \qquad (4.5.33)$$

4.5.2 Systems With Slowly Varying Parameters

In classical mechanics, whenever a system has a periodic motion, the action integral $\oint pdq$ taken over a period is a constant of motion. This is nearly so even when slow variations (compared with the period) occur and the motion then being not quite periodic; the approximate constant is then called an adiabatic invariant. More generally, a Hamiltonian system with constant parameters has certain constants of motion; when these parameters vary slowly these constants of motion break up but new approximate constants of motion, called adiabatic invariants, appear.

Consider a Hamiltonian system with a parameter a, with the Hamiltonian given by $H = H(p,q,a)$. By means of a generating function $F(q,\theta,a)$, ($F(q,\theta,a)$ is related to the usual generator $S = \int pdq$ by $F = S - J\theta$) one may go to an action/angle variable representation (see Appendix 2). The new Hamiltonian is given by

$$K(J,\theta,a) = H(J,a) + \frac{\partial F}{\partial t} = H(J,a) + \frac{\partial F}{\partial a}\dot{a}. \qquad (4.5.34)$$

Hamilton's equations in the new frame then give

$$\dot{J} = -\frac{\partial K}{\partial \theta} = -\dot{a}\frac{\partial}{\partial \theta}\left(\frac{\partial F}{\partial a}\right). \qquad (4.5.35)$$

If a is slowly-varying, $\dot{J}$ will be small, so in order to find its cumulative value over long times, one averages $\dot{J}$ over a period τ (treating $\dot{a}$ to be a constant in this interval),

$$\bar{\dot{J}} = -\frac{\dot{a}}{\tau}\oint \frac{\partial}{\partial \theta}\left(\frac{\partial F}{\partial a}\right)dt + 0\left(\ddot{a}, \dot{a}^2\right). \qquad (4.5.36)$$

From $S = \int pdq$ and $J = \frac{1}{2\pi}\oint pdq$, it is obvious that when θ increases by 2π, S will increase by $2\pi J$. Noting that $J\theta$ will also increase by $2\pi J$ during this period, we see that $F = S - J\theta$ is a periodic function of θ and thus can be represented by a Fourier series:

$$F = \sum_k A_k(J,a)\, e^{ik\theta}. \qquad (4.5.37)$$

We then have from equation (4.5.36),

$$\bar{\dot{J}} = -\frac{\dot{a}}{\tau}\int \sum_k ik\frac{\partial A_k}{\partial a}e^{ik\theta}dt + O\left(\ddot{a},\dot{a}^2\right) = O\left(\ddot{a},\dot{a}^2\right). \qquad (4.5.38)$$

Thus, $\bar{J}$ is zero to within $O(\ddot{a}, \dot{a}^2)$; J is an adiabatic invariant. An adiabatic invariant is a dynamical quantity which is approximately constant under slow or adiabatic changes in the Hamiltonian.

One may characterize slow variations by introducing $\tilde{t} = \varepsilon t$, $\varepsilon \ll 1$, thus $a = a(\tilde{t})$; then $\dot{a} = \varepsilon a'(\tilde{t})$.

Equation (4.5.38) then implies that $\bar{J}$ is zero within $O(\varepsilon^2)$.[2]

Example 6: Consider a simple pendulum with constant string length. Then, both energy E and frequency ω are constants. If the string length is now allowed to vary slowly, neither E nor ω will be constants; however, the ratio of E and ω, namely, $J = \dfrac{E}{\omega}$, will be an adiabatic invariant, i.e., the change in J will be of a higher order than the change in the parameter a.

A procedure to construct adiabatic invariants of nearly periodic systems with many degrees of freedom was also given by Gardner (1959) and Kruskal (1962). If the original system is Hamiltonian, then one can define the usual action integral

$$J = \oint p \cdot dq$$

taken around a closed curve. J remains constant in the course of time if the curve evolves in accordance with the equations of motion. This result is then used to reduce the given system to a system of one less degree of freedom by means of a canonical transformation to new coordinates, one of which is the constant J, in which case its conjugate coordinate, viz., the angle variable ϕ describing the phase around the closed curves, is ignorable. If the reduced system can itself be put in Hamiltonian form, and if its solutions are nearly periodic, the whole procedure can be reapplied.

---

[2] In studying adiabatic invariants using perturbation theory, it must be noted that a quantity might appear to be constant to all orders in some perturbation parameter ε without being only constant. For example, the change in the adiabatic invariant may be described by a term like $e^{-1/\varepsilon}$ which does not have a power series expansion in ε about $\varepsilon = 0$ (see Example 3 in Chapter 1).

As an illustration of the method of canonical transformations on the Hamiltonian of systems with slowly varying parameters, let us consider the motion in a system of nonlinearly-coupled oscillators with slowly varying individual frequencies.

For a system of nonlinearly coupled oscillators there is, in general, no effective energy sharing without an internal resonance in the system (see also Example 9, Chapter 6). The energy sharing is usually brought about by resonant interactions among the natural modes of the system. Mathematically, an internal resonance occurs if the uncoupled frequencies $\{\omega_k\}$ of the various oscillators are such that $\sum_{k=1}^{N} n_k \omega_k \approx 0$ for a set of integers $\{n_k\}$. Here, N is the number of oscillators in the system. However, if one allows the various system parameters, such as the uncoupled frequencies of the various oscillators, to vary slowly in time (the restriction to slow variation of the system parameters is crucial for the workability of an analytic perturbation theory), then an internal resonance can prevail only temporarily. One issue of interest in these problems is how the presence of nonlinear coupling affects the adiabatic invariance of the individual actions of the two oscillators. It turns out that for times t sufficiently far from the instant, say T, corresponding to the advent of an internal resonance, the two oscillators individually exhibit adiabatic action invariants. The latter have different values in the pre- and post-resonant regions. During the period when an internal resonance prevails, a certain linear combination of the various actions will be an adiabatic invariant (Kevorkian, 1980, Shivamoggi, 1987).

Example 7: Consider a system of two nonlinearly-coupled oscillators with slowly varying individual frequencies ω_k (ω_k's change very little during the individual periods $2\pi/\omega_k$) having the Hamiltonian

$$H = \frac{1}{2}\left(p_1^2 + p_2^2\right) + \frac{1}{2}\left(\omega_1^2 q_1^2 + \omega_2^2 q_2^2\right) + \varepsilon\left(q_1^2 q_2 - \frac{1}{3} q_2^3\right) \quad (4.5.39)$$

where the small parameter ε characterizes the weak coupling between the two oscillators. Let the parameters $\omega_1(\tilde{t})$ and $\omega_2(\tilde{t})$, where $\tilde{t} = \varepsilon t$, evolve in such a way that the system passes through a state of internal resonance. We now apply successive canonical transformations on the Hamiltonian to eliminate the resonant variables (when an internal resonance prevails) from the Hamiltonian.

Using a canonical transformation generated by a function

$$S(q_1, q_2, I_1, I_2, t) = \int_0^{q_1} \sqrt{2I_1\omega_1 - \omega_1^2 q_1^2}\ dq_1 + \int_0^{q_2} \sqrt{2I_2\omega_2 - \omega_2^2 q_2^2}\ dq_2 \qquad (4.5.40)$$

we have the following action-angle formulation

$$\left. \begin{array}{l} q_1 = \sqrt{\dfrac{2I_1}{\omega_1}}\ \sin\varphi_1, \qquad q_2 = \sqrt{\dfrac{2I_2}{\omega_2}}\ \sin\varphi_2 \\[2mm] p_1 = \sqrt{2I_1\,\omega_1}\ \cos\varphi_1, \quad p_2 = \sqrt{2I_2\,\omega_2}\ \cos\varphi_2 \end{array} \right\} \qquad (4.5.41)$$

where I_1 and I_2 are the actions and φ_1 and φ_2 are the corresponding angles of the two oscillators.

In the action-angle formulation, the new Hamiltonian $\hat{H}$ is then given by

$$\hat{H} = H + \frac{\partial S}{\partial t}$$

$$= (I_1\omega_1 + I_2\omega_2) + \varepsilon \left\{ \left[\frac{I_1}{\omega_1}\sqrt{\frac{2I_2}{\omega_2}} - \frac{1}{4}\left(\frac{2I_2}{\omega_2}\right)^{3/2} \right] \sin\varphi_2 \right.$$

$$\left. - \frac{I_1}{2\omega_1}\sqrt{\frac{2I_2}{\omega_2}} \left[\sin(\varphi_2 - 2\varphi_1) + \sin(\varphi_2 + 2\varphi_1) \right] + \frac{1}{12}\left(\frac{2I_2}{\omega_2}\right)^{3/2} \sin 3\varphi_2 \right\}$$

$$+ \varepsilon \left[\frac{\omega_1' J_1}{2\omega_1} \sin 2\varphi_1 + \frac{\omega_2'}{2\omega_2} \sin 2\varphi_2 \right]$$

$$(4.5.42)$$

where primes denote differentiation with respect to the argument.

Because of the nonlinear coupling between the two oscillators and the variations of the ω_k's (represented by the $0(\varepsilon)$ terms on the right in equation (4.5.42)) the actions I_k of the two oscillators will undergo slow variations. However, since the nonlinear coupling is weak and the variations in the ω_k's are slow, for times t sufficiently far from the instant T corresponding to the advent of an internal resonance, the changes in I_k become effective only when they accumulate over long times. Thus, averaging over φ_k, we obtain

$$\left\langle \hat{H} \right\rangle = I_1\omega_1 + I_2\omega_2 \qquad (4.5.43)$$

according to which

$$I_1 = \text{constant}, \quad I_2 = \text{constant}. \tag{4.5.44}$$

Thus, for times t sufficiently far from the instant T corresponding to the advent of an internal resonance, the individual actions of the two oscillators are adiabatic invariants.

In order to treat the situation for times t near the instant T when an internal resonance prevails, i.e., $\omega_2(\tilde{t}) = 2\omega_1(\tilde{t})$ (which can be recognized from the $0(\varepsilon)$ terms in $\hat{H}$), let us eliminate the resonant variables from $\hat{H}$ by a canonical transformation to a frame of reference that rotates with the resonant frequency. The new coordinates then describe the slow variation of the variables about their values at resonance. We then average over the rapidly rotating phase. The new canonical transformation is generated by the function

$$\tilde{S}(\tilde{I}_1, \tilde{I}_2, \varphi_1, \varphi_2) = (\varphi_2 - 2\varphi_1)\,\tilde{I}_2 + \varphi_1\,\tilde{I}_1 \tag{4.5.45}$$

where $\tilde{I}_k$ are the new "momenta". The new coordinates are then given by

$$\left.\begin{array}{l} \tilde{\varphi}_1 = \dfrac{\partial \tilde{S}}{\partial \tilde{I}_1} = \varphi_1 \\[3mm] \tilde{\varphi}_2 = \dfrac{\partial \tilde{S}}{\partial \tilde{I}_2} = \varphi_2 - 2\varphi_1 \end{array}\right\}. \tag{4.5.46}$$

On the other hand, the old "momenta" are given by

$$\left.\begin{array}{l} I_1 = \dfrac{\partial \tilde{S}}{\partial \varphi_1} = \tilde{I}_1 - 2\tilde{I}_2 \\[3mm] I_2 = \dfrac{\partial \tilde{S}}{\partial \varphi_2} = \tilde{I}_2 \end{array}\right\}. \tag{4.5.47}$$

Using (4.5.45)-(4.5.47), the new Hamiltonian $\hat{H}$ is given by

$$\hat{H} = \hat{H} + \frac{\partial \tilde{S}}{\partial t}$$

$$= \tilde{I}_1 \omega_1 + \tilde{I}_2 (\omega_2 - 2\omega_1) + \varepsilon \left\{ \left[\frac{\tilde{I}_1 - 2\tilde{I}_2}{\omega_1} \sqrt{\frac{2\tilde{I}_2}{\omega_2}} - \frac{1}{4}\left(\frac{2\tilde{I}_2}{\omega_2}\right)^{3/2} \right] \right.$$

$$\times \sin(\tilde{\varphi}_2 + 2\tilde{\varphi}_1) - \frac{\tilde{I}_1 - 2\tilde{I}_2}{2\omega_1}\sqrt{\frac{2\tilde{I}_2}{\omega_2}} \left[\sin \tilde{\varphi}_2 + \sin(\tilde{\varphi}_2 + 4\tilde{\varphi}_1) \right]$$

$$\left. +\frac{1}{12}\left(\frac{2\tilde{I}_2}{\omega_2}\right)^{3/2}\sin 3\left(\tilde{\varphi}_2+2\tilde{\varphi}_1\right)\right\}+\varepsilon\left[\frac{\omega_1'}{2\omega_1}\left(\tilde{I}_1+2\tilde{I}_2\right)\sin 2\tilde{\varphi}_1\right.$$

$$\left. +\frac{\omega_2'}{2\omega_2}\tilde{I}_2\sin 2\left(\tilde{\varphi}_2+2\tilde{\varphi}_1\right)\right]. \tag{4.5.48}$$

Note that near the internal resonance, $\omega_2=2\omega_1$, $\tilde{\varphi}_2$ is the slow variable. On averaging over the fast variable $\tilde{\varphi}_1$, we obtain

$$\langle\tilde{H}\rangle=\tilde{I}_1\omega_1+\tilde{I}_2\left(\omega_2-2\omega_1\right)-\varepsilon\left[\frac{\tilde{I}_1-2\tilde{I}_2}{2\omega_1}\sqrt{\frac{2\tilde{I}_2}{\omega_2}}\sin\tilde{\varphi}_2\right]. \tag{4.5.49}$$

Since, according to (4.5.49), $\tilde{\varphi}_1$ is now the ignorable coordinate, we have

$$\tilde{I}_1=\text{constant} \tag{4.5.50}$$

or, from (4.5.47),

$$I_1+2I_2=\text{constant}. \tag{4.5.51}$$

Thus, during the period an internal resonance prevails, a certain linear combination of the actions I_k is an adiabatic invariant, which of course signifies an exchange of energy between the two oscillators that occurs during an internal resonance.

4.6. Applications to Fluid Dynamics: Nonlinear Evolution of a Modulated Gravity Wavepacket on the Surface of a Fluid

Let us consider the nonlinear evolution of a wavepacket (for which the energy of the motion is concentrated in a narrow wavenumber band) propagating on deep water using the averaged Lagrangian method (Lighthill, 1967, Whitham, 1965, 1967, and Yuen and Lake, 1975).

Consider an initially quiescent semi-infinite fluid confined in the region $y<0$, and subjected to a gravitational field $\boldsymbol{g}=-g\boldsymbol{i}_y$, (see Figure 4.4). Let $y=\eta(x,t)$ denote the disturbed shape of the free surface (whose mean level is given by $y = 0$). The motion of this fluid system is described by the following Lagrangian (Luke, 1967) –

$$L = \int_{-\infty}^{\eta(x,t)} \left[\phi_t + \frac{1}{2} \left(\phi_x^2 + \phi_y^2 \right) + gy \right] dy \qquad (4.6.1)$$

where $\phi(x,y,t)$ is the velocity potential describing the flow, so the fluid velocity v is given by $v = \nabla\phi$.

The variational principle for the wavemotions on the free surface of a fluid is postulated to be

$$\delta \int_{t_1}^{t_2} \int_{x_1}^{x_2} L\, dx\, dt = 0 \qquad (4.6.2)$$

subject to the restrictions that $\delta\phi$, $\delta\eta = 0$ at $x = x_1, x_2$ and $t = t_1, t_2$.

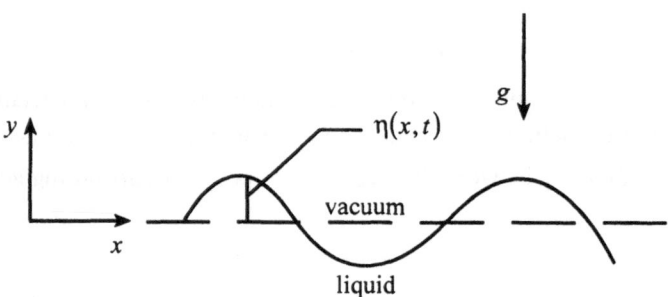

Figure 4.4. Deformed surface of fluid under gravity.

Following the usual procedure of calculus of variations, we obtain from (4.6.1) and (4.6.2):

$$y < \eta(x,t) : \phi_{xx} + \phi_{yy} = 0 \qquad (4.6.3)$$

$$y = \eta(x,t) : -\eta_t + \phi_y - \eta_x \phi_x = 0 \qquad (4.6.4)$$

$$\phi_t + \frac{1}{2} \left(\phi_x^2 + \phi_y^2 \right) + gy = 0 \qquad (4.6.5)$$

$$y \Rightarrow -\infty : \quad \phi_y \Rightarrow 0. \qquad (4.6.6)$$

Equation (4.6.4) describes the kinematic condition on the velocity field at the free surface. Equation (4.6.5) describes the dynamic condition of force balance at the free surface.

Let us now consider a finite-amplitude stationary gravity wave of frequency ω_0 and wavenumber k_0 propagating in the x-direction, and superpose on it a slowly-varying weak modulation, and study the evolution of such a modulation. If, following Whitham (1967), we assume that the wave can still be taken to be sinusoidal locally, i.e.,

$$\eta = a\cos\theta \qquad (4.6.7)$$

but with amplitude and phase varying slowly in x and t, i.e.,

$$\left.\begin{array}{l} a = a(x,t) \\ \theta = \theta(x,t) = k_0 x - \omega_0 t + \varphi(x,t) \end{array}\right\}, \qquad (4.6.8)$$

then we may introduce a generalized frequency ω and wavenumber k:

$$\left.\begin{array}{l} \omega = -\theta_t = \omega_0 - \varphi_t \\ k = \theta_x = k_0 + \varphi_x \end{array}\right\}. \qquad (4.6.9)$$

We have from (4.6.9), a compatibility condition:

$$\frac{\partial k}{\partial t} + \frac{\partial \omega}{\partial x} = 0. \qquad (4.6.10)$$

Using (4.6.3), (4.6.4), and (4.6.6), we obtain

$$\eta = a\cos\theta + k a^2 \cos 2\theta \qquad (4.6.11)$$

$$\phi = \left[\frac{\omega a}{k}\sin\theta + \frac{\omega a_x}{k^2}(1-ky)\cos\theta + \frac{1}{k}a_t\cos\theta\right]e^{ky} + \frac{\omega a^2}{2}\sin 2\theta\, e^{2ky}. \qquad (4.6.12)$$

In order to obtain the equations describing the long-time evolution of the wave, we use (4.6.11) and (4.6.12) in (4.6.1), and calculate the averaged Lagrangian $\mathcal{L}$:

$$\mathcal{L} \equiv \frac{1}{2\pi}\int_0^{2\pi} L\, d\theta = \left(-\frac{\omega^2}{4k} + \frac{g}{4}\right)a^2 + \frac{a_t^2}{4k} + \frac{a\, a_{tt}}{2k}$$

$$+ \frac{\omega a_x a_t}{4k^2} + \frac{3}{8}\frac{\omega^2 a\, a_{xx}}{k^3} + \frac{3}{4}\frac{\omega a\, a_{xt}}{k^2} + \frac{\omega^2 a_x^2}{8k^3} + \frac{1}{8}\omega^2 k a^4. \qquad (4.6.13)$$

Variation of $\mathcal{L}$ with respect to θ gives

$$\frac{\partial}{\partial t}\left(\frac{\partial \mathcal{L}}{\partial \omega}\right) - \frac{\partial}{\partial x}\left(\frac{\partial \mathcal{L}}{\partial k}\right) + \frac{\partial \mathcal{L}}{\partial \theta} = 0 \qquad (4.6.14)$$

which, on using (4.6.10), gives the well-known equation –

$$\frac{\partial}{\partial t}\left(a^2\right)+\frac{\partial}{\partial x}\left(\frac{d\omega}{dk}\,a^2\right)=0. \tag{4.6.15}$$

Next, variation of $\mathcal{L}$ with respect to a gives

$$\frac{\partial \mathcal{L}}{\partial a}=0 \tag{4.6.16}$$

which, on using (4.6.15), gives

$$\omega=\omega_0\left(1+\frac{1}{2}\,k^2a^2-\frac{1}{2\omega_0}\frac{d^2\omega_0}{dk^2}\frac{a_{xx}}{a}\right), \tag{4.6.17}$$

where

$$\omega_0=\sqrt{g\,k_0}\,. \tag{4.6.18}$$

For weak modulations, using (4.6.17), we may write

$$\omega=\omega_0+\frac{d\omega_0}{dk_0}\cdot\left(k-k_0\right)+\frac{1}{2}\frac{d^2\omega_0}{dk_0^2}\cdot\left[\left(k-k_0\right)^2-\frac{a_{xx}}{a}\right]+\frac{\partial\omega}{\partial a_0^2}\cdot\left(a^2-a_0^2\right). \tag{4.6.19}$$

Using (4.6.9), we obtain from (4.6.19) and (4.6.15),

$$\varphi_t+\frac{d\omega_0}{dk_0}\,\varphi_x+\frac{1}{2}\frac{d^2\omega_0}{dk_0^2}\left[\varphi_x^2-\frac{a_{xx}}{a}\right]+\frac{1}{2}\,\omega_0 k_0^2\left(a^2-a_0^2\right)=0 \tag{4.6.20}$$

$$\left(a^2\right)_t+\frac{d\omega_0}{dk_0}\left(a^2\right)_x+\frac{d^2\omega_0}{dk_0^2}\left(\varphi_x a^2\right)_x=0. \tag{4.6.21}$$

If we put

$$\chi=a\,e^{i\varphi} \tag{4.6.22}$$

then we may combine equations (4.6.20) and (4.6.21) into the Schrödinger equation:

$$i\left(\frac{\partial\chi}{\partial t}+\frac{d\omega_0}{dk_0}\frac{\partial\chi}{\partial x}\right)+\frac{1}{2}\frac{d^2\omega_0}{dk_0^2}\frac{\partial^2\chi}{\partial x^2}=\frac{1}{2}\,\omega_0 k_0^2\left(|\chi|^2-|\chi_0|^2\right). \tag{4.6.23}$$

In order to investigate the stability of the modulation governed by equation (4.6.23), we put

$$a=\sqrt{\rho\left(\xi,t\right)}, \tag{4.6.24}$$

where

$$\xi\equiv x-\frac{d\omega_0}{dk_0}\,t. \tag{4.6.25}$$

Then, equations (4.6.20) and (4.6.21) give

$$\rho_t + \frac{d^2\omega_0}{dk_0^2}(\rho\varphi)_\xi = 0 \tag{4.6.26}$$

$$-\frac{1}{2}\omega_0 k_0^2(\rho - \rho_0) - \varphi_t + \frac{1}{4\rho}\frac{d^2\omega_0}{dk_0^2}\left(\rho_{\xi\xi} - \frac{1}{2\rho}\rho_\xi^2 - 2\rho\varphi_\xi^2\right) = 0. \tag{4.6.27}$$

Let us put

$$\rho = \rho_0 + \rho_1(\xi, t), \quad \varphi = \varphi_1(\xi, t) \tag{4.6.28}$$

and assume

$$\rho_1, \varphi_1 \sim e^{i(\kappa\xi - \Omega t)} \tag{4.6.29}$$

and linearize in ρ_1 and φ_1. We then obtain from equations (4.6.26) and (4.6.27):

$$-i\Omega\rho_1 + \frac{\rho_0\omega_0}{4k_0^2}\kappa^2\varphi_1 = 0 \tag{4.6.30}$$

$$\left(-\frac{1}{2}\omega_0 k_0^2 + \frac{\omega_0\kappa^2}{16k_0^2\rho_0}\right)\rho_1 + i\Omega\varphi_1 = 0, \tag{4.6.31}$$

from which we obtain for Ω:

$$\Omega = \pm\left[\frac{\omega_0^2\kappa^2}{8k_0^2}\left(\frac{\kappa^2}{8k_0^2} - k_0^2|\chi_0|^2\right)\right]^{1/2}. \tag{4.6.32}$$

Equation (4.6.32) shows that the modulation is unstable $(\Omega^2 < 0)$ if $|\kappa| < \sqrt{8}\, k_0^2 |\chi_0|$.

4.7. Exercises

1. Use the Krylov-Bogoliubov method of averaging to construct an approximate solution of

$$\ddot{u} + (\delta + \varepsilon\cos 2t)\, u = 0, \quad \varepsilon \ll 1.$$

2. Use the Krylov-Bogoliubov-Mitropolski method to find the limit cycle of

$$\ddot{u} + \varepsilon(u^2 - 1)\,\dot{u} + u - \varepsilon u^3 = 0, \quad \varepsilon \ll 1.$$

3. Solve the forced Duffing equation –

$$\frac{d^2 u}{dt^2} + \omega_0^2 u = \varepsilon u^3 + \varepsilon F_0 \cos \lambda t$$

using the Krylov-Bogoliubov-Mitropolski method for $\omega_0 \neq \lambda$ and $\omega_0 \approx \lambda$ separately.

4. Use the Krylov-Bogoliubov-Mitropolski method to construct an approximate solution for a system of two nonlinearly-coupled oscillators

$$\ddot{x} + \omega_1^2 x = -\varepsilon \cdot 2xy,$$

$$\ddot{y} + \omega_2^2 y = \varepsilon(y^2 - x^2), \quad \varepsilon \ll 1.$$

Show that this solution breaks down for an internal resonance $\omega_2 = 2\omega_1$. Construct a solution that deals with this case.

5. Show that the Boussinesq equation

$$\phi_{tt} - \phi_{xx} - \phi_{xxxx} - (\phi^2)_{xx} = 0$$

has the variational characterization

$$\delta \iint L\,dx\,dt = 0$$

where the Lagrangian is

$$L = \frac{1}{2}\,\psi_{tt}\psi_{xx} - \frac{1}{2}\,\psi_{xx}^2 + \frac{1}{2}\,\psi_{xxx}^2 + \frac{1}{3}\,\psi_{xx}^3, \quad \phi = \psi_{xx}.$$

Find the averaged Lagrangian $\mathcal{L}$ and the variational equations.

6. Find the first-order corrections to the motion of the linear oscillator with the Hamiltonian

$$H_0(q,p) = \frac{1}{2}\left(p^2 + q^2\right)$$

perturbed by

$$H_1(q,p) = \frac{1}{2}\,q^4.$$

7. A particle of mass m slides under the influence of gravity on a smooth rigid wire in the shape $z = \frac{1}{2}\,\alpha^2 x^2$, where the z-axis is vertically upwards. Find the Hamiltonian for the system and show that an approximation to this Hamiltonian, correct to $O(p^2 x^2)$, is

$$H = \frac{p^2}{2m} + \frac{mg}{2}\,\alpha^2 x^2 - \frac{\alpha^4}{2m}\,x^2 p^2.$$

Find an approximation to the motion which is correct to $O\left(p^2 x^2\right)$.

4.8. Appendix 1

Review of Calculus of Variations

A1.1 Functionals with Second-Order Derivatives

Calculus of variations is a generalization of the theory of extremals of point functions. The problem is to find a function $y = \phi(x)$ from a suitable class Γ of admissible functions, which are single-valued with continuous first and second derivatives on the interval of interest, such that the functional

$$I(y) = \int_{x_1}^{x_2} F(x, y, y')\, dx \tag{A1.1}$$

is an extremal. A necessary condition for the functional I to have an extremal is the vanishing of its first variation.

In this context, it is useful to briefly review first the extrema of point functions.

If a point function $f(x, y)$ is defined throughout a region R and if the partial derivatives $\dfrac{\partial f}{\partial x}$ and $\dfrac{\partial f}{\partial y}$ exist and are continuous in R, then *necessary* conditions that f possess an extremal at some point (x_0, y_0) in R are that $\partial f / \partial x = \partial f / \partial y = 0$ at the point (x_0, y_0) or that the differential df satisfies

$$df = \frac{\partial f}{\partial x}\, dx + \frac{\partial f}{\partial y}\, dy = 0 \tag{A1.2}$$

at the point (x_0, y_0). Let us now generalize the concept of differential (A1.2) so that we can treat functionals in a similar manner.

If we write

$$y = \phi(x) + \varepsilon\eta(x) \tag{A1.3}$$

the change $\varepsilon\eta(x)$ is called the variation of y,

$$\delta y \equiv \varepsilon\eta(x). \tag{A1.4}$$

Here, δy plays a role similar to that of the differential df in the discussion of point functions. However, from a geometrical viewpoint, the differential df of a point function $f(x)$ of one variable is a first-order approximation to the change in that function *along a particular curve*. On the other hand, the variation δy represents a first-order approximation to the change in y from *curve to curve*. More generally, we have the following definition for a variation of functional.

Definition: *The first variation of the functional I is*

$$\delta I \equiv \varepsilon \lim_{\varepsilon \to 0} \frac{d}{d\varepsilon} I(\phi + \varepsilon \eta)$$

whenever this limit exists for all η in the class Γ.

In order to calculate the first variation of the functional I, one compares the values of I that correspond to neighboring curves $y(x)$ anchored on the same end points at $x = x_1$ and x_2 (see Figure 4.5).

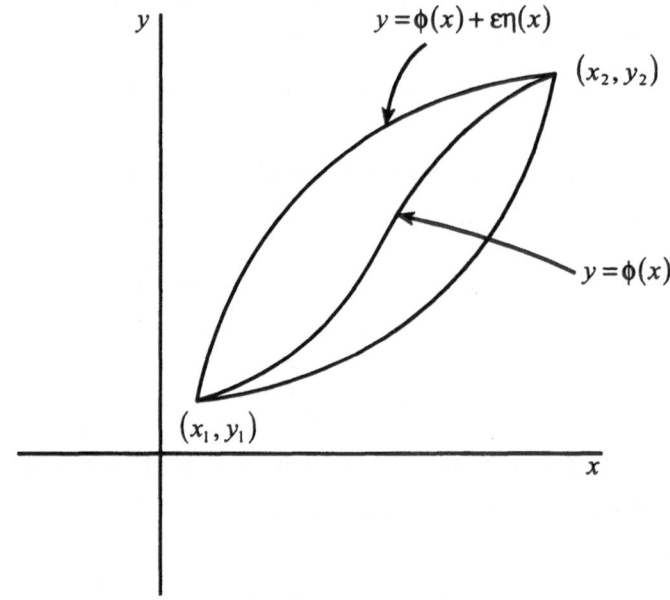

Figure 4.5. Variations about the optimal curve.

Applying the variational operator δ to the functional –

$$I(y) = \int_{x_1}^{x_2} F(x, y, y') \, dx, \tag{A1.5}$$

we have

$$\delta I(y) = \int\limits_{x_1}^{x_2} \delta F(x,y,y')\, dx \tag{A1.6}$$

under suitable continuity requirements on F.

Note,

$$\delta F = \varepsilon \lim_{\varepsilon \to 0} \frac{d}{d\varepsilon}\, F\big(x, \phi + \varepsilon\eta,\ \phi' + \varepsilon\eta'\big)$$

$$= \varepsilon \left[\eta(x)\, \frac{\partial F}{\partial y} + \eta'(x)\, \frac{\partial F}{\partial y'} \right].$$

Setting

$$\varepsilon\eta = \delta y \quad \text{and} \quad \varepsilon\eta' = \delta y',$$

we have

$$\delta F = \frac{\partial F}{\partial y}\, \delta y + \frac{\partial F}{\partial y'}\, \delta y'. \tag{A1.7}$$

Therefore,

$$\delta I = \int\limits_{x_1}^{x_2} \left(\frac{\partial F}{\partial y}\, \delta y + \frac{\partial F}{\partial y'}\, \delta y' \right) dx. \tag{A1.8}$$

Noting the commutativity of the two operators δ and d/dx,

$$\delta y' \equiv \varepsilon\eta' = (\varepsilon\eta)' = (\delta y)'$$

and integrating by parts, we have

$$\delta I = \frac{\partial F}{\partial y'}\, \delta y \bigg]_{x_1}^{x_2} + \int\limits_{x_1}^{x_2} \left[\frac{\partial F}{\partial y} - \frac{d}{dx}\left(\frac{\partial F}{\partial y'} \right) \right] \delta y\, dx. \tag{A.19}$$

We have the forced boundary conditions

$$\left. \begin{array}{l} x = x_1 : y = y_1 \\ x = x_2 : y = y_2 \end{array} \right\} \tag{A1.10a}$$

so,

$$x = x_1 \ \text{and} \ x_2 : \delta y = 0$$

or

$$x = x_1 \ \text{and} \ x_2 : \eta = 0. \tag{A1.10b}$$

Lemma: *If $g(x)$ is a continuous function in $x_1 \leq x \leq x_2$, and if*

$$\int_{x_1}^{x_2} \eta(x)\, g(x)\, dx = 0$$

for arbitrary η in class Γ for which $\eta(x_1) = 0$ and $\eta(x_2) = 0$, then $g(x) \equiv 0$ in $x_1 \leq x \leq x_2$.

Using the above lemma, an extremal of $I(y)$, which corresponds to $\delta I = 0$, with δI given by (A.19), is given by the Euler-Lagrange equation –

$$\frac{\partial F}{\partial y} - \frac{d}{dx}\left(\frac{\partial F}{\partial y'}\right) = 0. \tag{A1.11}$$

A1.2 Functionals with Higher-Order Derivatives

Consider extremals of functionals of the form

$$I(y) = \int_{x_1}^{x_2} F(x, y, y', y'')\, dx \tag{A1.12}$$

where F satisfies general continuity and differentiability requirements in each of its arguments x, y, y', y''. We seek extremals of (A1.12) subject to the forced boundary conditions –

$$\left.\begin{array}{l} y(x_1) = y_1, \quad y(x_2) = y_2 \\ y'(x_1) = y_1', \quad y'(x_2) = y_2' \end{array}\right\}. \tag{A1.13}$$

A necessary condition for I to have an extremal is the vanishing of its first variation. In accordance with the variational notation introduced above, we have

$$\delta I = \int_{x_1}^{x_2}\left(\frac{\partial F}{\partial y}\,\delta y + \frac{\partial F}{\partial y'}\,\delta y' + \frac{\partial F}{\partial y''}\,\delta y''\right) dx. \tag{A1.14}$$

Replacing $\delta y'$ by $(\delta y)'$ and $\delta y''$ by $(\delta y)''$, and then integrating by parts we obtain for an extremal of I,

$$\delta I = \int_{x_1}^{x_2}\left[\frac{\partial F}{\partial y} - \frac{d}{dx}\left(\frac{\partial F}{\partial y'}\right) + \frac{d^2}{dx^2}\left(\frac{\partial F}{\partial y''}\right)\right]\delta y\, dx$$

$$+ \left[\frac{\partial F}{\partial y'} - \frac{d}{dx}\left(\frac{\partial F}{\partial y''}\right)\right]\delta y\,\Bigg|_{x_1}^{x_2} + \frac{\partial F}{\partial y''}\,\delta y'\,\Bigg|_{x_1}^{x_2} = 0 \tag{A1.15}$$

from which, invoking the Lemma, we deduce the Euler-Lagrange equation

$$\frac{\partial F}{\partial y} - \frac{d}{dx}\left(\frac{\partial F}{\partial y'}\right) + \frac{d^2}{dx^2}\left(\frac{\partial F}{\partial y''}\right) = 0. \tag{A1.16}$$

Because of the forced boundary conditions (A1.13), we have $\delta y = 0$ and $\delta y' = 0$ at $x = x_1, x_2$, and no natural boundary conditions are required. On the other hand, if no forced boundary conditions are specified, then in order to satisfy (A1.15) we must prescribe the natural boundary conditions

$$\left[\frac{\partial F}{\partial y'} - \frac{d}{dx}\left(\frac{\partial F}{\partial y''}\right)\right]\Bigg|_{x=x_1} = 0, \quad \frac{\partial F}{\partial y''}\Bigg|_{x=x_1} = 0 \tag{A1.17}$$

$$\left[\frac{\partial F}{\partial y'} - \frac{d}{dx}\left(\frac{\partial F}{\partial y''}\right)\right]\Bigg|_{x=x_2} = 0, \quad \frac{\partial F}{\partial y''}\Bigg|_{x=x_2} = 0. \tag{A1.18}$$

One may generalize the above results to functionals of the more general form

$$I(y) = \int_{x_1}^{x_2} F\left(x, y, y', \ldots, y^{(n)}\right) dx \tag{A1.19}$$

where n is a fixed integer satisfying $n > 2$. For extremals of such functionals, it can be shown that the corresponding Euler-Lagrange equation has the form

$$\frac{\partial F}{\partial y} - \frac{d}{dx}\left(\frac{\partial F}{\partial y'}\right) + \frac{d^2}{dx^2}\left(\frac{\partial F}{\partial y''}\right) - \cdots + (-1)^n \frac{d^n}{dx^n}\left(\frac{\partial F}{\partial y^{(n)}}\right) = 0. \tag{A1.20}$$

A1.3 Functionals with Several Independent Variables

Let us now consider a functional that is a function of a point function u of two independent variables x and y. We assume that u is continuous with continuous first and second partial derivatives in some bounded, simply-connected region R with smooth boundary curve C and assumes prescribed boundary values on C but is otherwise arbitrary. The functional that we shall study is then defined by

$$I(u) = \iint_R F\left(x, y, u, u_x, u_y\right) dx\, dy \tag{A1.21}$$

where u_x, u_y and F are continuous and F has continuous first and second partial derivatives in each of its arguments.

In considering functionals of the type (A1.21), it proves to be useful to note Green's Theorem in the Plane:

Theorem: *If R is a bounded, simply-connected region bounded by a simple closed curve[3] C in the x,y-plane, and if* $P(x,y)$ *and* $Q(x,y)$ *together with their first partial derivatives are continuous in R, then*

$$\iint_R \left(\frac{\partial Q}{\partial x} - \frac{\partial P}{\partial y} \right) dx\, dy = \oint_C \left[P(x,y)\, dx + Q(x,y)\, dy \right].$$

If we perturb the solution u about the extremal solution $\phi(x,y)$ according to

$$u = \phi(x,y) + \varepsilon\eta(x,y) = \phi(x,y) + \delta u \tag{A1.22}$$

then we have for the first variation of I,

$$\delta I = \iint_R \left(\frac{\partial F}{\partial u}\, \delta u + \frac{\partial F}{\partial u_x}\, \delta u_x + \frac{\partial F}{\partial u_y}\, \delta u_y \right) dx\, dy. \tag{A1.23}$$

However, if we make use of the identity –

$$\frac{\partial}{\partial x}\left(\frac{\partial F}{\partial u_x}\, \delta u \right) + \frac{\partial}{\partial y}\left(\frac{\partial F}{\partial u_y}\, \delta u \right)$$

$$= \left[\frac{\partial}{\partial x}\left(\frac{\partial F}{\partial u_x} \right) + \frac{\partial}{\partial y}\left(\frac{\partial F}{\partial u_y} \right) \right] \delta u + \frac{\partial F}{\partial u_x}\, \delta u_x + \frac{\partial F}{\partial u_y}\, \delta u_y \tag{A1.24}$$

then we can rewrite (A1.22) as

$$\delta I = \iint_R \left[\frac{\partial F}{\partial u} - \frac{\partial}{\partial x}\left(\frac{\partial F}{\partial u_x} \right) - \frac{\partial}{\partial y}\left(\frac{\partial F}{\partial u_y} \right) \right] \delta u\, dx\, dy$$

$$+ \iint_R \left[\frac{\partial}{\partial x}\left(\frac{\partial F}{\partial u_x}\, \delta u \right) + \frac{\partial}{\partial y}\left(\frac{\partial F}{\partial u_y}\, \delta u \right) \right] dx\, dy. \tag{A1.25}$$

By applying Green's Theorem in the Plane to the second integral in (A1.25), we obtain

$$\delta I = \iint_R \left[\frac{\partial F}{\partial u} - \frac{\partial}{\partial x}\left(\frac{\partial F}{\partial u_x} \right) - \frac{\partial}{\partial y}\left(\frac{\partial F}{\partial u_y} \right) \right] \delta u\, dx\, dy$$

$$+ \oint_C \left[-\frac{\partial F}{\partial u_y}\, dx + \frac{\partial F}{\partial u_x}\, dy \right] \delta u. \tag{A1.26}$$

---

[3] A simple closed curve is a curve consisting of a finite number of arcs with continuously turning tangents, joined at the end points, but not crossing over itself.

Here,

$$\frac{\partial}{\partial x}\, G\!\left(x,y,u,u_x,u_y\right) = \frac{\partial G}{\partial x} + \frac{\partial G}{\partial u}\, u_x + \frac{\partial G}{\partial u_x}\, u_{xx} + \frac{\partial G}{\partial u_y}\, u_{xy}.$$

For stationary values of $I(u)$, we have $\delta I = 0$, which upon invoking a two-dimensional form of the lemma, leads to the two-dimensional Euler-Lagrange equation

$$\frac{\partial F}{\partial u} - \frac{\partial}{\partial x}\left(\frac{\partial F}{\partial u_x}\right) - \frac{\partial}{\partial y}\left(\frac{\partial F}{\partial u_y}\right) = 0 \qquad\qquad \text{(A1.27)}$$

along with either the forced boundary condition –

$$u = f(x,y) \quad \text{on} \quad C \qquad\qquad \text{(A1.28)}$$

in which case $\delta u = 0$ on C or the natural boundary condition –

$$\frac{\partial F}{\partial u_y} - \frac{\partial F}{\partial u_x}\frac{dy}{dx} = 0 \quad \text{on} \quad C \qquad\qquad \text{(A1.29)}$$

in order that the line integral in (A1.26) vanishes.

Example 1 Consider the functional

$$I(u) = \iint_R \left(u_x^2 + u_y^2\right) dx\, dy$$

with $u = f(x,y)$ on C.

We have for this functional –

$$\frac{\partial F}{\partial u} = 0, \quad \frac{\partial F}{\partial u_x} = 2u_x, \quad \frac{\partial F}{\partial u_y} = 2u_y.$$

The Euler-Lagrange equation (A1.27) associated with the stationary condition of the above functional then reduces to the Laplace equation

$$u_{xx} + u_{yy} = 0.$$

The solution of this equation in a particular region R of the xy-plane, subject to the forced boundary condition $u = f(x,y)$ on C, is the *Dirichlet problem*.

4.9. Appendix 2

Hamilton-Jacobi Theory[4]

A2.1 Hamilton's Equations

Consider a system of n degrees of freedom which is described by a Lagrangian L^5 which is a function of the generalized coordinates q_j, $(j = 1, 2, \ldots, n)$, the generalized velocities $\dot{q}_j$ $(j = 1, 2, \ldots, n)$ and the time t –

$$L = L\left(q_j, \dot{q}_j, t\right). \tag{A2.1}$$

Then, the actual motion of the system from a position $q_j^{(1)}$ at time t_1 to a position $q_j^{(2)}$ at time t_2 can be determined from Hamilton's principle of least action which requires that the integral of the Lagrangian function takes the minimum possible value between the initial time t_1 and the final time t_2 for the actual path. More precisely, the actual path in configuration space between two configurations $q_j(t_1)$ and $q_j(t_2)$ at times t_1 and t_2, respectively, is that which makes the time integral of the Lagrangian function stationary with respect to variations δq_j of the path which vanish at the end points, i.e.,

$$\delta \Phi = \delta \int_{t_1}^{t_2} L\left(q_j, \dot{q}_j, t\right) dt = 0. \tag{A2.2}$$

The functional Φ is called Hamilton's principal function (or the action integral) for the path $q(t)$.

Using the techniques of variational calculus, (A2.2) can be expressed as

$$\delta \Phi = \sum_{j=1}^{n} \int_{t_1}^{t_2} \left[\frac{\partial L}{\partial q_j} - \frac{d}{dt}\left(\frac{\partial L}{\partial \dot{q}_j} \right) \right] \delta q_j \, dt = 0, \quad j = 1, 2, \ldots, n. \tag{A2.3}$$

Because of the arbitrariness of δq_j between $t = t_1$ and $t = t_2$, equation (A2.3) leads to Lagrange's equations

---

[4] See Shivamoggi (1997) for a more detailed discussion.

[5] A Lagrangian of a system is typically the difference between the kinetic energy and the potential energy of the system.

$$\frac{\partial L}{\partial q_j} - \frac{d}{dt}\left(\frac{\partial L}{\partial \dot{q}_j}\right) = 0, \quad j = 1, 2, \ldots, n. \tag{A2.4}$$

If we introduce generalized momenta according to

$$p_j = \frac{\partial L}{\partial \dot{q}_j}, \quad j = 1, 2, \ldots, n, \tag{A2.5}$$

Lagrange's equations (A2.4) become

$$\dot{p}_j = \frac{\partial L}{\partial q_j}, \quad j = 1, 2, \ldots, n. \tag{A2.6}$$

Let us now use q_j, p_j, t as the independent variables (q_j and p_j are called the conjugate variables) rather than $q_j, \dot{q}_j, t$. Noting that

$$\begin{aligned}
\delta L &= \sum_j \frac{\partial L}{\partial q_j}\,\delta q_j + \sum_j \frac{\partial L}{\partial \dot{q}_j}\,\delta \dot{q}_j + \frac{\partial L}{\partial t}\,\delta t \\
&= \sum_j \dot{p}_j\,\delta q_j + \sum_j p_j\,\delta \dot{q}_j + \frac{\partial L}{\partial t}\,\delta t \\
&= \delta\left(\sum_j p_j \dot{q}_j\right) + \sum_j \dot{p}_j\,\delta q_j - \sum_j \dot{q}_j\,\delta p_j + \frac{\partial L}{\partial t}\,\delta t,
\end{aligned} \tag{A2.7}$$

we have

$$\delta\left(\sum_j p_j \dot{q}_j - L\right) = -\sum_j \dot{p}_j\,\delta q_j + \sum_j \dot{q}_j\,\delta p_j - \frac{\partial L}{\partial t}\,\delta t \tag{A2.8}$$

which describes a Legendre transformation from L to the Hamiltonian H,[6]

$$H = H(q_j, p_j, t) = \sum_j p_j \dot{q}_j - L. \tag{A2.9}$$

We have from (A2.9)

$$\delta H = \sum_j \frac{\partial H}{\partial q_j}\,\delta q_j + \sum_j \frac{\partial H}{\partial p_j}\,\delta p_j + \frac{\partial H}{\partial t}\,\delta t \tag{A2.10}$$

which, on comparing with equation (A2.8), leads to Hamilton's equations –

---

[6] H is typically the energy of the system.

$$\left. \begin{array}{l} \dot{q}_j = \dfrac{\partial H}{\partial p_j} \\[2ex] \dot{p}_j = -\dfrac{\partial H}{\partial q_j} \\[2ex] \dfrac{\partial H}{\partial t} = -\dfrac{\partial L}{\partial t} \end{array} \right\}_{j=1,2,\ldots,n} \qquad (A2.11)$$

Observe that if a particular coordinate is cyclic, i.e., it does not appear explicitly in H, the corresponding momentum is a constant of the motion. Similarly, if L does not depend on t explicitly, H is a constant of motion and we have conservation of energy.

A2.2 Canonical Transformations

Canonical transformations can be of practical use in simplifying the integration of Hamilton's equations.

Let $L\left(q_j,\dot{q}_j,t\right)$ and $\overline{L}\left(\overline{q}_j,\dot{\overline{q}}_j,t\right)$ be two Lagrangian functions involving the same number of degrees of freedom. Let q_i and $\overline{q}_i$ be related so that a path in the q_i-space for which the time integral of L is stationary corresponds to a path in $\overline{q}_i$-space for which the time integral of $\overline{L}$ is stationary. Then these two Lagrangians provide two different descriptions of the same system. This implies that

$$\overline{L}\left(\overline{q}_j,\dot{\overline{q}}_j,t\right) = L\left(q_j,\dot{q}_j,t\right) - \frac{d}{dt}\,\psi\left(q_j,\overline{q}_j,t\right) \qquad (A2.12)$$

because L is unique to within an additive total derivative of a scalar function ψ.[7] Thus, we have from (A2.12),

$$\delta\,\overline{L}\left(\overline{q}_j,\dot{\overline{q}}_j,t\right) = \delta\,L\left(q_j,\dot{q}_j,t\right) - \frac{d}{dt}\,\delta\psi\left(q_j,\overline{q}_j,t\right)$$

or

---

[7] The total time derivative cannot contribute to the variation of the time integral in Hamilton's principle (A2.2).

$$\sum_i \frac{\partial \overline{L}}{\partial \overline{q}_i}\, \delta \overline{q}_i + \sum_i \overline{p}_i\, \delta \dot{\overline{q}}_i + \frac{\partial \overline{L}}{\partial t}\, \delta t$$

$$= \sum_i \frac{\partial L}{\partial q_i}\, \delta q_i + \sum_i p_i\, \delta \dot{q}_i + \frac{\partial L}{\partial t}\, \delta t$$

$$- \frac{d}{dt}\left(\sum_i \frac{\partial \psi}{\partial q_i}\, \delta q_i + \sum_i \frac{\partial \psi}{\partial \overline{q}_i}\, \delta \overline{q}_i + \frac{\partial \psi}{\partial t}\, \delta t \right). \qquad \text{(A2.13)}$$

Choosing

$$p_i = \frac{\partial \psi}{\partial q_i}, \quad \overline{p}_i = -\frac{\partial \psi}{\partial \overline{q}_i}; \quad i = 1, 2, \ldots, n, \qquad \text{(A2.14)}$$

we find, from (A2.13) that the Lagrange equations remain form invariant under a group of canonical transformations generated by the function $\psi\left(q_j, \overline{q}_j, t\right)$.[8]

Using (A2.12) and (A2.14), the new Hamiltonian is then given by

$$\overline{H} = \sum_i \overline{p}_i\, \dot{\overline{q}}_i - \overline{L} = \sum_i p_i \dot{q}_i - L + \frac{\partial \psi}{\partial t}$$

$$\qquad \text{(A2.15)}$$

$$= H\left(q_i, p_i, t\right) + \frac{\partial}{\partial t}\, \psi\left(q_i, \overline{q}_i, t\right).$$

ψ is called the generating function, and we have from (A2.14) and (A2.15),

$$d\psi = \sum_i p_i dq_i - \sum_i \overline{p}_i\, d\overline{q}_i - \left(H - \overline{H}\right) dt. \qquad \text{(A2.16)}$$

One may find other types of generating functions by making Legendre transformations. Thus, letting

$$\psi\left(q_j, \overline{q}_j, t\right) = \tilde{\psi}\left(q_j, \overline{p}_j, t\right) - \sum_j \overline{p}_j\, \overline{q}_j \qquad \text{(A2.17)}$$

we find that $\tilde{\psi}\left(q_j, \overline{p}_j, t\right)$ generates a canonical transformation according to

$$p_j = \frac{\partial \tilde{\psi}\left(q_j, \overline{p}_j, t\right)}{\partial q_j}, \quad \overline{q}_j = \frac{\partial \tilde{\psi}\left(q_j, \overline{p}_j, t\right)}{\partial \overline{p}_j}. \qquad \text{(A2.18)}$$

Using (A2.15) and (A2.18), we have for the new Hamiltonian

$$\overline{H}\left(q_j, \overline{p}_j, t\right) = H\left(q_j, p_j, t\right) + \frac{\partial \tilde{\psi}\left(q_j, \overline{p}_j, t\right)}{\partial t}. \qquad \text{(A2.19)}$$

---

[8] Note that the coordinates and momenta do not remain necessarily distinct under such transformations.

A2.3 Hamilton-Jacobi Equation

Let us choose a canonical transformation so that $\overline{H}$=constant. Then, we may take $\overline{H} = 0$, which trivially integrates the canonical equations to give $\overline{p}_i$=constant and $\overline{q}_i$=constant.

Noting (A2.19), the Hamilton-Jacobi equation for the generating function $\tilde{\psi}$ is then given by

$$\frac{\partial \tilde{\psi}}{\partial t} + H\left(q_j, \frac{\partial \tilde{\psi}}{\partial q_j}, t\right) = 0; \quad j = 1,\ldots,n. \tag{A2.20}$$

If H is not an explicit function of t, then

$$H(q_i, p_i) = \text{constant} = \alpha_1, \text{ say}. \tag{A2.21}$$

We now seek a canonical transformation generated by $\tilde{\psi} = S(q_i, \alpha_i)$ such that all the new momenta $\overline{p}_i$ are constants, say α_j. The new Hamiltonian $\overline{H}$ will then be equal to H or α_1 and will be cyclic in all the new coordinates $\overline{q}_j$. The new equations of motion are

$$\left.\begin{aligned}
\dot{\overline{q}}_i &= \frac{\partial \overline{H}}{\partial \overline{p}_i} = \frac{\partial \overline{H}}{\partial \alpha_i} = \begin{cases} 1, & i = 1 \\ 0, & i \neq 1 \end{cases} \\
\dot{\overline{p}} &= -\frac{\partial \overline{H}}{\partial \overline{q}_i} = 0; \quad i = 1,\ldots,n
\end{aligned}\right\}. \tag{A2.22}$$

The generating function $S(q_i, \alpha_i)$ produces the following transformation:

$$p_i = \frac{\partial S}{\partial q_i}, \quad \overline{q}_i = \frac{\partial S}{\partial \overline{p}_i} = \frac{\partial S}{\partial \alpha_i} \tag{A2.23}$$

and satisfies the following equation:

$$H\left(q_i, \frac{\partial S}{\partial q_i}\right) = \alpha_1. \tag{A2.24}$$

α_1 is called an isolating integral (Whittaker, 1964) since it isolates one degrees of freedom from the other $(n-1)$ degrees of freedom.

In order to see the physical significance of S, note that

$$\frac{dS}{dt} = \sum_i \frac{\partial S}{\partial q_i} \dot{q}_i = \sum_i p_i \dot{q}_i \tag{A2.25}$$

so that

$$S = \sum_i \int p_i \dot{q}_i dt = \sum_i \int p_i dq_i \qquad \text{(A2.26)}$$

which is simply the action integral!

A2.4 Action-Angle Variables

In cases where the motion is periodic, one may be interested in some average characteristics of the motion rather than the details of the motion. Toward this objective, one modifies the Hamilton-Jacobi approach slightly so that the integration constants α_i are chosen to define a set of n constants called action variables J_i.

Consider, for the sake of illustration, a system with one degree of freedom described by a Hamiltonian $H = H(q, p)$. Apply a canonical transformation generated by $\tilde{S}(q, \bar{p})$ so that

$$p = \frac{\partial \tilde{S}}{\partial q}, \quad \bar{q} = \frac{\partial \tilde{S}}{\partial \bar{p}}, \quad \bar{H} = H. \qquad \text{(A2.27)}$$

Hamilton's equations in the new coordinates are then

$$\dot{\bar{p}} = -\frac{\partial \bar{H}}{\partial \bar{q}}, \quad \dot{\bar{q}} = \frac{\partial \bar{H}}{\partial \bar{p}}. \qquad \text{(A2.28)}$$

Let $\bar{p}$=constant=J, say. Then $\bar{q} = \theta$ is a cyclic coordinate, and the Hamilton-Jacobi equation (A2.20) for the generating function $\tilde{S}$ becomes

$$\bar{H} = \bar{H}(J) = H(q, p) = H\left(q, \frac{\partial \tilde{S}}{\partial q}\right) \qquad \text{(A2.29)}$$

with

$$\theta = \frac{\partial \tilde{S}}{\partial J}. \qquad \text{(A2.30)}$$

On the other hand, from equations (A2.28), we have

$$\dot{\theta} = \frac{d\bar{H}}{dJ} = \omega, \text{ say}, \qquad \text{(A2.31)}$$

from which,

$$\theta = \omega t + \delta. \qquad \text{(A2.32)}$$

Note that if J has the dimension of action, θ is dimensionless and is called the angle variable.

Let us suppose now that q and p are periodic functions of t, and let us take J to be the action evaluated over one period

$$J = \frac{1}{2\pi} \oint p\,dq. \tag{A2.33}$$

Note that during one period of the motion, θ increases by an amount

$$\Delta\theta = \oint d\theta = \oint \frac{\partial}{\partial q}\left(\frac{\partial \tilde{S}}{\partial J}\right)dq = \frac{\partial}{\partial J}\oint \frac{\partial \tilde{S}}{\partial q}dq = \frac{\partial}{\partial J}\oint p\,dq = 2\pi \tag{A2.34}$$

where we have use (A2.27), (A2.30), and (A2.33). (A2.34) shows that ω is the frequency of the periodic motion. Since the integral $\oint p\,dq$ represents the area enclosed by an orbit of energy $H = E$ in the phase plane, each orbit is uniquely labelled by J, which is constant along every orbit. Each point on an orbit is labelled by a single-valued function of θ. Thus, the action-angle formulation enables us to calculate the frequencies of the periodic motions directly without finding the variations of the coordinates with time.

On using (A2.27), (A2.33) shows that

$$J = \frac{1}{2\pi}\oint\left(\frac{\partial \tilde{S}}{\partial q}\right)dq. \tag{A2.35}$$

So, the change in S during one period is given by

$$\Delta\tilde{S} \equiv \oint d\tilde{S} = 2\pi J \tag{A2.36}$$

The relation (see (A2.17))

$$\tilde{S}(q,\bar{p}) = S(q,\theta) + J\theta \tag{A2.37}$$

then shows that $S(q,\theta)$ is a periodic function θ with period 2π.

Chapter 5

The Method of Matched Asymptotic Expansions

5.1. Introduction

In cases where a small parameter multiplies the highest derivative in a differential equation, there occurs a sharp change in the dependent variable in a certain region of the domain of the independent variable. In constructing a solution to the differential equation through uniformly-valid expansions, one characterizes the sharp changes by a magnified scale that is different from the scale characterizing the behavior of the dependent variable outside the "boundary-layer" regions. In other words, one represents the solution by two different asymptotic expansions using the independent variables x and x/ε say. Since they are different asymptotic representations of the same function, they should be related to each other in a rational manner in an overlapping region where both are valid (Friedrichs, 1955); this leads to the asymptotic matching principle (the latter makes the two representations completely determinate).

5.2. Physical Motivation

In order to physically motivate this method, consider the response of a linear spring-mass-damping system initially at rest to an impulse I_0. Denoting the mass by m, the damping constant by β, and the spring constant by k, we have the initial-value problem (Cole, 1968):

$$\left.\begin{array}{l} m\ddot{y} + \beta\dot{y} + ky = I_0\,\delta(t) \\ y(0^-) = 0, \quad \dot{y}(0^-) = 0 \end{array}\right\}. \tag{5.2.1}$$

For very small values of m, the initial-value problem (5.2.1) may be approximated by

$$\left.\begin{array}{l} \beta\dot{y}^{(0)} + ky^{(0)} = I_0\,\delta(t) \\ y^{(0)}(0^-) = 0 \end{array}\right\}. \tag{5.2.2}$$

Since the order of the differential equation in (5.2.2) has decreased from 2 to 1, one of the two initial conditions in (5.2.1) must be dropped. This implies the existence of a boundary layer at $t = 0$, and the above truncated equation is valid only for t away from $t = 0$. Noting, from (5.2.2), that

$$y^{(0)}(0^+) = \frac{I_0}{\beta},$$

we have

$$y^{(0)} = \frac{I_0}{\beta}\,e^{-\frac{kt}{\beta}}. \tag{5.2.3}$$

However, in a short interval near $t = 0$, the displacement is produced infinitely rapidly from 0 to $\frac{I_0}{\beta}$. In order to describe the motion during this initial interval, we note that inertia is dominant at $t = 0$ (impulse-momentum balance) and that due to the large initial velocity, damping becomes effective immediately, but the spring does not, since the deflection is small. Thus, for $t \approx 0$, the initial-value problem (5.2.1) may be approximated by

$$\left.\begin{array}{l} m\ddot{y}^{(i)} + \beta\dot{y}^{(i)} = I_0\,\delta(t) \\ y^{(i)}(0^-) = \dot{y}^{(i)}(0^-) = 0 \end{array}\right\}. \tag{5.2.4}$$

Integrating with respect to t from 0^- to t, we have from (5.2.4),

$$m\dot{y}^{(i)} + \beta y^{(i)} = I_0 \tag{5.2.5}$$

from which we obtain

$$y^{(i)} = \frac{I_0}{\beta} + c\,e^{-\frac{\beta}{m}t} \tag{5.2.6}$$

where c is an arbitrary constant.

Noting from equation (5.2.5) that

$$\dot{y}^{(i)}(0^+) = \frac{I_0}{m} \tag{5.2.7}$$

we have, on using (5.2.6),

$$c = -\frac{I_0}{\beta}. \tag{5.2.8}$$

Using (5.2.8), (5.2.6) becomes

$$y^{(i)} = \frac{I_0}{\beta}\left[1 - e^{-\frac{\beta}{m}t}\right]. \tag{5.2.9}$$

Equation (5.2.3) and (5.2.9) appear to imply that

$$y^{(i)}(\infty) = y^{(0)}(0^+). \tag{5.2.10}$$

Equation (5.2.10) embodies the general idea behind the asymptotic matching principle.

The solutions (5.2.3) and (5.2.9) are sketched in Figure 5.1 along with the exact solution of (5.2.1). Observe that the exact solution looks like $y^{(1)}$ near $t = 0$ and like $y^{(0)}$ as t becomes large.

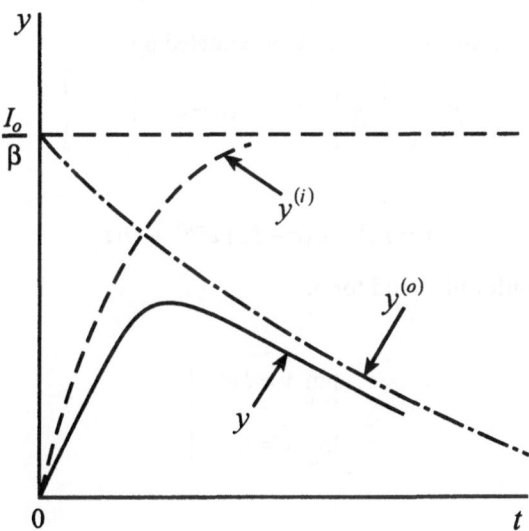

Figure 5.1. Inner, outer, and exact solutions.

5.3. The Inner and Outer Expansions

Let us now illustrate the method of matched asymptotic expansions through an example.

Example 1: Consider the boundary-value problem:

$$\left. \begin{array}{l} \varepsilon y'' + y' + y = 0, \quad 0 \le x \le 1 \\ y(0) = a, \ y(1) = b \end{array} \right\}. \tag{5.3.1}$$

The exact solution of the boundary-value problem (5.3.1) is

$$y^{(e)} = \frac{\left(ae^{s_2} - b\right) e^{s_1 x} + \left(b - ae^{s_1}\right) e^{s_2 x}}{e^{s_2} - e^{s_1}} \tag{5.3.2}$$

where

$$s_{1,2} = \frac{-1 \pm \sqrt{1 - 4\varepsilon}}{2\varepsilon} \approx -1, -\frac{1}{\varepsilon} + 1.$$

The above exact solution may be approximated by

$$y \approx \frac{1}{\left(-e^{-1}\right)} \left[-be^{-x} + \left(b - ae^{-1}\right) e^{x - \frac{x}{\varepsilon}} \right]$$

or

$$y = be^{1-x} + (a - be) e^{-x/\varepsilon} + O(\varepsilon) \tag{5.3.3}$$

which shows a uniformly valid form.

Note that

$$\left. \begin{array}{l} \lim_{\varepsilon \to 0} y = be^{1-x} \\ \\ \lim_{x \to 0} y = a \end{array} \right\} \tag{5.3.4}$$

from which

$$\lim_{\varepsilon \to 0} \left(\lim_{x \to 0} y \right) = a \ne \lim_{x \to 0} \left(\lim_{\varepsilon \to 0} y \right) = be \tag{5.3.5}$$

so that the two limits are non-commutable.

Note that the expansion (5.3.3) cannot be obtained by keeping either x or x/ε fixed. In the former case, one obtains

$$y^{(0)} = be^{1-x} + O(\varepsilon) \tag{5.3.6}$$

which is not valid in the boundary layer near $x = 0$ since

$$y^{(0)}(0) = be \neq a. \tag{5.3.7}$$

In the latter case, one obtains, on the other hand,

$$y^{(i)} = be + (a - be)\, e^{-x/\varepsilon} + O(\varepsilon) \tag{5.3.8}$$

which is not valid, as $x \to 1$, since

$$y^{(i)}(1) = be \neq b. \tag{5.3.9}$$

This suggests that we represent the solution by using two different asymptotic expansions using the variables x and x/ε. The occurrence of the term $e^{-x/\varepsilon}$ makes this problem singular because there is no expansion of $e^{-x/\varepsilon}$, which is valid in a neighborhood of $\varepsilon = 0$.

Let us seek, for the boundary-value problem (5.3.1), an outer expansion (valid away from $x = 0$) of the form –

$$y^{(0)}(x;\varepsilon) \sim \sum_{n=0}^{N-1} \varepsilon^n y_n^{(0)}(x) + O(\varepsilon^N). \tag{5.3.10}$$

We then obtain, from equation (5.3.1), to various orders in ε,

$$\left. \begin{array}{l} O(1): {y_0'^{(0)}}' + y_0^{(0)} = 0 \\[2mm] O(\varepsilon): {y_1'^{(0)}}' + y_1^{(0)} = -{y_0''^{(0)}}'' \end{array} \right\} \tag{5.3.11}$$

etc.

Note that in the outer expansion the order of equation (5.3.1) is reduced. Therefore, the outer equations (5.3.11) cannot take on both of the boundary conditions in (5.3.1), and one of these boundary conditions, viz., $y(0) = a$, must be dropped. This means that $y^{(0)}$ is valid everywhere, except in the region $x = O(\varepsilon)$. Thus, we have the following boundary conditions on $y^{(0)}$ –

$$y_0^{(0)}(1) = b, \quad y_1^{(0)}(1) = 0, \text{ etc.} \tag{5.3.12}$$

Using these conditions, we obtain, for equations (5.3.11), the following solutions –

$$y_0^{(0)} = be^{1-x}$$

$$\left. y_1^{(0)} = b(1-x)\, e^{1-x} \right\}. \qquad (5.3.13)$$

etc.

Thus, the outer expansion is given by

$$y^{(0)} = b\left[1 + \varepsilon(1-x)\right] e^{1-x} + O\left(\varepsilon^2\right). \qquad (5.3.14)$$

For small ε, $y^{(0)}$ is close to $y^{(e)}$ everywhere except in a small interval at $x = 0$, where $y^{(e)}$ changes rapidly in order to retrieve the boundary condition there which is about to be lost (see Figure 5.2).

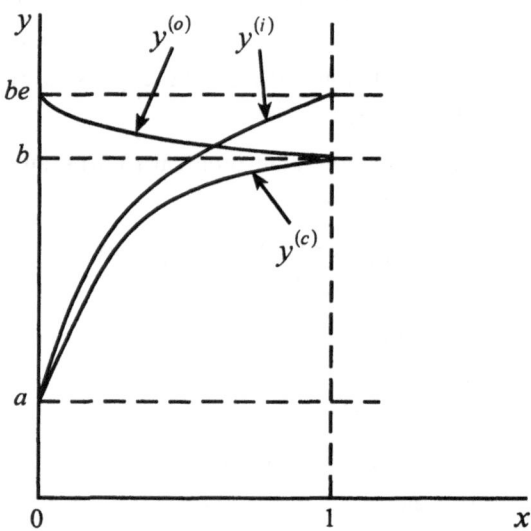

Figure 5.2. Inner, outer, and composite expansions for equation (5.3.1).

To determine an inner expansion which is valid in the boundary layer in $x = O(\varepsilon)$, we introduce a new independent variable $\xi = x/\varepsilon$. This allows the width of the boundary layer region to become independent of ε as $\varepsilon \Rightarrow 0$ and enables us to retain the highest derivative in the given equation when $\varepsilon \to 0$. This is essential to representing the rapid variation of y in the boundary layer. We then have

$$\frac{d^2 y}{d\xi^2} + \frac{dy}{d\xi} + \varepsilon y = 0. \tag{5.3.15}$$

Let us seek an inner expansion of the following form –

$$y^{(i)}(\xi;\varepsilon) \sim \sum_{n=0}^{N-1} \varepsilon^n y_n^{(i)}(\xi) + O(\varepsilon^N). \tag{5.3.16}$$

We then obtain, from equation (5.3.15), to various orders in ε –

$$\left. \begin{array}{l} O(1): \dfrac{d^2 y_0^{(i)}}{d\xi^2} + \dfrac{dy_0^{(i)}}{d\xi} = 0 \\[4mm] O(\varepsilon): \dfrac{d^2 y_1^{(i)}}{d\xi^2} + \dfrac{dy_1^{(i)}}{d\xi} = -y_0^{(i)} \\[4mm] \text{etc.} \end{array} \right\}. \tag{5.3.17}$$

Noting that $y^{(i)}$ is valid only in the region $x = O(\varepsilon)$, we have the following boundary conditions on $y^{(i)}$ –

$$\left. \begin{array}{l} y_0^{(i)}(0) = a \\[2mm] y_1^{(i)}(0) = 0 \\[2mm] \text{etc.} \end{array} \right\}. \tag{5.3.18}$$

Using these conditions, we have, for equations (5.3.17), the following solutions -

$$O(1): \ y_0^{(i)} = a - A_0 \left(1 - e^{-\xi}\right) \tag{5.3.19}$$

$$O(\varepsilon): \ y_1^{(i)} = A_1 \left(1 - e^{-\xi}\right) - \left[a - A_0 \left(1 + e^{-\xi}\right)\right] \xi \tag{5.3.20}$$

etc.

Thus, the inner expansion is given by

$$y^{(i)} = a - A_0 \left(1 - e^{-\xi}\right) + \varepsilon \left\{ A_1 \left(1 - e^{-\xi}\right) - \left[a - A_0 \left(1 + e^{-\xi}\right)\right] \xi \right\} + O(\varepsilon^2). \tag{5.3.21}$$

In order to relate $y^{(i)}$ to $y^{(0)}$ in an overlapping domain of validity, let us use the following asymptotic matching principle (Shivamoggi, 1978b)[1] –

---

[1] There are other types of asymptotic matching principles available in the literature. One such type is due to Kaplun (1957) which involves intermediate limits and the other type

"The n - term *formal* Laurent series
expansion of the outer expansion about   =   " The n - term *formal* outer
the inner boundary written in terms   limit of the inner expansion."
of the inner variable."

Thus, near $x = 0$, let us write

$$y^{(0)}(x) = y^{(0)}(0) + xy^{(0)'}(0) + 0(x^2)$$

$$= y_0^{(0)}(0) + \varepsilon \left[\frac{x}{\varepsilon} y_0^{(0)'}(0) + y_1^{(0)}(0) \right] + O(\varepsilon^2) \qquad (5.3.22)$$

$$= y_0^{(0)}(0) + \varepsilon \left[\xi y_0^{(0)'}(0) + y_1^{(0)}(0) \right] + O(\varepsilon^2).$$

We then have according to the above matching principle

$$be + \varepsilon [be - be\xi] + O(\varepsilon^2) = (a - A_0) + \varepsilon \left[A_1 - (a - A_0)\xi \right] + O(\varepsilon^2), \qquad (5.3.23)$$

from which

$$A_0 = a - be, \quad A_1 = be, \text{ etc.} \qquad (5.3.24)$$

Using (5.3.24), the inner expansion (5.3.21) becomes

$$y^{(i)} = be + (a - be) \, e^{-\xi} + \varepsilon \left\{ be(1 - e^{-\xi}) - [be - (a - be) \, e^{-\xi}] \, \xi \right\} + O(\varepsilon^2). \quad (5.3.25)$$

$y^{(0)}$ is valid everywhere except in a small interval of $O(\varepsilon)$ near the origin while $y^{(i)}$ is valid only in a small interval of $O(\varepsilon)$ near the origin. Although $y^{(0)}$ and $y^{(i)}$ have overlapping domains, one needs to switch from one expansion to the other if a numerical solution is desired over the whole interval. However, the switching location is not known precisely. This difficulty can be circumvented by combining both expansions into a single composite expansion $y^{(c)}$ –

$$y^{(c)} = y^{(0)} + y^{(i)} - y^{(0)_i} \left(\text{or } y^{(i)_0} \right), \qquad (5.3.26)$$

where $y^{(0)_i}$ represents the inner limit of the outer expansion and $y^{(i)_0}$ represents the outer limit of the inner expansion.

Note,

---

is due to Van Dyke (1975). (See Fraenkel, 1969 and Eckaus, 1979 for a critical assessment of the various asymptotic matching principles.)

$$y^{(c)_o} = y^{(0)_o} + y^{(i)_o} - y^{(i)_{oo}} = y^{(0)} + y^{(i)_o} - y^{(i)_o} = y^{(0)}$$

and

$$y^{(c)_i} = y^{(0)_i} + y^{(i)_i} - y^{(0)_{ii}} = y^{(0)_i} + y^{(i)} - y^{(0)_i} = y^{(i)}$$

so that $y^{(c)}$ reproduces $y^{(0)}$ in the outer domain while it reproduces $y^{(i)}$ in the inner domain. Therefore, $y^{(c)}$ is valid everywhere.

For the present example, on using (5.3.14) and (5.3.25), we have for the composite expansion –

$$y^{(c)} = b\left[1 + \varepsilon(1-x)\right] e^{1-x} + O(\varepsilon^2) + be + (a-be)e^{-\xi}$$
$$+ \varepsilon\left\{be\left(1 - e^{-\xi}\right) - \left[be - (a-be)e^{-\xi}\right]\xi\right\}$$
$$+ O(\varepsilon^2) - \left[be + \varepsilon(be - be\xi)\right] + O(\varepsilon^2) \qquad (5.3.27)$$

or

$$y^{(c)} = b\left[1 + \varepsilon(1-x)\right] e^{1-x} + \left[(a-be)(1+x) - \varepsilon be\right] e^{-\frac{x}{\varepsilon}} + O(\varepsilon^2).$$

Here, note that

$$e^{-\frac{x}{\varepsilon}} = \begin{cases} o(\varepsilon^n) & \text{as } \varepsilon \Rightarrow 0, \ \forall n, \ \text{if } x = O(1), \\ O(1) & \text{as } \varepsilon \Rightarrow 0, \ \text{if } x = O(\varepsilon). \end{cases} \qquad (5.3.28)$$

Let us now consider the determination of the location of boundary layers in a singular-perturbation problem.

Example 2: Consider the boundary-value problem

$$\left.\begin{array}{l} \varepsilon y'' + a(x) y' + b(x) y = 0 \\ y(0) = \alpha, \ y(1) = \beta \end{array}\right\}. \qquad (5.3.29)$$

In the limit $\varepsilon \to 0$, equation (5.3.29) gives

$$a(x) y' + b(x) y = 0 \qquad (5.3.30)$$

whose solution cannot satisfy both boundary conditions, and one of them must be dropped as a consequence. The boundary condition that must be dropped depends on the sign of $a(x)$ in the interval $(0,1)$. In general, if the outer limit of the inner solution diverges and does not exist, the boundary layer does not arise at the assumed location. Thus, the location of the boundary layer is determined so as to have the resulting inner expansion possess a proper outer limit. If $a(x) > 0$, $y(0) = \alpha$ must be dropped, and an inner expansion near $x = 0$ must be developed and matched with the outer solution. If $a(x) < 0$, $y(1) = \beta$ must be dropped, and an inner expansion near $x = 1$ must be obtained and matched with the outer expansion. If $a(x)$ changes sign in $(0,1)$, y may change from oscillatory to monotonic across the zeros of $a(x)$ (called the turning points – see Section 5.9).

Let us consider here the case $a(x) > 0$. Then, we have for the outer solution $y_0^{(0)}$ –

$$\left. \begin{aligned} a(x)\, y_0^{(0)'} + b(x)\, y_0^{(0)} = 0 \\ y_0^{(0)}(1) = \beta \end{aligned} \right\} \tag{5.3.31}$$

from which we have

$$y_0^{(0)} = \beta \, \exp\left[-\int_1^x \frac{b(t)}{a(t)}\, dt \right]. \tag{5.3.32}$$

In order to obtain the inner solution, introduce a new independent variable

$$\xi = x/\varepsilon \tag{5.3.33}$$

so we have, in the limit $x \to 0$,

$$\left. \begin{aligned} \frac{d^2 y_0^{(i)}}{d\xi^2} + a(0)\, \frac{dy_0^{(i)}}{d\xi} = 0 \\ y_0^{(i)}(0) = \alpha \end{aligned} \right\} \tag{5.3.34}$$

from which we obtain

$$y_0^{(i)} = \alpha - B + b e^{-a(0)\xi} \tag{5.3.35}$$

where B is an arbitrary constant.

Matching asymptotically $y_0^{(0)}$ with $y_0^{(i)}$, we obtain

$$B = \alpha - \beta \exp\left[\int_0^1 \frac{b(t)}{a(t)}\, dt \right]. \tag{5.3.36}$$

Using (5.3.32), (5.3.35), and (5.3.36), the composite expansion is then given by

$$y^{(c)} = y^{(0)} + y^{(i)} - y^{(0)},$$

$$= \beta \exp\left[-\int_1^x \frac{b(t)}{a(t)} \, dt \right] + \left\{ \alpha - \beta \exp\left[\int_0^1 \frac{b(t)}{a(t)} \, dt \right] \right\} e^{-\frac{a(0)x}{\varepsilon}} + O(\varepsilon). \qquad (5.3.37)$$

Example 3: Consider the boundary-value problem

$$\left. \begin{array}{l} \varepsilon y'' + (2x+1)\, y' + 2y = 0 \\ y(0) = \alpha, \quad y(1) = \beta \end{array} \right\}. \qquad (5.3.38)$$

Look for an outer expansion of the form –

$$y^{(0)} = \sum_{n=0}^{\infty} \varepsilon^n y_n^{(0)}(x). \qquad (5.3.39)$$

We then obtain from the boundary-value problem (5.3.38)

$$\left. \begin{array}{l} (2x+1)\, y_0^{(0)\,'} + 2y_0^{(0)} = 0 \\ y_0^{(0)}(1) = \beta \end{array} \right\} \qquad (5.3.40)$$

etc.

On solving these problems, we obtain

$$\left. \begin{array}{l} y_0^{(0)} = \dfrac{3\beta}{2x+1} \end{array} \right\}. \qquad (5.3.41)$$

etc.

Look for an inner expansion of the form –

$$y^{(i)} = \sum_{n=0}^{\infty} \varepsilon^n y_n^{(i)}(\xi), \quad \xi \equiv \frac{x}{\varepsilon}. \qquad (5.3.42)$$

We then obtain from the boundary-value problem (5.3.38)

$$\left. \begin{array}{l} y_0^{(i)\,''} + y_0^{(i)\,'} = 0 \\ y_0^{(i)}(0) = \alpha \end{array} \right\}. \qquad (5.3.43)$$

etc.

On solving these problems, we obtain

$$\left. \begin{array}{l} y_0^{(i)} = \alpha - A\left(1 - e^{-\xi}\right) \end{array} \right\} \qquad (5.3.44)$$

etc.

where A is an arbitrary constant.

Asymptotic matching between $y^{(0)}$ and $y^{(i)}$ then gives

$$\lim_{x \Rightarrow 0} y_0^{(0)} = \lim_{\xi \Rightarrow \infty} y_0^{(i)} \qquad (5.3.45)$$

from which, we have on using (5.3.41) and (5.3.44),

$$3\beta = \alpha - A$$

or

$$A = \alpha - 3\beta. \qquad (5.3.46)$$

Using (5.3.46), (5.3.44) becomes

$$y_0^{(i)} = \alpha - (\alpha - 3\beta)\left(1 - e^{-\xi}\right). \qquad (5.3.47)$$

The composite expansion is then given by

$$y^{(c)} = y^{(0)} + y^{(i)} - y^{(i)_0}$$

$$= \frac{3\beta}{2x+1} + (\alpha - 3\beta)\, e^{-\frac{x}{\varepsilon}} + O(\varepsilon). \qquad (5.3.48)$$

5.4. Hyperbolic Equations

Consider the linear hyperbolic partial differential equation (Cole, 1968) –

$$\varepsilon\left(\frac{\partial^2 v}{\partial x^2} - \frac{\partial^2 v}{\partial t^2}\right) = a\,\frac{\partial v}{\partial x} + b\,\frac{\partial v}{\partial t} \qquad (5.4.1)$$

which has real characteristics (see Figure 5.3)

$$r = t - x, \quad s = t + x. \qquad (5.4.2)$$

The characteristics serve to define the region of influence, propagating into the future, of a disturbance at a point Q (see Figure 5.3). The manner of specification of boundary conditions on an arc for a fully-posed boundary-value problem depends on the nature of the arc with respect to the characteristic directions of propagation. One boundary condition is specified on the time-like arc (see Figure 5.4) corresponding to one characteristic leading into the adjacent region in which the solution is defined. Two boundary conditions are given on the space-like arc (see Figure 5.4) corresponding to the two characteristics leading into the adjacent domain. When the boundary curves are along the characteristic curves, only one condition can be prescribed, and the characteristic relations must hold. The characteristic-initial-value problem describes one condition each on AB and on AC to define the solution in $ABCD$ (see Figure 5.5).

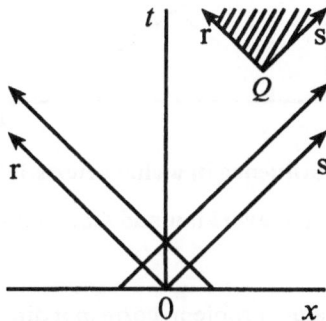

Figure 5.3. Characteristics and region of influence

(from Kevorkian and Cole, 1996).

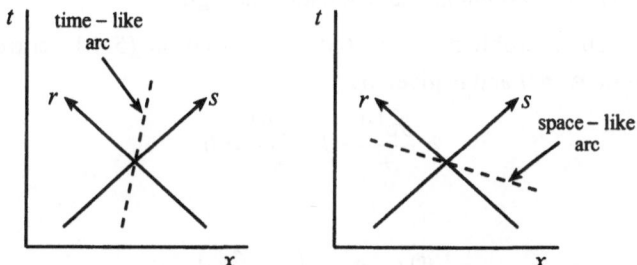

Figure 5.4. Time-like arc and space-like arc

(from Kevorkian and Cole, 1996).

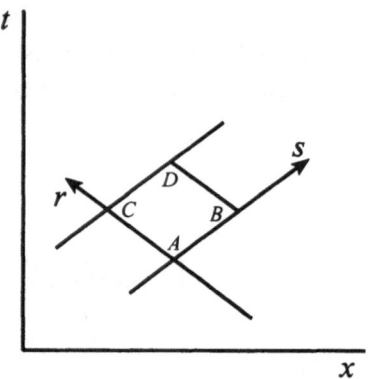

Figure 5.5. Domain of influence in a characteristic initial-value problem

(from Kevorkian and Cole, 1996).

Consider the initial-value problem corresponding to equation (5.4.1) in $-\infty < x < \infty$ with

$$t = 0 : v = F(x), \quad v_t = G(x).$$ (5.4.3)

According to the general theory of characteristics, the solutions at a point $P(x,t)$ (see Figure 5.6) can depend only on that part of the initial data which can send a signal to P. This is part $(x_1 < x < x_2)$ of the initial line contained between the backward running characteristics through P.

Now, the outer problem associated with equation (5.4.1) corresponds to taking the limit $\varepsilon \Rightarrow 0$ and is given by

$$a \frac{\partial v^{(0)}}{\partial x} + b \frac{\partial v^{(0)}}{\partial t} = 0$$ (5.4.4)

from which,

$$v^{(0)}(x,t) = f\left(x - \frac{a}{b} t\right).$$ (5.4.5)

In the limit $\varepsilon \Rightarrow 0$ the solution $v(x,t)$ depends only on the data connected to P along a subcharacteristic of equation (5.4.1) given by

$$bx - at = \text{const.}$$ (5.4.6)

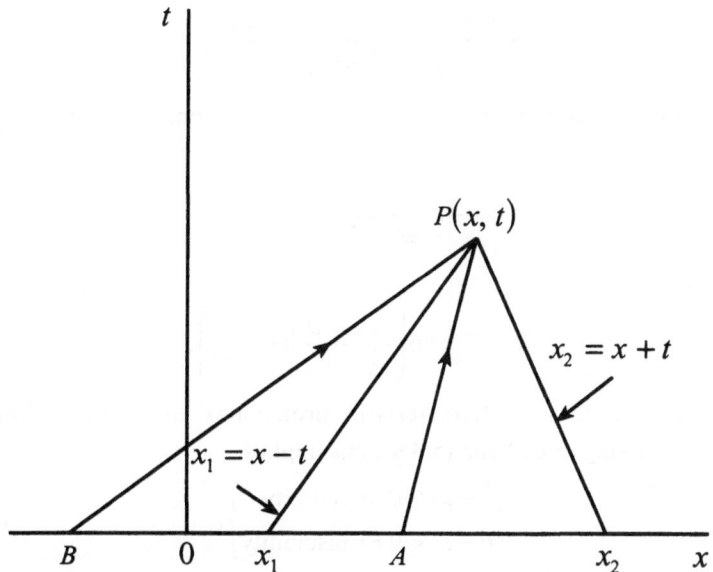

**Figure 5.6. Domain of influence and time-like and space-like
subcharacteristics (from Kevorkian and Cole, 1996).**

Now the subcharacteristic, reaching P, originated at point A between x_1, x_2 if $|b/a| > 1$ i.e., if it is time-like. Then the limit $\varepsilon \Rightarrow 0$ preserves the domain of influence. However if $|b/a| < 1$ the subcharacteristic reaching P is space-like and lies outside the domain of influence, originating at B (see Figure 5.6). In this case, the limit $\varepsilon \Rightarrow 0$ increases the domain of influence and implies that the outer solution cannot be obtained by taking the outer limit of the solution of the full problem for equation (5.4.1), contrary to what should be the case!

It turns out that even the issue of stability of the solution $v(x,t)$ (here stability/instability refers to the exponential decay/growth along the characteristic of the discontinuities in the derivatives of the solution across characteristic) is related to whether the subcharacterstics are time-like or space-like. In order to see that, note that in terms of the characteristic coordinates (5.4.2), equation (5.4.1) becomes

$$-4\varepsilon \frac{\partial^2 v}{\partial r \partial s} = (b-a)\frac{\partial v}{\partial r} + (b+a)\frac{\partial v}{\partial s}. \qquad (5.4.7)$$

Consider the propagation of a jump in $\partial v/\partial r$ along $r = r_0 = $ const. Let,

$$K \equiv \left[\frac{\partial v}{\partial r}\right]_{r=r_0} \equiv \left(\frac{\partial v}{\partial r}\right)_{r_0^+} - \left(\frac{\partial v}{\partial r}\right)_{r_0^-} . \tag{5.4.8}$$

Assuming that v itself is continuous across $r = r_0$ one finds from equation (5.4.7)

$$-4\varepsilon\frac{\partial K}{\partial s} = (b - a)\,K ,$$

from which

$$K = K_0 \exp\left[-\left(\frac{b-a}{4\varepsilon}\right)(s - s_0)\right]. \tag{5.4.9}$$

Now, a jump across a characteristic propagates to infinity along that characteristic. Using (5.4.8) and (5.4.9), this implies

$$\left.\begin{array}{l} (b - a) > 0 \Rightarrow \text{stability} \\ (b - a) < 0 \Rightarrow \text{instability} \end{array}\right\}. \tag{5.4.10}$$

Similarly, a consideration of a jump in $\partial v/\partial s$ across a characteristic $s = s_0$ gives

$$\left.\begin{array}{l} (b + a) > 0 \Rightarrow \text{stability} \\ (b + a) < 0 \Rightarrow \text{instability} \end{array}\right\}. \tag{5.4.11}$$

From (5.4.10) and (5.4.11) one obtains,

$$|b/a| > 1 \quad \text{for stability.} \tag{5.4.12}$$

We restrict further discussion to the stable case.

Now, note that the solution $v^{(0)}(x,t)$ given in (5.4.5) can only satisfy one initial condition, so that one may expect the existence of a boundary layer on the line $t = 0$.

Assume an initially-valid expansion

$$v^{(i)}\left(x,\tilde{t};\varepsilon\right) \sim v_0^{(i)}\left(x,\tilde{t}\right) + \beta_1(\varepsilon)\,v_1^{(i)}\left(x,\tilde{t}\right) + \cdots \tag{5.4.13}$$

where, $\tilde{t}$ is the time scale relevant for $t \approx 0$,

$$\tilde{t} = \frac{t}{\delta(\varepsilon)}$$

and

$$\beta_1, \delta \Rightarrow 0 \quad \text{as} \quad \varepsilon \Rightarrow 0$$

with the associated inner limit process $\varepsilon \Rightarrow 0$, $x, \tilde{t}$ held fixed. Taking

$$\beta_1(\varepsilon) = \delta(\varepsilon), \tag{5.4.14}$$

the initial conditions (5.4.3) give

$$\left.\begin{array}{l} \tilde{t} = 0 : v_0^{(i)} = F(x), \quad v_n^{(i)} = 0 \quad \text{for} \quad n > 0 \\[2mm] \tilde{t} = 0 : \dfrac{\partial v_0^{(i)}}{\partial \tilde{t}} = 0, \quad \dfrac{\partial \tilde{v}_1^{(i)}}{\partial \tilde{t}} = G(x), \\[3mm] \qquad\qquad \dfrac{\partial v_n^{(i)}}{\partial \tilde{t}} = 0 \quad \text{for} \quad n > 1 \end{array}\right\}. \tag{5.4.15}$$

Choosing $\delta(\varepsilon) = \varepsilon$, and substituting (5.4.13) and (5.4.14), equation (5.4.1) gives

$$O(1) : \frac{\partial^2 v_0^{(i)}}{\partial \tilde{t}^2} + b\,\frac{\partial v_0^{(i)}}{\partial \tilde{t}} = 0 \tag{5.4.16}$$

$$O(\varepsilon) : \frac{\partial^2 v_1^{(i)}}{\partial \tilde{t}^2} + b\,\frac{\partial v_1^{(i)}}{\partial \tilde{t}} = -a\,\frac{\partial v_0^{(i)}}{\partial x} \tag{5.4.17}$$

etc.

Notice that the boundary-layer equations (5.4.16) and (5.4.17) are ordinary-differential equations, which is a feature of the boundary layers not occurring on a subcharacteristic. This is true for any hyperbolic-initial value problem, since a space-like arc can never be a subcharacteristic.

Using the initial conditions (5.4.15), equations (5.4.16) and (5.4.17) give

$$v_0^{(i)}(x, \tilde{t}) = F(x) \tag{5.4.18}$$

$$v_1^{(i)}(x, \tilde{t}) = \left[G(x) + \frac{a}{b} F'(x) \right]\left[1 - e^{-\tilde{t}} \right] - \frac{a}{b} \tilde{t} F'(x) \tag{5.4.19}$$

so that

$$v^{(i)}(x, \tilde{t}; \varepsilon) = F(x) + \varepsilon\left[\left\{ G(x) + \frac{a}{b} F'(x) \right\}\left\{ 1 - e^{-\tilde{t}} \right\} - \frac{a}{b} \tilde{t} F'(x) \right] + \cdots. \tag{5.4.20}$$

Note that (5.4.20) possesses terms that persist in the limit $t \Rightarrow \infty$, as well as terms that decay in time which are typical of a boundary layer.

Next, construct an outer expansion, with the associated outer limit process, $\varepsilon \Rightarrow 0$, x and t held fixed,

$$v^{(0)}(x, t, \varepsilon) \sim v_0^{(0)}(x, t) + \varepsilon\, v_1^{(0)}(x, t) + \cdots \tag{5.4.21}$$

so that equation (5.4.1) gives

$$O(1): \quad a\, \frac{\partial v_0^{(0)}}{\partial x} + b\, \frac{\partial v_0^{(0)}}{\partial t} = 0 \tag{5.4.22}$$

$$O(\varepsilon): \quad a\, \frac{\partial v_1^{(0)}}{\partial x} + b\, \frac{\partial v_1^{(0)}}{\partial t} = \left(\frac{\partial^2 v_0^{(0)}}{\partial x^2} - \frac{\partial^2 v_0^{(0)}}{\partial t^2} \right) \tag{5.4.23}$$

etc.

One obtains from equation (5.4.22)

$$v_0^{(0)} = f(\xi), \quad \xi = x - \frac{a}{b}\, t. \tag{5.4.24}$$

Using (5.4.22), equation (5.4.23) becomes

$$a\, \frac{\partial v_1^{(0)}}{\partial x} + b\, \frac{\partial v_1^{(0)}}{\partial t} = \left(1 - \frac{a^2}{b^2} \right) f''(\xi) \tag{5.4.25}$$

from which

$$v_1^{(0)} = \frac{a}{b^2} \frac{b^2 - a^2}{b^2 + a^2} \left(x + \frac{b}{a}\, t \right) f''(\xi) + f_1(\xi) \tag{5.4.26}$$

so that

$$v^{(0)}(x,t;\varepsilon) = f(\xi) + \varepsilon \left[f_1(\xi) + \frac{a}{b^2} \frac{b^2 - a^2}{b^2 + a^2} \left(x + \frac{b}{a}\, t \right) f''(\xi) \right] + \cdots. \tag{5.4.27}$$

The asymptotic matching between $v^{(i)}$ and $v^{(0)}$ requires

$$v^{(0)}(x,0;\varepsilon) + \varepsilon \tilde{t}\, v_1^{(0)}(x,0;\varepsilon) + \cdots = v^{(i)}(x,\infty;\varepsilon). \tag{5.4.28}$$

Using (5.4.20) and (5.4.27), (5.4.28) gives

$$f(x) = F(x),$$

$$f_1(x) + \frac{a}{b^2} \frac{b^2 - a^2}{b^2 + a^2}\, x\, f''(x) = G(x) + \frac{a}{b}\, F'(x) \tag{5.4.29}$$

so that (5.4.27) becomes

$$v^{(0)}(x,t;\varepsilon) \sim F\left(x - \frac{a}{b}\, t \right) + \varepsilon \left[\frac{b^2 - a^2}{b^3}\, t\, F''\left(x - \frac{a}{b}\, t \right) \right.$$

$$\left. + \frac{a}{b}\, F'\left(x - \frac{a}{b}\, t \right) + G\left(x - \frac{a}{b}\, t \right) \right] + \cdots. \tag{5.4.30}$$

Consider, next, a radiation problem in which boundary conditions are prescribed on a time-like arc and propagate into the quiescent medium in $x > 0$,

(Figure 5.7). When the boundary condition is prescribed for instance, at $x = 0$, one has to distinguish two cases depending on whether the subcharacterstics run into or out of the boundary $x = 0$. Recall, from (5.4.6), that the subcharacterstics are given by

$$\xi = x - \frac{a}{b} t = \text{const.} \tag{5.4.6}$$

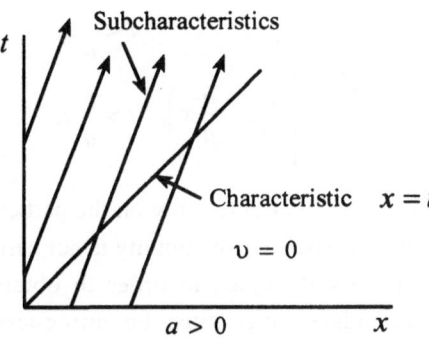

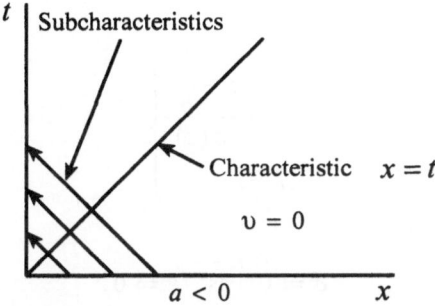

Figure 5.7. Radiation problem with boundary conditions prescribed on a finite portion of the boundary (from Kevorkian and Cole, 1996).

Note, that the characteristics are incoming or outgoing according as $a \lessgtr 0$. Let the boundary condition be

$$x = 0 : \upsilon = F(t), \quad t > 0. \tag{5.4.31}$$

Outgoing Characteristics: Assume an outer solution,

$$v^{(0)}(x,t;\varepsilon) \sim v_0^{(0)}(x,t) + \varepsilon\, v_1^{(0)}(x,t) + \cdots \qquad (5.4.32)$$

where

$$v_0^{(0)} = f(\zeta), \quad \zeta = t - \frac{b}{a}x. \qquad (5.4.5)$$

Substituting (5.4.5) into the boundary condition (5.4.31), we obtain

$$v_0^{(0)} = \begin{cases} 0, & t < \dfrac{b}{a}x, \\[2mm] F\!\left(t - \dfrac{b}{a}x\right), & t > \dfrac{b}{a}x. \end{cases} \qquad (5.4.33)$$

This solution obviously has a discontinuity on the particular subcharacteristic through the origin. However, such a discontinuity is not permitted in the solution to equation (5.4.1) with $\varepsilon \neq 0$. Thus, in order to obtain a uniformly-valid solution, a suitable boundary layer must be introduced on the particular subcharacteristic $\zeta = 0$ which supports the discontinuity in the outer solution $v^{(0)}$. Assume an inner expansion

$$v^{(i)}(\tilde{x},t;\varepsilon) \sim v_0^{(i)}(\tilde{x},\tilde{t}) + \mu(\varepsilon)\, v_1^{(i)}(\tilde{x},\tilde{t}) + \cdots$$

where,

$$\left. \begin{aligned} \tilde{x} &= \frac{x - \dfrac{a}{b}t}{\delta(\varepsilon)} \\[2mm] \tilde{t} &= t \end{aligned} \right\} \qquad (5.4.34)$$

and

$$\delta \Rightarrow 0 \quad \text{as} \quad \varepsilon \Rightarrow 0,$$

with an associated inner limit process $\varepsilon \Rightarrow 0$, $\tilde{x}$ and $\tilde{t}$ held fixed. Choosing

$$\delta = \sqrt{\varepsilon}$$

and substituting (5.4.34), equation (5.4.1) gives in the limit $\varepsilon \Rightarrow 0$,

$$K\frac{\partial^2 v_0^{(i)}}{\partial \tilde{x}^2} = \frac{\partial v_0^{(i)}}{\partial \tilde{t}} \qquad (5.4.35)$$

where

$$K \equiv \frac{1 - \dfrac{a^2}{b^2}}{b} > 0$$

which ensures that $\tilde{t} = t$ is a positive time-like variable so that equation (5.4.35) is a diffusion equation that describes the spreading of the discontinuity in the outer expansion $v^{(0)}$ on the subcharacteristic $\zeta = 0$. Matching $v^{(i)}$ to $v^{(0)}$ asymptotically, as before, one obtains

$$v_0^{(i)}\left(\tilde{x}, \tilde{t}\right) = \frac{F\left(0^+\right)}{2}\, erfc\left(\frac{\tilde{x}}{2\sqrt{K\tilde{t}}}\right). \tag{5.4.36}$$

Incoming Characteristics: Assume an outer expansion

$$v^{(0)}\left(x, t; \varepsilon\right) \sim v_0^{(0)}\left(x, t\right) + \varepsilon\, v_1^{(0)}\left(x, t\right) + \cdots. \tag{5.4.37}$$

Since the disturbances now propagate along the subcharacteristics from the quiescent region to the boundary, one has

$$v^{(0)} \equiv 0. \tag{5.4.38}$$

Then the discontinuity in $v^{(0)}$ occurs at the boundary $x = 0$ so that one has a boundary layer at $x = 0$. Since the line $x = 0$ is not a subcharacteristic, the boundary layer equations should now be ordinary differential equations. Assume an inner expansion

$$v^{(i)}\left(\tilde{x}, \tilde{t}; \varepsilon\right) \sim v_0^{(i)}\left(\tilde{x}, \tilde{t}\right) + v_1\left(\varepsilon\right) v_1^{(i)}\left(\tilde{x}, \tilde{t}\right) + \cdots \tag{5.4.39}$$

where

$$\tilde{x} = \frac{x}{\delta(\varepsilon)}, \quad \tilde{t} = t \quad \text{and} \quad v_1 \Rightarrow 0 \quad \text{as} \quad \varepsilon \Rightarrow 0 \tag{5.4.40}$$

with an associated inner limit process, $\varepsilon \Rightarrow 0$, $\tilde{x}$ and $\tilde{t}$ held fixed. If one chooses $\delta = \varepsilon$, substituting (5.4.39) into equation (5.4.1), we obtain

$$\frac{\partial^2 v_0^{(i)}}{\partial \tilde{x}^2} = a \frac{\partial v_0^{(i)}}{\partial \tilde{x}}. \tag{5.4.41}$$

On using the boundary condition (5.4.31), one obtains from equation (5.4.41)

$$v_0^{(i)}\left(\tilde{x}, \tilde{t}\right) = F\left(\tilde{t}\right) \exp\left(a\tilde{x}\right), \quad a < 0. \tag{5.4.42}$$

5.5. Elliptic Equations

Consider an elliptic equation of the form

$$\varepsilon \left[\alpha_{11} \frac{\partial^2 u}{\partial x^2} + 2\alpha_{12} \frac{\partial^2 u}{\partial x \partial y} + \alpha_{22} \frac{\partial^2 u}{\partial y^2} \right] = a \frac{\partial u}{\partial x} + b \frac{\partial u}{\partial y} \qquad (5.5.1)$$

where α's, a, and b are constants, and $\varepsilon \ll 1$, and

$$\alpha_{12}^2 - \alpha_{11}\alpha_{22} < 0.$$

In order to determine a solution $u(x,y;\varepsilon)$ to equation (5.5.1) uniquely, it is sufficient to prescribe one boundary condition on u or its normal derivative, or a combination on a closed boundary.

Consider an interior boundary-value problem with $u = u_B(x_B)$ prescribed on a closed boundary curve, (see Figure 5.8).

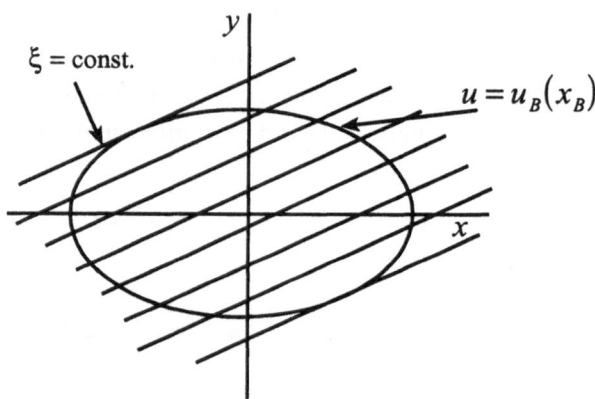

Figure 5.8. The subcharacteristics.

The curves

$$\xi = bx - ay = \text{const.} \qquad (5.5.2)$$

are the characteristics of equation (5.5.1), in the limit $\varepsilon \Rightarrow 0$, and are the subcharacteristics of equation (5.5.1).

Introducing another independent variable

$$\eta = ax + by, \qquad (5.5.3)$$

transforming the independent variables x, y to ξ, η, equation (5.5.1) becomes

$$\varepsilon\left[A_{11}u_{\xi\xi} + 2A_{12}u_{\xi\eta} + A_{22}u_{\eta\eta}\right] = u_{\eta} \tag{5.5.4}$$

where

$$A_{11} = \frac{\alpha_{11}b^2 - 2\alpha_{12}ab + \alpha_{22}a^2}{a^2 + b^2},$$

$$A_{12} = \frac{\alpha_{11}ab + \alpha_{12}\left(b^2 - a^2\right) - \alpha_{22}ab}{a^2 + b^2},$$

$$A_{22} = \frac{\alpha_{11}a^2 + 2\alpha_{12}ab + \alpha_{22}b^2}{a^2 + b^2}.$$

Now, in equation (5.5.4), let

$$\lim_{\substack{\varepsilon \Rightarrow 0 \\ \xi, \eta \text{ fixed}}} u\left(\xi, \eta; \varepsilon\right) \Rightarrow u^{(0)}\left(\xi\right) \tag{5.5.5}$$

where the boundary condition on one side of the domain (see Figure 5.9) is sufficient to determine $u^{(0)}(\xi)$ uniquely in the whole domain, but $u^{(0)}(\xi)$ does not, in general, satisfy the boundary condition on the other side of the domain, so that one may expect a boundary layer to arise there. In order to study the latter region, we introduce a new independent variable

$$\eta^* = \frac{\eta - \eta_B(\xi)}{\delta(\varepsilon)} \tag{5.5.6}$$

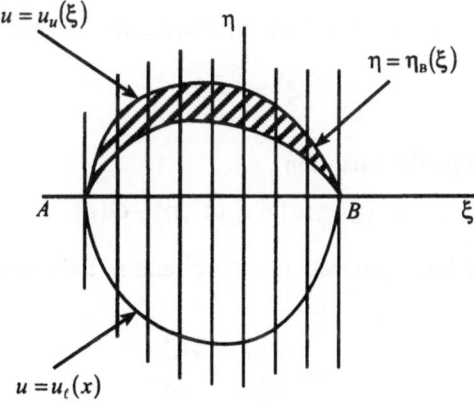

Figure 5.9. Production of a boundary layer on $u = u_u\left(\xi\right)$

(from Kevorkian and Cole, 1996).

with an associated limit process $\varepsilon \Rightarrow 0$, η^* and ξ held fixed. The retention of the highest-order derivatives in equation (5.5.4) then requires $\delta(\varepsilon) = \varepsilon$. Seeking an asymptotic solution of the form

$$u^*\left(\xi, \eta^*; \varepsilon\right) \sim u_0^*\left(\xi, \eta\right) + O(\varepsilon), \qquad (5.5.7)$$

equation (5.5.4) gives, in the limit $\varepsilon \Rightarrow 0$,

$$\kappa(\xi)\, u_{0\eta^*\eta^*} = u_{0\eta^*}^*. \qquad (5.5.8)$$

where

$$\kappa(\xi) \equiv A_{11}\eta_{B\xi}^2 - 2A_{12}\eta_{B\xi} + A_{22}.$$

The elliptic nature of equation (5.5.4) ensures that $\kappa(\xi) > 0$.

We obtain from equation (5.5.8),

$$u_0^*\left(\xi, \eta^*\right) = A(\xi) + B(\xi)\, \exp\!\left(\frac{\eta^*}{\kappa(\xi)}\right). \qquad (5.5.9)$$

Equation (5.5.9) shows that the boundary layer occurs on the upper boundary. Matching (5.5.9) asymptotically to the interior solution $u_0^{(0)}(\xi)$, we obtain

$$u_0^*\left(\xi, \eta^*\right) = u_0^{(0)}\left(\xi\right) + \left[u_U(\xi) - u_0^{(0)}\left(\xi\right)\right] \exp\!\left(\frac{\eta^*}{\kappa(\xi)}\right)$$

where the subscript U refers to values on $\eta = \eta_B(\xi)$.

This solution breaks down for the case when the boundary is a subcharacteristic, say $\xi = \xi_s$. We then introduce a new independent variable

$$\xi^* = \frac{\xi - \xi_s}{\sqrt{\varepsilon}} \qquad (5.5.10)$$

and assume an asymptotic expansion

$$u^*\left(\xi^*, \eta; \varepsilon\right) = u_0^*\left(\xi^*, \eta\right) + O(\varepsilon) \qquad (5.5.11)$$

with an associated limit process $\varepsilon \Rightarrow 0$, ξ^* and η held fixed. Then equation (5.5.4) gives

$$A_{11}\frac{\partial^2 u_0^*}{\partial \xi^{*2}} = \frac{\partial u_0^*}{\partial \eta}. \qquad (5.5.12)$$

Since $A_{11} > 0, \eta$ is a time-like coordinate, which means that one requires the prescription

$$\xi^* = \xi_s^* : u = u_s(\eta). \tag{5.5.13}$$

Further, the requirement of matching of u_0^* with the interior solution $u_0^{(0)}$ gives

$$\xi^* \Rightarrow -\infty : u_0^*(\xi^*, \eta) \Rightarrow u_0^{(0)}(\xi). \tag{5.5.14}$$

Thus, the boundary layers arising on the subcharacteristics are characterized by a diffusion-like behavior, which we also saw in Section 5.4.

If $u_s(\eta) = \text{const.}$, then we have

$$u_0^*(\xi^*, \eta) = u_s + \left[u_0^{(0)}(\xi) - u_s\right] erf\left(\frac{\xi^*}{2\sqrt{\eta}}\right). \tag{5.5.15}$$

5.6. Parabolic Equations

Consider a nonlinear diffusion equation (also called the Burgers equation (Burgers, 1948)) –

$$u_t + uu_x = \varepsilon u_{xx}, \quad -\infty < x < \infty, \quad t > 0 \tag{5.6.1}$$

$$u(x,0) = \phi(x). \tag{5.6.2}$$

It is assumed here that $\phi(x)$ is smooth and bounded except for a jump discontinuity at $x = 0$. Moreover, $\phi'(x) \geq 0$ for $x \neq 0$, and $\phi(0^-) > \phi(0^+)$.

Look for an outer solution of the form –

$$u^{(0)}(x,t) \sim u_0^{(0)}(x,t) + \varepsilon u_1^{(0)}(x,t) + \cdots. \tag{5.6.3}$$

Substituting (5.6.3) into the initial-value problem (5.6.1) and (5.6.2), we obtain

$$u_{0,t}^{(0)} + u_0^{(0)} u_{0,x}^{(0)} = 0 \tag{5.6.4}$$

$$u_0^{(0)}(x,0) = \phi(x). \tag{5.6.5}$$

The characteristics of this equation are given by (see Appendix 2)

$$x = x_0 + \phi(x_0)t. \tag{5.6.6}$$

Therefore, the solution of the above equation is given by

$$u_0^{(0)}(x,t) = \phi(x_0), \tag{5.6.7}$$

where x_0 is given implicitly in terms of x and t via (5.6.6). This solution is valid as long as the characteristics do not intersect.

Let us assume that there is a single (smooth) curve $x = s(t)$ – the shock (see Figure 5.10), where the characteristics intersect, and hence, the outer solution becomes discontinuous across such a curve. In order to develop a proper description in the transition layer around this curve, introduce a new coordinate

$$\xi = \frac{x - s(t)}{\varepsilon} \tag{5.6.8}$$

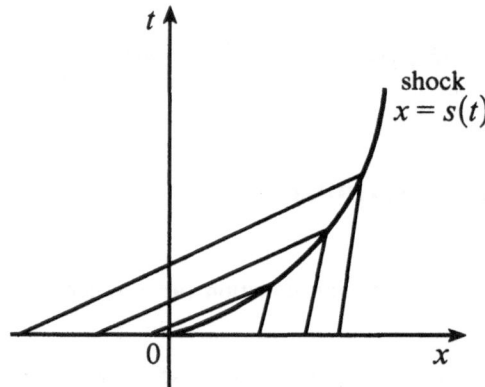

Figure 5.10. The shock curve.

and assume an inner expansion

$$u^{(1)}(\xi, t) = u_0^{(i)}(\xi, t) + \varepsilon u_1^{(i)}(\xi, t) + \cdots. \tag{5.6.9}$$

Substituting (5.6.9) into equation (5.6.1), we obtain to $0(1)$,

$$-s'(t) u_{0_\xi}^{(i)} + u_0^{(i)} u_{0_\xi}^{(i)} = u_{0_{\xi\xi}}^{(i)} \tag{5.6.10}$$

$$\xi \Rightarrow \pm\infty : u_0^{(i)} = u_0^{(0)\pm} \tag{5.6.11}$$

where,

$$u_0^{(0)\pm} = \lim_{x \Rightarrow s(t)^\pm} u_0^{(0)}(x, t).$$

We obtain from equation (5.6.10), on one integration,

$$u_{0_\xi}^{(i)} = \frac{1}{2} u_0^{(i)^2} - s'(t) u_0^{(i)} + A(t). \tag{5.6.12}$$

The boundary conditions (5.6.11) then give

$$A(t) = -\frac{1}{2}\left(u_0^{(0)-}\right)^2 + s'(t)\, u_0^{(0)-}.$$

(5.6.13)

This leads to

$$s'(t) = \frac{1}{2}\left(u_0^{(0)+} + u_0^{(0)-}\right),$$

(5.6.14)

which determines the position of the shock.

Using these expressions, another integration of equation (5.6.12) gives

$$u_0^{(i)}(\xi,t) = \frac{u_0^{(0)+} + u_0^{(0)-}\, B(t)\, e^{-\frac{1}{2}\left(u_0^{(0)-}-u_0^{(0)+}\right)\xi}}{1 + B(t)\, e^{-\frac{1}{2}\left(u_0^{(0)-}-u_0^{(0)+}\right)\xi}}$$

(5.6.15)

where $B(t)$ is an arbitrary function in the zeroth order problem. The solution (5.6.15) is sketched in Figure 5.11.

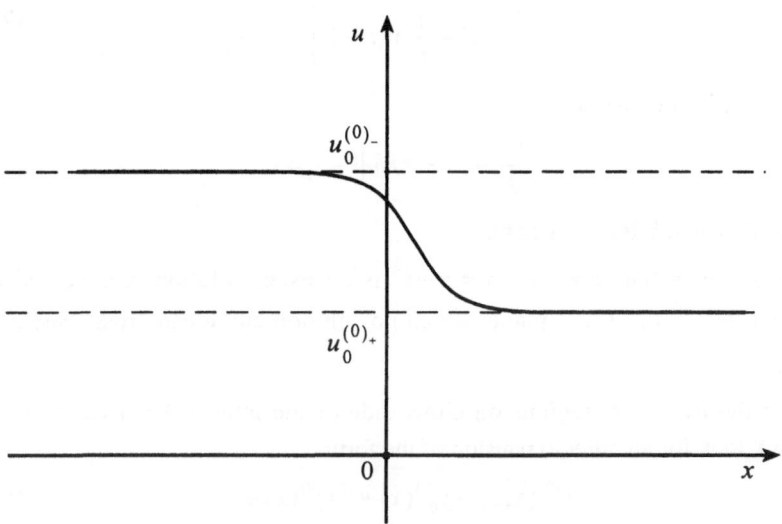

Figure 5.11. The shock structure.

5.7. Interior Layers

The rapid variations in the solution that are typical of a boundary layer do not have to occur only at the boundary. When this happens, the problems turn out to be a little bit more complicated because the location of the layer is usually not known until the expansions have been matched.

Example 4: Consider the nonlinear differential equation (Lagerstrom, 1988) –

$$\varepsilon y'' - yy' + y = 0, \quad 0 < x < 1 \tag{5.7.1}$$

$$x = 0 : y = 1 \tag{5.7.2}$$

$$x = 1 : y = -1. \tag{5.7.3}$$

Equation (5.7.1) was advanced by Lagerstrom (1988) to model shock layers in gas dynamics.

Equation (5.7.1) may be rewritten as

$$\left. \begin{array}{l} y' = z \\ z' = \dfrac{1}{\varepsilon} y(z-1) \end{array} \right\} \tag{5.7.4}$$

which admits an integral –

$$\frac{y^2}{2} - \varepsilon z + \varepsilon \ell n \, |1 - z| = C, \tag{5.7.5}$$

where C is an arbitrary constant.

This shows that $z = 1$ or $y = x + C$ is an exact solution represented by a straight line in the (x, y)-plane, which no solution curve can cross. (see Figure 5.12).

For the two outer regions on either side of the interior layer (to be located below), look for an outer expansion of the form

$$y^{(0)}(x;\varepsilon) \sim y_0^{(0)}(x) + \varepsilon y_1^{(0)}(x) + \cdots. \tag{5.7.6}$$

Equation (5.7.1) then leads to

$$y_0^{(0)} y_0^{(0)'} - y_0^{(0)} = 0, \tag{5.7.7}$$

which gives

$$y_0^{(0)} = x + \alpha \tag{5.7.8}$$

where α is an arbitrary constant to be determined by imposing the boundary conditions (5.7.2) and (5.7.3).

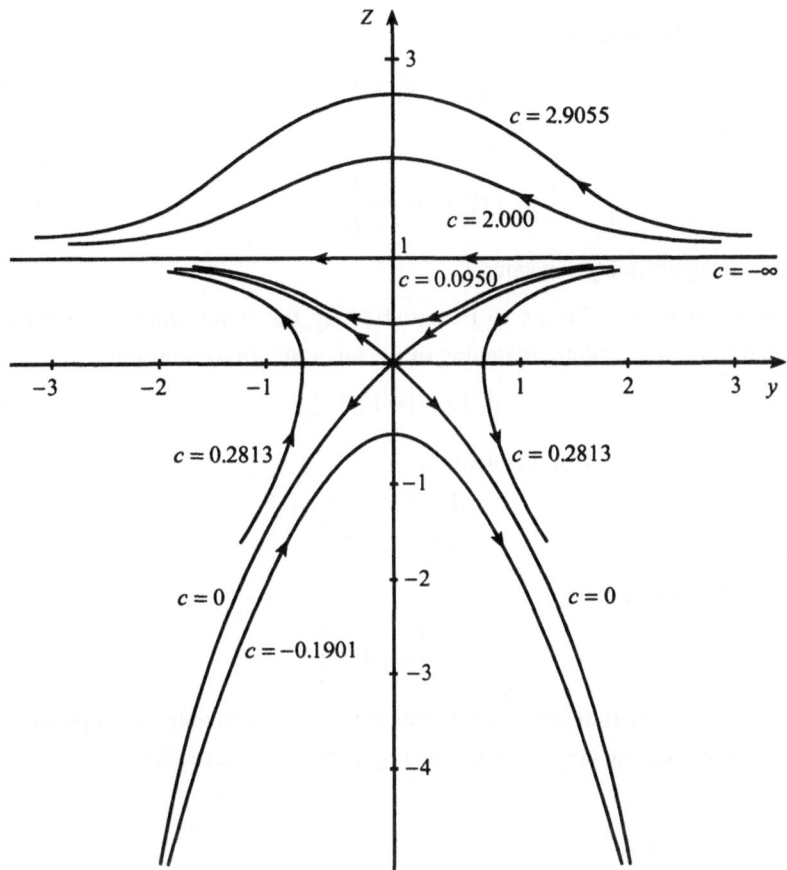

Figure 5.12. First integral: $\left(\dfrac{y^2}{2} - \epsilon z + \epsilon \, \ell n \, |1 - z| = C\right)$ **curves**

(from Lagerstrom, 1988).

If there is an interior layer at $x = x_0$, then we have

$$y_0^{(0)} = \begin{cases} x+1, & 0 \le x < x_0 \\ x-2, & x_0 < x \le 1. \end{cases} \tag{5.7.9}$$

In order to determine the solution that is valid in the interior layer, introduce

$$\xi = \frac{x - x_0}{\varepsilon} \tag{5.7.10}$$

and look for an inner expansion of the form

$$y^{(i)}(\xi,\varepsilon) \sim y_0^{(i)}(\xi) + \varepsilon\, y_1^{(i)}(\xi) + \cdots. \tag{5.7.11}$$

Equation (5.7.1) then leads to

$$y_0^{(i)''} - y_0^{(i)} y_0^{(i)'} = 0 \tag{5.7.12}$$

which gives

$$y_0^{(i)} = A\, \frac{1 - Be^{A\xi}}{1 + Be^{A\xi}}, \tag{5.7.13}$$

A and B being arbitrary constants.

The boundary conditions to be satisfied by the inner solution (5.7.13) are provided by the asymptotic matching of the outer and inner solutions

$$y_0^{(i)}(\pm\infty) = y_0^{(0)}(x_0^{\pm}). \tag{5.7.14}$$

Using (5.7.9) and (5.7.13), this leads to

$$\left.\begin{array}{l} A = x_0 + 1 \\ -A = x_0 - 2 \end{array}\right\} \tag{5.7.15}$$

from which we obtain

$$x_0 = \frac{1}{2}, \quad A = \frac{3}{2}. \tag{5.7.16}$$

In order to determine the other constant B, one reasonable assumption is that $y^{(i)}$ is anti-symmetric about $\xi = 0$ (see Figure 5.13). This leads to

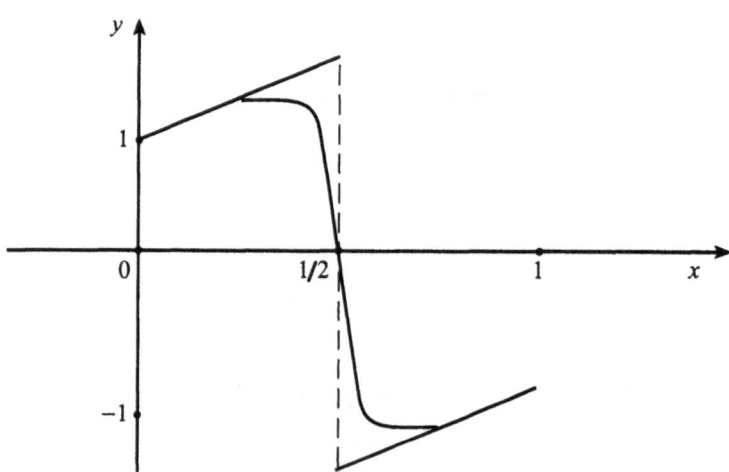

Figure 5.13. Interior layer transition.

$$B = 1. \tag{5.7.17}$$

The composite expansion for the region $0 < x < 1$ is then given by using (5.7.9), (5.7.13), (5.7.16), and (5.7.17),

$$y \sim x + 1 - \frac{3}{1 + e^{-\frac{3}{2}\left(x - \frac{1}{2}\right)/\epsilon}}, \quad 0 < x < 1 \tag{5.7.18}$$

which is shown in Figure 5.14.

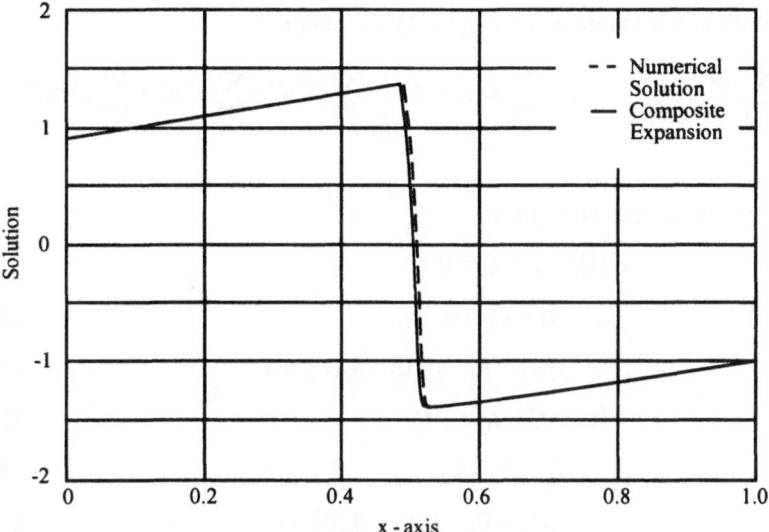

Figure 5.14. Comparison between the composite expansion given in (5.7.18) and the numerical solution of (5.7.1). In the calculations $\epsilon = 10^{-2}$ (from Holmes, 1995).

5.8. Latta's Method of Composite Expansions

Rather than determining outer and inner expansions, matching them, and then forming a composite expansion, as in the above, Latta (1951) suggested that, one may instead start with a solution which has the form of the composite expansion.

Example 5: Consider the boundary-value problem

$$\varepsilon y'' + y' + y = 0, \quad 0 \le x \le 1 \tag{5.8.1}$$

$$y(0) = a, \quad y(1) = b. \tag{5.8.2}$$

Let us look for a solution of the form

$$y \sim \sum_{n=0}^{\infty} \varepsilon^n f_n(x) + e^{-\frac{x}{\varepsilon}} \sum_{n=0}^{\infty} \varepsilon^n h_n(x). \tag{5.8.3}$$

Substituting (5.8.3) into equation (5.8.1), we obtain

$$\varepsilon \sum_{n=0}^{\infty} \varepsilon^n f_n'' + \sum_{n=0}^{\infty} \varepsilon^n f_n' + \sum_{n=0}^{\infty} \varepsilon^n f_n + e^{-\frac{x}{\varepsilon}} \left(\varepsilon \sum_{n=0}^{\infty} \varepsilon^n h_n'' - \sum_{n=0}^{\infty} \varepsilon^n h_n' + \sum_{n=0}^{\infty} \varepsilon^n h_n \right) = 0.$$

$$\tag{5.8.4}$$

This gives at various order in ε,

$$O(1) : f_0' + f_0 = 0 \tag{5.8.5}$$

$$h_0' - h_0 = 0 \tag{5.8.6}$$

$$f_1(1) = b, \quad f_0(0) + h_0(0) = a \tag{5.8.7}$$

$$O(\varepsilon) : f_1' + f_1 = -f_0'' \tag{5.8.8}$$

$$h_1' - h_1 = h_0'' \tag{5.8.9}$$

$$f_1(1) = 0, \quad f_1(0) + h_1(0) = 0 \tag{5.8.10}$$

etc.

On solving these problems, we obtain

$$O(1) : \left. \begin{array}{l} f_0 = b e^{1-x} \\ h_0 = (a - be) e^x \end{array} \right\} \tag{5.8.11}$$

$$O(\varepsilon) : \left. \begin{array}{l} f_1 = b(1-x) e^{1-x} \\ h_1 = \left[-be + (a - be) x \right] e^x \end{array} \right\} \tag{5.8.12}$$

etc.

Substituting (5.8.11) and (5.8.12), into (5.8.3), we obtain

$$y = \left[b e^{1-x} + \varepsilon b(1-x) \, e^{1-x} \right] + e^{-x/\varepsilon} \big\{ (a - be) e^x +$$

$$+ \varepsilon \left[-be + (a - be) x \right] e^x \big\} + O(\varepsilon^2)$$

or

$$y \sim b\left[1+\varepsilon(1-x)\right] e^{1-x} + e^{-\frac{x}{\varepsilon}}\left\{(a-be)(1+x)-\varepsilon\, be\right\} + O\!\left(\varepsilon^2\right) \qquad (5.8.13)$$

in agreement with the results in Example 1.

Example 6: Consider the boundary-value problem (Nayfeh, 1973)

$$\varepsilon y'' + (2x+1)y' + 2y = 0, \quad 0 \le x \le 1 \qquad (5.8.14)$$

$$y(0) = a, \quad y(1) = b. \qquad (5.8.15)$$

Let us look for a solution of the form

$$y \sim \sum_{n=0}^{\infty} \varepsilon^n f_n(x) + e^{-\frac{g(x)}{\varepsilon}} \sum_{n=0}^{\infty} \varepsilon^n h_n(x) \qquad (5.8.16)$$

where $g(x)$ is another function to be determined. Substituting (5.8.16), equation (5.8.14) gives

$$\varepsilon \sum_{n=0}^{\infty} \varepsilon^n f_n'' + (2x+1)\sum_{n=0}^{\infty} f_n' + 2\sum_{n=0}^{\infty} \varepsilon^n f_n$$

$$+ \varepsilon\left[\left(-\frac{g''}{\varepsilon}+\frac{g'^2}{\varepsilon^2}\right)\sum_{n=0}^{\infty} \varepsilon^n h_n - \frac{2g'}{\varepsilon}\sum_{n=0}^{\infty} \varepsilon^n h_n' + \sum_{n=0}^{\infty} \varepsilon^n h_n''\right] \qquad (5.8.17)$$

$$+ (2x+1)\left[-\frac{g'}{\varepsilon}\sum_{n=0}^{\infty} \varepsilon^n h_n + \sum_{n=0}^{\infty} \varepsilon^n h_n'\right] + 2\sum_{n=0}^{\infty} \varepsilon^n h_n = 0.$$

This gives at various orders in ε,

$$O(1/\varepsilon): \ h_0 g'\left[g' - (2x+1)\right] = 0 \qquad (5.8.18)$$

$$O(1): \ (2x+1)f_0' + 2f_0 = 0 \qquad (5.8.19)$$

$$(-2g' + 2x + 1)\, h_0' + (2 - g'')\, h_0 = 0 \qquad (5.8.20)$$

etc.

Substituting (5.8.16), the boundary conditions (5.8.15) give

$$g(0) = 0, \quad f_0(1) = b, \quad f_0(0) + h_0(0) = a. \qquad (5.8.21)$$

On solving these problems, we obtain

$$
\left.
\begin{array}{l}
O(1/\varepsilon): \ g = x^2 + x \\[2mm]
O(1): \ f_0 = \dfrac{3b}{2x+1} \\[2mm]
h_0 = a - 3b \\[2mm]
\text{etc.}
\end{array}
\right\}.
\tag{5.8.22}
$$

Substituting (5.8.22) into (5.8.16), we obtain

$$
y \sim \frac{3b}{2x+1} + (a - 3b)\, e^{-\frac{x^2+x}{\varepsilon}} + O(\varepsilon).
\tag{5.8.23}
$$

Example 7: Consider now a boundary-value problem for a diffusion equation (Keller, 1968) –

$$
\varepsilon u_\tau = u_{xx}, \quad 0 \le x \le b(\tau), \quad \tau = \varepsilon t
\tag{5.8.24}
$$

$$
\left.
\begin{array}{l}
u(0,\tau) = \varphi(\tau), \quad u\big[b(\tau),\tau\big] = 0 \\[2mm]
u(x,0) = \psi(x)
\end{array}
\right\}.
\tag{5.8.25}
$$

Let us look for a solution of the form

$$
u = \sum_{n=0}^{\infty} \varepsilon^n f_n(x,\tau) + e^{-\frac{g(\tau)}{\varepsilon}} \sum_{n=0}^{\infty} \varepsilon^n h_n(x,\tau).
\tag{5.8.26}
$$

Substituting (5.8.26) into the boundary-value problem (5.8.24) and (5.8.25), we obtain to various orders in ε:

$$
0(1): \ f_{0_{xx}} = 0
\tag{5.8.27}
$$

$$
f_0(0) = \varphi(\tau)
\tag{5.8.28}
$$

$$
f_0\big(b(\tau),\tau\big) = 0
\tag{5.8.29}
$$

$$
h_{0_{xx}} + g' h_0 = 0
\tag{5.8.30}
$$

$$
h_0(0,\tau) = 0
\tag{5.8.31}
$$

$$
h_0\big(b(\tau),\tau\big) = 0
\tag{5.8.32}
$$

$$
f_0(x,0) + h_0(x,0) = \psi(x)
\tag{5.8.33}
$$

$$
O(\varepsilon): \ f_{1_{xx}} = f_{0_\tau}
\tag{5.8.34}
$$

$$
f_1(0,\tau) = f_1\big[b(\tau),\tau\big] = 0
\tag{5.8.35}
$$

$$h_{1_{xx}} + g'h_1 = h_{0_\tau} \tag{5.8.36}$$

$$h_1(0,\tau) = h_1\big[b(\tau),\tau\big] = 0 \tag{5.8.37}$$

$$f_1(x,0) + h_1(x,0) = 0 \tag{5.8.38}$$

etc.

Solving the $0(1)$ problem, we obtain

$$f_0 = \varphi(\tau)\left[1 - \frac{x}{b(\tau)}\right]. \tag{5.8.39}$$

Since the boundary conditions (5.8.31) and (5.8.32) on h_0 are homogeneous, equation (5.8.30) has a nontrivial solution if and only if g' is one of the eigenvalues

$$g'_\kappa = \left[\frac{\kappa\pi}{b(\tau)}\right]^2; \quad \kappa = 1,2,\ldots . \tag{5.8.40}$$

The corresponding eigenfunctions are

$$\chi_\kappa = \left[\frac{2}{b(\tau)}\right]^{1/2} \sin\frac{\kappa\pi x}{b(\tau)}. \tag{5.8.41}$$

Thus,

$$h_0 = a_0(\tau)\,\chi_\kappa(x,\tau). \tag{5.8.42}$$

Using (5.8.39)-(5.8.42), the $O(\varepsilon)$ problem becomes

$$h_{1_{xx}} + g'_\kappa h_1 = a'_0\,\chi_\kappa + a_0\,\chi_{\kappa\tau} \tag{5.8.43}$$

$$h_1(0,\tau) = h_1\big[b(\tau),\tau\big] = 0. \tag{5.8.44}$$

Expanding h_1 in terms of the eigenfunctions χ_s,

$$h_1 = \sum_{s=1}^{\infty} C_s(\tau)\,\chi_s(x,\tau),$$

we obtain from equation (5.8.43)

$$\sum_{s=1}^{\infty} (g'_\kappa - g'_s)\,C_s\,\chi_s = a'_0\,\chi_\kappa + a_0\,\chi_{\kappa_\tau}. \tag{5.8.45}$$

Multiplying equation (5.8.45) by χ_κ, and integrating over $\big(0,b(\tau)\big)$ we have

$$\int_0^{b(\tau)} \left(a_0' \, \chi_\kappa^2 + a_0 \, \chi_{\kappa_\tau} \, \chi_\kappa \right) dx = 0 \tag{5.8.46}$$

which is just the solvability condition for equation (5.8.43).

From the normalization condition,

$$\int_0^{b(\tau)} \chi_\kappa^2 \, dx = 1, \tag{5.8.47}$$

we have

$$\frac{d}{d\tau} \int_0^{b(\tau)} \chi_\kappa^2 \, dx = 2 \int_0^{b(\tau)} \chi_\kappa \, \chi_{\kappa_\tau} dx = 0. \tag{5.8.48}$$

Using (5.8.48), we have from the solvability condition (5.8.46)

$$a_0' = 0 \quad \text{or} \quad a_0 = \text{constant}. \tag{5.8.49}$$

Using (5.8.39)-(5.8.42), and (5.8.49), we have finally

$$u \sim \varphi(\tau) \left[1 - \frac{x}{b(\tau)} \right] + \sum_{\kappa=1}^{\infty} a_\kappa \left[\frac{2}{b(\tau)} \right]^{1/2} \sin \frac{\kappa \pi x}{b(\tau)} \exp \left[-\frac{\kappa^2 \pi^2}{\varepsilon} \int_0^{\tau} \frac{d\xi}{b^2(\xi)} \right] + O(\varepsilon) \tag{5.8.50}$$

where

$$a_\kappa = \left[\frac{2}{b(0)} \right]^{1/2} \int_0^{b(0)} \left\{ \psi(x) - \varphi(0) \left[1 - \frac{x}{b(0)} \right] \right\} \sin \frac{\kappa \pi x}{b(0)} \, dx.$$

5.9. Turning-Point Problems

5.9.1 JWKB Approximation

In this procedure, one assumes an exponential dependence for the fast variation while the slow variation is described via a polynomial correction. With this assumption, the first-term approximation of the solution is determined by solving two first-order differential equations. One of these equations, called the eikonal equation, is nonlinear and determines the fast variation in the solution. The other, called the transport equation, is linear and determines the slow variation. It turns out that the higher-order terms in the JWKB (Jeffreys, 1924,

Wentzel, 1926, Kramers, 1926, Brillouin, 1926) expansion are linear and can be determined in principle conveniently.

Consider an equation of the form

$$\frac{d^2y}{dx^2} + \lambda^2 q(x) y = 0 \tag{5.9.1}$$

where λ is a constant.

If $\lambda \gg 1$, then between two successive zeros of $y(x)$, $q(x)$ is nearly constant. This suggests that we look for a solution of the form

$$y(x, \lambda) = e^{\lambda g(x, \lambda)}. \tag{5.9.2}$$

Equation (5.9.1) then becomes

$$\frac{1}{\lambda} g_{xx} + g_x^2 + q = 0. \tag{5.9.3}$$

Setting

$$g_x = h \quad \text{or} \quad g = \int h \, dx, \tag{5.9.4}$$

equation (5.9.3) becomes

$$\frac{1}{\lambda} h_x + h^2 + q = 0. \tag{5.9.5}$$

Expanding h in powers of $1/\lambda$,

$$h(x, \lambda) \sim h_0(x) + \frac{1}{\lambda} h_1(x) + \cdots, \tag{5.9.6}$$

equation (5.9.5) gives to various orders in $\dfrac{1}{\lambda}$:

$$O(1): \ h_0^2 + q = 0, \quad \text{eikonal equation} \tag{5.9.7}$$

$$O(1/\lambda): \ 2 h_0 h_1 + h_0' = 0 \tag{5.9.8}$$

etc.

We have, on solving equations (5.9.7) and (5.9.8),

$$O(1): \ h_0(x) = \pm i \sqrt{q(x)} \tag{5.9.9}$$

$$O(1/\lambda): \ h_1(x) = -\frac{1}{4} \left[\ell n \, q(x) \right]' \tag{5.9.10}$$

etc.

Consequently, we have from (5.9.2), (5.9.4), (5.9.6), (5.9.9), and (5.9.10)

$$y(x,\lambda) \sim \frac{1}{[q(x)]^{1/4}} e^{\left[\lambda\left[\pm i\int_{}^{x}\sqrt{q(t)}\,dt + O(1/\lambda^2)\right]\right]}.$$ (5.9.11)

Thus,

$$y \sim \frac{a_1 \cos\left[\lambda\int\sqrt{q(x)}\,dx\right] + b_1 \sin\left[\lambda\int\sqrt{q(x)}\,dx\right]}{\sqrt[4]{q(x)}}, \quad q > 0$$ (5.9.12a)

$$y \sim \frac{a_2 \exp\left[\lambda\int\sqrt{-q(x)}\,dx\right] + b_2 \exp\left[-\lambda\int\sqrt{-q(x)}\,dx\right]}{\sqrt[4]{-q(x)}}, \quad q < 0.$$ (5.9.12b)

This is called the JWKB approximation and is valid as long as x is away from the zeros of $q(x)$. The latter is called a turning point because y is oscillatory on one side of a zero of $q(x)$ while it is monotonic on the other side.

Example 8: Consider the boundary-value problem (Holmes, 1995),

$$y'' + \lambda^2 e^{2x} y = 0, \quad 0 < x < 1$$ (5.9.13)

$$x = 0 : y = a$$ (5.9.14)

$$x = 1 : y = b.$$ (5.9.15)

The solution (5.9.12) then becomes

$$y \sim e^{-\frac{x}{2}}\left[\frac{be^{1/2}\sin\lambda(e^x - 1) - a\sin\lambda(e^x - e)}{\sin\lambda(e - 1)}\right].$$ (5.9.16)

The exact solution for this problem is

$$y = C_1 J_0(\lambda e^x) + C_2 Y_0(\lambda e^x)$$ (5.9.17)

where,

$$C_1 \equiv \frac{1}{\Delta}\left[b\,Y_0(\lambda) - a\,Y_0(\lambda e)\right],$$

$$C_2 \equiv \frac{1}{\Delta}\left[a\,J_0(\lambda e) - b\,J_0(\lambda)\right],$$

$$\Delta \equiv J_0(\lambda e)\,Y_0(\lambda) - Y_0(\lambda e)\,J_0(\lambda).$$

Figure 5.15 shows that (5.9.16) and (5.9.17) are in good agreement.

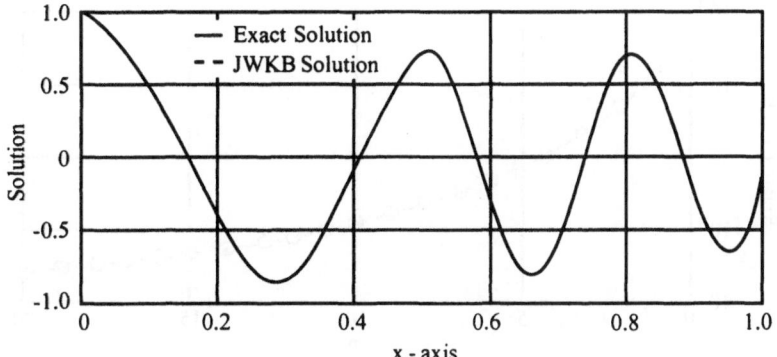

Figure 5.15. Comparison between the exact solution (5.9.17) and the JWKB approximation given in (5.9.16). In the calculations, $\epsilon = 10^{-1}$, $\alpha = 1$, and $b = 0$. (The two curves are so close they are essentially indistinguishable from each other.) (From Holmes, 1995).

Example 9: Consider the eigenvalue problem (Holmes, 1995),

$$y'' + \lambda^2 e^{2x} y = 0, \quad 0 < x < 1 \tag{5.9.18}$$

$$x = 0 \text{ and } 1 : y = 0. \tag{5.9.19}$$

From the solution (5.9.16), the eigenvalues are given by

$$\sin \lambda (e - 1) \approx 0$$

or

$$\lambda (e - 1) \approx n\pi$$

or

$$\lambda \approx \frac{n\pi}{e - 1}. \tag{5.9.20}$$

For the exact solution, on the other hand, the eigenvalues are given by

$$J_0 (\lambda e) \, Y_0 (\lambda) - Y_0 (\lambda e) \, J_0 (\lambda) = 0. \tag{5.9.21}$$

The comparison between these two sets of numbers (see Figure 5.16) shows that the JWKB approximation does well even for the first few eigenvalues.

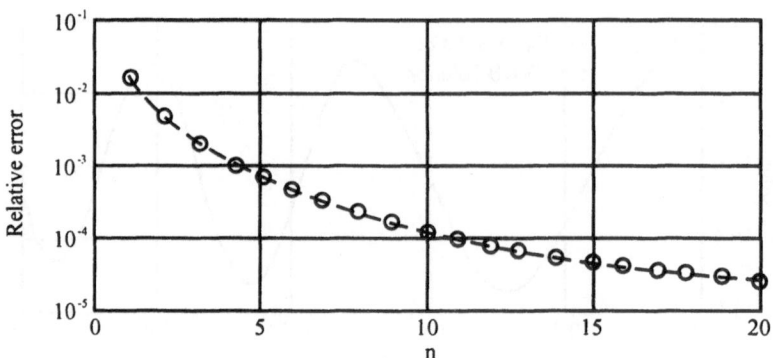

Figure 5.16. Relative error, in absolute value, between the JWKB approximation (5.9.20) of the eigenvalues and the values obtained by solving (5.9.18) numerically. Here $\epsilon = 10^{-1}$ (from Holmes, 1995).

5.9.2 Solution Near the Turning Point

Let us write, for $x \approx \mu$,

$$q(x) = (x - \mu) f(x), \quad f(x) > 0. \tag{5.9.22}$$

The JWKB solutions (5.9.12) are the outer expansions valid for $x > \mu$ and $x < \mu$ which break down for $x \approx \mu$. In order to determine the inner expansion valid near $x = 0$, we introduce

$$\xi = (x - \mu) \lambda^{2/3} \tag{5.9.23}$$

so equation (5.9.1) becomes

$$\frac{d^2 y}{d\xi^2} + \xi f(\mu) y = 0. \tag{5.9.24}$$

Putting

$$z = -\xi \left(f(\mu) \right)^{1/3} \tag{5.9.25}$$

equation (5.9.24) becomes

$$\frac{d^2 y}{dz^2} - zy = 0, \tag{5.9.26}$$

from which

$$y^{(i)} = a_3 \, Ai(z) + b_3 \, Bi(z)^{\,2} \qquad (5.9.27)$$

where $Ai(z)$ and $Bi(z)$ are Airy functions, which have the following asymptotic values for large z:

$$
\left.
\begin{aligned}
Ai(z) &\approx \frac{1}{2\sqrt{\pi}} \, z^{-1/4} \, e^{-\zeta} \\[4pt]
Ai(-z) &\approx \frac{1}{\sqrt{\pi}} \, |z|^{-1/4} \, \sin(\zeta + \pi/4) \\[4pt]
Bi(z) &\approx \frac{1}{\sqrt{\pi}} \, z^{-1/4} \, e^{\zeta} \\[4pt]
Bi(-z) &\approx \frac{1}{\sqrt{\pi}} \, |z|^{-1/4} \, \cos(\zeta + \pi/4)
\end{aligned}
\right\}
\qquad (5.9.28)
$$

where

$$\zeta = \frac{2}{3} \, |z|^{3/2} .$$

In order to match the inner solution (5.9.27) with the outer solution (5.9.12a) for $\xi > 0$, we note first that

$$\lambda \int_\mu^x \sqrt{q(\tau)} \; d\tau = \lambda \int_\mu^x \sqrt{\tau - \mu} \; \sqrt{f(\tau)} \; d\tau = \frac{2}{3} \sqrt{f(\mu)} \; \xi^{3/2} + O(\lambda^{-2/3}) \qquad (5.9.29)$$

so that (5.9.12a) becomes, for $x \approx \mu$ and $x > \mu$,

$$y^{(0)} \sim \frac{\lambda^{1/6}}{\sqrt[4]{\xi f(\mu)}} \left[a_1 \cos\left(\frac{2}{3} \sqrt{f(\mu)} \; \xi^{3/2} \right) + b_1 \sin\left(\frac{2}{3} \sqrt{f(\mu)} \; \xi^{3/2} \right) \right] + \cdots, \quad \xi > 0 .$$

$$(5.9.30)$$

---

[2] The solution of equation (5.9.26) is given by

$$y = \sqrt{z} \left[C_1 \, I_{1/3}\!\left(\frac{2}{3} z^{3/2} \right) + C_2 \, I_{-1/3}\!\left(\frac{2}{3} z^{3/2} \right) \right].$$

Putting

$$Ai(z) = \frac{1}{3} \sqrt{z} \left[I_{-1/3}\!\left(\frac{2}{3} z^{3/2} \right) - I_{1/3}\!\left(\frac{2}{3} z^{3/2} \right) \right],$$

$$Bi(z) = \frac{1}{3} \sqrt{z} \left[I_{-1/3}\!\left(\frac{2}{3} z^{3/2} \right) + I_{1/3}\!\left(\frac{2}{3} z^{3/2} \right) \right],$$

the above solution becomes

$$y = C_1 \, Ai(z) + C_2 \, Bi(z) .$$

Further, we note that for large $\xi > 0$, we have from (5.9.27) and (5.9.28),

$$y^{(i)} \sim \frac{\xi^{-1/4} f^{-1/12}}{\sqrt{\pi}} \left[a_3 \sin\left(\frac{2}{3}\sqrt{f(\mu)}\xi^{3/2} + \pi/4\right) + b_3 \cos\left(\frac{2}{3}\sqrt{f(\mu)}\xi^{3/2} + \pi/4\right) \right] + \cdots,$$

$$\xi > 0. \tag{5.9.31}$$

Matching $y^{(0)}$ with $y^{(i)}$, we obtain

$$a_1 = \frac{\lambda^{-1/6} f^{1/6}}{\sqrt{\pi}} \left[a_3 \sin\frac{\pi}{4} + b_3 \cos\frac{\pi}{4} \right],$$

$$b_1 = \frac{\lambda^{-1/6} f^{1/6}}{\sqrt{\pi}} \left[a_3 \cos\frac{\pi}{4} - b_3 \sin\frac{\pi}{4} \right]. \tag{5.9.32}$$

Next, in order to match the inner solution (5.9.27) with the outer solution (5.9.12b) for $\xi < 0$, note

$$\lambda \int_x^\mu \sqrt{-q(\tau)} \, d\tau = \lambda \int_x^\mu \sqrt{(\mu - \tau) f(\tau)} \, d\tau$$

$$= \frac{2}{3} \sqrt{f(\mu)} \, (-\xi)^{3/2} + O(\lambda^{-2/3}) \tag{5.9.33}$$

so that (5.9.12b) becomes for $x \approx \mu$, $x < \mu$,

$$y^{(0)} \sim \frac{\lambda^{1/6}}{\sqrt[4]{-f(\mu)\xi}} \left\{ a_2 \exp\left[\frac{2}{3}\sqrt{f(\mu)}(-\xi)^{3/2}\right] + b_2 \exp\left[-\frac{2}{3}\sqrt{f(\mu)}(-\xi)^{3/2}\right] \right\} + \cdots,$$

$$\xi < 0. \tag{5.9.34}$$

Also for large $|\xi|$, $\xi < 0$, we have from (5.9.27) and (5.9.28),

$$y^{(i)} \sim \frac{(-\xi)^{-1/4} f^{-1/12}}{\sqrt{\pi}} \left\{ \frac{1}{2} a_3 \exp\left[-\frac{2}{3}\sqrt{f(\mu)}(-\xi)^{3/2}\right] \right.$$

$$\left. + b_3 \exp\left[\frac{2}{3}\sqrt{f(\mu)}(-\xi)^{3/2}\right] \right\} + \cdots, \quad \xi < 0. \tag{5.9.35}$$

Matching $y^{(0)}$ with $y^{(i)}$, we obtain

$$a_2 = \frac{\lambda^{-1/6} f^{1/6}}{\sqrt{\pi}} b_3, \quad b_2 = \frac{\lambda^{-1/6} f^{1/6}}{2\sqrt{\pi}} a_3. \tag{5.9.36}$$

Thus, an approximate solution to an equation with a turning point at $x = \mu$ is given by three separate expansions (Rayleigh, 1912) –

(i)   one near $x \approx \mu$;

(ii)   one for $x < \mu$;

(iii)   one for $x > \mu$.

5.9.3 Langer's Method

One may seek to describe the behavior of the solution of a turning-point problem of the form

$$\varepsilon^2 y'' - f(x) y = 0 \tag{5.9.37}$$

over the given entire domain in terms of the Airy functions. For this purpose, we introduce another independent variable (Langer, 1931, 1935, Wasow, 1965),

$$\xi = \varepsilon^{-2/3} g(x) \quad \text{with} \quad g(0) = 0, \tag{5.9.38}$$

and look for a solution of the form

$$y = y(x, \xi; \varepsilon). \tag{5.9.39}$$

Noting that

$$\left. \begin{array}{l} y' = y_x + \dfrac{1}{\varepsilon^{2/3}} g' y_\xi \\[2mm] y'' = y_{xx} + \dfrac{2}{\varepsilon^{2/3}} g' y_{x\xi} + \dfrac{1}{\varepsilon^{2/3}} g'' y_\xi + \dfrac{1}{\varepsilon^{4/3}} g'^2 y_{\xi\xi} \end{array} \right\} \tag{5.9.40}$$

we obtain from equation (5.9.37)

$$gg'^2 y_{\xi\xi} + \varepsilon^{2/3} g(2g' y_{x\xi} + g'' y_\xi) + \varepsilon^{4/3} gy_{xx} - \xi f(x)y = 0. \tag{5.9.41}$$

If, for convenience, we set

$$gg'^2 = f, \tag{5.9.42}$$

equation (5.9.41) becomes

$$y_{\xi\xi} + \varepsilon^{2/3} \left(\frac{2}{g'} y_{x\xi} + \frac{g''}{g'^2} y_\xi \right) + \varepsilon^{4/3} \left(\frac{1}{g'^2} y_{xx} \right) - \xi y = 0. \tag{5.9.43}$$

We have from equation (5.9.42)

$$g(x) = \operatorname{sgn}(x) \left[\frac{3}{2} \int_0^x |f(t)|^{1/2} \, dt \right]^{2/3}. \tag{5.9.44}$$

Expanding the solution as

$$y(x,\xi,\varepsilon) \sim y(x,\xi) + \varepsilon^{2/3} y_1(x,\xi) + \cdots \tag{5.9.45}$$

we obtain from equation (5.9.43), to various orders in ε,

$$O(1): y_{0_{\xi\xi}} - \xi y_0 = 0 \tag{5.9.46}$$

$$O(\varepsilon): y_{1_{\xi\xi}} - \xi y_1 = -\frac{2}{g'} y_{0_{x\xi}} - \frac{g''}{g'^2} y_{0_\xi}, \tag{5.9.47}$$

etc.

We have from equation (5.9.46)

$$y_0(x,\xi) = C_0(x) \, Ai(\xi). \tag{5.9.48}$$

Using (5.9.48), equation (5.9.47) becomes

$$y_{1_{\xi\xi}} - \xi y_1 = -\left[\frac{2}{g'} C_0' + \frac{g''}{g'^2} C_0\right] Ai'(\xi). \tag{5.9.49}$$

Removal of the secular term in equation (5.9.49) requires

$$\frac{2}{g'} C_0' + \frac{g''}{g'^2} C_0 = 0 \tag{5.9.50}$$

from which

$$C_0(x) = \frac{C_0}{[g'(x)]^{1/4}} = C_0 \left(\frac{g}{f}\right)^{1/4}, \tag{5.9.51}$$

where C_0 is an arbitrary constant.

Using (5.9.51), (5.9.48) becomes

$$y(x) \sim C_0 \left[\frac{g(x)}{f(x)}\right]^{1/4} Ai\left[\varepsilon^{-2/3} g(x)\right] + O(\varepsilon^{2/3}). \tag{5.9.52}$$

Note that as $x \to +\infty$, (5.9.52) gives

$$g \sim \frac{C_0 \, \varepsilon^{1/6}}{2\sqrt{\pi} \, [f(x)]^{1/4}} e^{-\frac{1}{\varepsilon} \int\limits^x \sqrt{f(t)}\, dt} \tag{5.9.53}$$

while as $x \to -\infty$, (5.9.52) gives

$$y \sim \frac{C_0 \, \varepsilon^{1/6}}{\sqrt{\pi} \, |f(x)|^{1/4}} \sin\left[\frac{1}{\varepsilon} \int\limits^x |f(t)|^{1/2} \, dt + \pi/4\right] \tag{5.9.54}$$

in agreement with the WKBJ approximation discussed in the previous section.

5.10. Applications to Fluid Dynamics: Boundary Layer Flow Past a Flat Plate

For flows past streamlined bodies at large Reynolds numbers (i.e., in fluids of small viscosity), Prandtl (1904) proposed that one need recognize the effects of viscosity only in a thin layer - boundary layer adjacent to the body and the rest of the flow may be considered inviscid. As a first approximation, the inviscid-flow equations are solved with appropriate boundary conditions, ignoring the presence of the boundary layer. However, in general, the inviscid flow will not satisfy the condition of the no-slip of the fluid at the body, and it is necessary to introduce a boundary layer between the inviscid flow and the body to adjust the inviscid solution toward satisfying this condition on the body.

Lagerstrom and Cole (1955) pointed out that Prandtl's boundary-layer theory can be embedded in a systematic scheme of successive approximations via the method of matched asymptotic expansions. However, this solution, which is represented by an infinite series in powers of $R_E^{-1/2}$ (R_E being the Reynolds number, $R_E \equiv UL/\upsilon$, U being a reference velocity, say, the velocity of the fluid in the free stream far away from the body, L being a characteristic length of the body, and υ being the kinematic viscosity of the fluid) turns out to be asymptotic.

We consider here the problem of two-dimensional viscous incompressible flow past a flat plate. It turns out that the external inviscid flow is associated with an outer limit process, and the boundary layer with an inner-limit process. The order of the differential equations is lowered in the outer limit, and the boundary condition of no-slip of the flow at the plate is lost so that the problem under consideration is one of singular-perturbation type.

5.10.1 The Outer Expansion

Let x measure the distance along the plate from the leading edge and y the distance normal to the plate.

The equations governing this flow are –

(i)   conservation of mass –

$$\nabla \cdot \boldsymbol{v} = 0 , \tag{5.10.1}$$

(ii) conservation of momentum –

$$\frac{\partial v}{\partial t} + (v \cdot \nabla) \, v = -\frac{1}{\rho} \, \nabla p + \upsilon \nabla^2 v, \tag{5.10.2}$$

where v is the fluid velocity, p is the fluid pressure, and ρ is the fluid density.

Eliminating the pressure p, equations (5.10.1) and (5.10.2) lead to

$$\frac{\partial \omega}{\partial t} + (v \cdot \nabla) \, \omega = \upsilon \nabla^2 \omega \tag{5.10.3}$$

where ω is the vorticity

$$\omega \equiv \nabla \times v \cdot \hat{i}_z. \tag{5.10.4}$$

Introducing the stream function ψ according to

$$v = \nabla \Psi \times \hat{i}_z, \tag{5.10.5}$$

equations (5.10.3) and (5.10.4) become

$$\frac{\partial}{\partial t} \nabla^2 \Psi + \left(\frac{\partial \Psi}{\partial y} \frac{\partial}{\partial x} - \frac{\partial \Psi}{\partial x} \frac{\partial}{\partial y} \right) \nabla^2 \Psi = \upsilon \nabla^4 \Psi \tag{5.10.6}$$

and

$$\omega = -\nabla^2 \Psi. \tag{5.10.7}$$

The boundary conditions on this flow are

$$y = 0 : v = 0 \tag{5.10.8}$$

$$\text{upstream} : v = U \hat{i}_x. \tag{5.10.9}$$

Equation (5.10.8) describes the no-slip condition at the plate.

Assuming the flow to be steady, the boundary-value problem in nondimensionalized flow variables is given by

$$\left(\Psi_y \frac{\partial}{\partial x} - \Psi_x \frac{\partial}{\partial y} - \frac{1}{R_E} \nabla^2 \right) \nabla^2 \Psi = 0 \tag{5.10.10}$$

$$\left. \begin{array}{l} y = 0 : \Psi = 0, \quad \Psi_y = 0, \quad 0 < x < 1 \quad \text{or} \quad \infty \\[2mm] \text{upstream} : \Psi \sim y \end{array} \right\}. \tag{5.10.11}$$

If the plate is semi-infinite there is no natural length in the problem; however, the apparent difficulty may be circumvented by choosing some reference length L.

Let us seek a straightforward asymptotic expansion, as $R_E \to \infty$, of the form

$$\Psi\left(x,y;R_E\right) \sim \psi_1^{(0)}(x,y) + \delta_2^{(0)}\left(R_E\right)\psi_2^{(0)}(x,y) + \cdots \qquad (5.10.12)$$

with the associated outer limit process $R_E \to \infty$, and x,y fixed. Then, equation (5.10.10) gives

$$\left(\psi_{1y}^{(0)}\frac{\partial}{\partial x} - \psi_{1x}^{(0)}\frac{\partial}{\partial y}\right)\nabla^2\psi_1^{(0)} = 0 \qquad (5.10.13)$$

from which

$$\nabla^2\psi_1^{(0)} = -\omega_1^{(0)}\left(\psi_1^{(0)}\right). \qquad (5.10.14)$$

If the oncoming stream is irrotational, equation (5.10.14) gives

$$\nabla^2\psi_1^{(0)} = 0. \qquad (5.10.15)$$

The boundary-conditions (5.10.11) give

$$\left.\begin{array}{l} y = 0 : \psi_1^{(0)} = 0 \\[2mm] \text{upstream} : \psi_1^{(0)} \sim y \end{array}\right\}. \qquad (5.10.16)$$

Note that the no-slip condition

$$y = 0 : \psi_{1y}^{(0)} = 0$$

has been dropped since, in the outer limit, the order of the differential equation (5.10.10) has dropped.

From (5.10.15) and (5.10.16), we obtain

$$\psi_1^{(0)}(x,y) = y \qquad (5.10.17)$$

so that, in the limit $R_E \to \infty$, a flat plate causes no disturbance.

5.10.2 The Inner Expansion

Because of the loss of the no-slip condition, the basic inviscid solution is not valid close to the flat plate. Therefore, assume an inner expansion valid within the boundary layer

$$\psi^{(i)}\left(x,y;R_E\right) \sim \delta_1^{(i)}\left(R_E\right)\psi_1^{(i)}(x,Y) + \delta_2^{(i)}\left(R_E\right)\psi_2^{(i)}(x,Y) + \cdots \qquad (5.10.18)$$

with the associated inner limit process $R_E \to \infty, x, Y$ fixed, where

$$Y = \frac{y}{\delta_1^{(i)}\left(R_E\right)}. \qquad (5.10.19)$$

Then, equation (5.10.10) gives

$$\left(\psi_{1Y}^{(i)}\frac{\partial}{\partial x}-\psi_{1y}^{(i)}\frac{\partial}{\partial Y}\right)\psi_{1YY}^{(i)}=\lim_{R_E\to\infty}\left[\frac{1}{R_E\delta_1^{(i)2}(R_E)}\right]\psi_{1YYYY}^{(i)} \qquad (5.10.20)$$

so that the retention of the highest derivative (the term on the right-hand side in (5.10.20)) requires

$$\delta_1^{(i)}(R_E)=\frac{1}{\sqrt{R_E}}. \qquad (5.10.21)$$

Thus, we have from (5.10.19),

$$Y=\sqrt{R_E}\ y \qquad (5.10.22)$$

and equation (5.10.20) becomes

$$\left(\frac{\partial^2}{\partial Y^2}-\psi_{1Y}^{(i)}\frac{\partial}{\partial x}+\psi_{1x}^{(i)}\frac{\partial}{\partial Y}\right)\psi_{1YY}^{(i)}=0$$

or

$$\frac{\partial}{\partial Y}\left(\psi_{1YYY}^{(i)}+\psi_{1x}^{(i)}\ \psi_{1YY}^{(i)}-\psi_{1Y}^{(i)}\ \psi_{1xY}^{(i)}\right)=0 \qquad (5.10.23)$$

from which we have

$$\psi_{1YYY}^{(i)}+\psi_{1x}^{(i)}\ \psi_{1YY}^{(i)}-\psi_{1Y}^{(i)}\ \psi_{1xY}^{(i)}=f(x) \qquad (5.10.24)$$

where $f(x)$ is proportional to the pressure gradient impressed on the plate by the inviscid flow. Note that this implies that the pressure is almost constant across the boundary layer.

The asymptotic matching between the outer and the inner solutions gives

$$\psi_1^{(i)}(x,\infty)=Y\psi_{1y}^{(i)}(x,0)+\cdots$$

from which

$$\psi_{1Y}^{(i)}(x,\infty)=\psi_{1y}^{(0)}(x,0) \qquad (5.10.25)$$

which simply implies that the tangential velocity of the boundary-layer flow approaches at the outer edge of the boundary layer, $y\to\infty$, the inviscid flow-velocity $\psi_{1y}^{(0)}(x,0)$.

Using the outer inviscid solution, equation (5.10.24) becomes

$$\psi_{1YYY}^{(i)}+\psi_{1x}^{(i)}\ \psi_{1YY}^{(i)}-\psi_{1Y}^{(i)}\ \psi_{1xY}^{(i)}=-\psi_{1y}^{(0)}(x,0)\ \psi_{1xy}^{(0)}(x,0). \qquad (5.10.26)$$

Notice that the boundary-layer equation is parabolic with x acting as a time-like variable – although the original Navier-Stokes equations were elliptic. This is in agreement with the result in Section 5.4 that the boundary layers arising on the subcharacteristics are characterized by a diffusion-like behavior. This means that the upstream influence is lost so that the first-order boundary-layer solution on a flat plate is not affected by the trailing edge (if the plate is finite) and the wake beyond.

For a flat plate, equation (5.10.26) becomes

$$\psi^{(i)}_{1YYY} + \psi^{(i)}_{1x}\,\psi^{(i)}_{1YY} - \psi^{(i)}_{1Y}\,\psi^{(i)}_{1xY} = 0, \tag{5.10.27}$$

the boundary conditions being

$$\left. \begin{array}{l} Y = 0 : \psi^{(i)}_1 = 0, \quad \psi^{(i)}_{1Y} = 0 \quad \text{for} \quad 0 < x < 1 \text{ or } \infty \\[2mm] Y \to \infty : \psi^{(i)}_Y = 1 \end{array} \right\}. \tag{5.10.28}$$

Now, the problem given by (5.10.27), (5.10.28) is invariant under the transformation

$$\psi^{(i)}_1 \to c\psi^{(i)}_1, \quad x \to c^2 x, \quad Y \to cY$$

so that it is possible to look for solutions in which $\psi^{(i)}_1$, x and y occur in certain combinations – self-similar solutions. This also reduces equation (5.10.27) to an ordinary differential equation. Thus, putting

$$\psi^{(i)}_1(x, Y) = \sqrt{2x}\; f_1(\eta), \quad \eta = \frac{Y}{\sqrt{2x}} \tag{5.10.29}$$

equation (5.10.27) gives the Blasius equation –

$$f'''_1 + f_1 f''_1 = 0 \tag{5.10.30}$$

while the boundary conditions (5.10.28) give

$$\left. \begin{array}{l} \eta = 0 : f_1 = 0, \quad f'_1 = 0 \\[2mm] \eta \to \infty : f'_1 \to 1 \end{array} \right\}. \tag{5.10.31}$$

Here the primes denote differentiation with respect to η. Note that the flow velocity components are given by

$$u = f'_1(\eta), \quad v = (\eta f'_1 - f_1)$$

which implies that the velocity profile in the boundary layer is the same for all x except for the change of scale.

In order to find the asymptotic behavior of the solution of (5.10.30) and (5.10.31), note first that equation (5.10.30) admits a solution of the form

$$f_1 \sim \frac{1}{\eta}. \tag{5.10.32}$$

This implies that equation (5.10.30) has the scaling group

$$\bar{f}_1 = \alpha^{-1} f_1, \quad \bar{\eta} = \alpha \eta. \tag{5.10.33}$$

We may, therefore, introduce the following canonical coordinates –

$$s = f_1 \eta, \quad t = \eta^2 \frac{df_1}{d\eta}. \tag{5.10.34}$$

The transformation from (s,t) to (f,η) is given differentially by

$$\frac{ds}{t+s} = \frac{d\eta}{\eta}. \tag{5.10.35}$$

The transformation rules of the various derivatives are

$$\frac{d^2 f_1}{d\eta^2} = -\frac{2}{\eta^3} t + \frac{1}{\eta^3} \frac{dt}{ds}(t+s),$$

$$\frac{d^3 f_1}{d\eta^3} = \frac{6}{\eta^4} t - \frac{5}{\eta^4}(t+s)\frac{dt}{ds} + \frac{1}{\eta^4}\frac{d^2 t}{ds^2}(t+s)^2 + \frac{1}{\eta^4}\left(\frac{dt}{ds}\right)(t+s)\left(\frac{dt}{ds}+1\right). \tag{5.10.36}$$

In terms of the new coordinates (s,t), the boundary-value problem (5.10.30) and (5.10.31) becomes

$$(t+s)^2 \frac{d^2 t}{ds^2} + (t+s)\left(\frac{dt}{ds}+1\right)\frac{dt}{ds} - 5(t+s)\frac{dt}{ds} + 6t + s\left[-2t + (t+s)\frac{dt}{ds}\right] = 0 \tag{5.10.37}$$

$$\left.\begin{array}{l} s = 0 : t = 0 \\ s \to \infty : t \to \infty \end{array}\right\}. \tag{5.10.38}$$

Near $s = 0$, equation (5.10.37) shows that

$$t = \lambda s \tag{5.10.39}$$

with

$$\lambda(\lambda+1)^2 - 5\lambda(\lambda+1) + 6\lambda = 0 \tag{5.10.40a}$$

or

$$\lambda = 1, 2. \tag{5.10.40b}$$

$\lambda = 1$ turns out to be the spurious root. For $\lambda = 2$, we obtain from equation (5.10.35),

$$\frac{ds}{3s} = \frac{d\eta}{\eta}$$

(5.10.41)

from which we have

$$s = \eta^3.$$

(5.10.42)

Hence, we have from (5.10.34),

$$\eta \to 0 : f_1 \approx \eta^2.$$

(5.10.43)

Next, near $s \to \infty$, equation (5.10.37) shows that

$$t = \lambda s$$

(5.10.44)

with

$$-2\lambda + \lambda(\lambda + 1) = 0$$

(5.10.45a)

or

$$\lambda = 1.$$

(5.10.45b)

We then obtain from equation (5.10.35),

$$\frac{ds}{2s} \approx \frac{d\eta}{\eta}$$

(5.10.46)

from which we have

$$s \approx \eta^2.$$

(5.10.47)

Hence, we have from (5.10.34),

$$\eta \to \infty : f_1 \approx \eta - \beta_1$$

(5.10.48)

as expected! Here, β_1 is an arbitrary constant.

Figure 5.17 shows the numerical solution of equation (5.10.30) due to Schlichting (1972) compared with the experimental data for several values of Reynolds numbers. The agreement between the two shows the validity of the various approximations and assumptions made in the boundary layer theory. Besides, the velocity profile is seen to preserve its shape as one moves downstream despite the fact that the boundary layer thickness is changing.

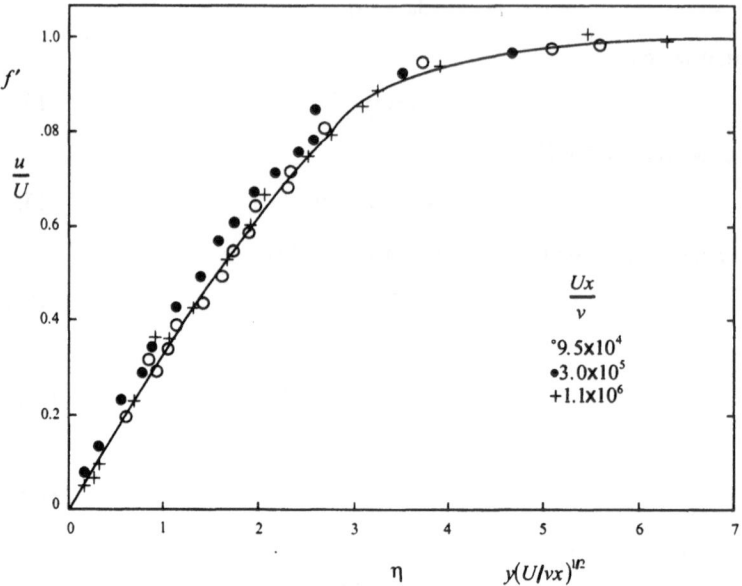

Figure 5.17. Theoretical Blasius profile (Schlichting, 1972) and experimental confirmation (Dhawan, 1952), plotted by Tritton, 1988.

5.10.3 Flow Due to Displacement Thickness

Using (5.10.12), (5.10.17), (5.10.18), (5.10.21), (5.10.27), (5.10.43), and (5.10.48), the asymptotic matching gives

$$0 + \delta_2^{(0)}\left(R_e\right)\psi_2^{(0)}(x,0) + \cdots + \frac{Y}{\sqrt{R_E}} + \cdots$$

$$= \frac{1}{\sqrt{R_E}}\;\psi_1^{(i)}(x,\infty) + \cdots$$

$$= \frac{1}{\sqrt{R_E}}\left[\sqrt{2x}\;f_1\left(\frac{Y}{\sqrt{2x}}\right)\right] + \cdots, \quad \text{as } Y \to \infty$$

$$= \frac{Y}{\sqrt{R_E}} - \frac{1}{\sqrt{2R_E}}\;\beta_1\,\sqrt{2x} + \cdots, \quad \text{as } Y \to \infty. \qquad (5.10.49)$$

Thus,

$$\delta_2^{(0)}(R_E) = \frac{1}{\sqrt{R_E}} \Bigg\}$$

$$y = 0 : \psi_2^{(0)} = -\beta_1 \sqrt{x} \Bigg]$$

(5.10.50)

so that $\psi^{(0)} = 0$ at $y = \left(1/\sqrt{R_E}\right)\beta_1\sqrt{x}$, which implies that the presence of a boundary layer endows a certain thickness to the plate, which then displaces the outer inviscid flow like a solid parabola of nose radius β_1^2/R_E.

For a semi-infinite flat plate, we then have

$$\nabla^2 \psi_2^{(0)} = 0 \qquad\qquad (5.10.51)$$

$$y = 0 : \psi_2^{(0)} = 0, \quad x < 0 \Bigg]$$

$$\psi_2^{(0)} = -\beta_1\sqrt{2x}, \quad x > 0 \Bigg\}$$

$$\text{upstream} : \psi_2^{(0)} = o(y) \Bigg]$$

(5.10.52)

which corresponds to linearized flow for a body given by $y = \beta_1\left(\sqrt{2x/R_e}\right)$, and we have

$$\psi_2^{(0)} = -\beta_1 \,\text{Re}\left[\sqrt{2(x + iy)}\right]. \qquad\qquad (5.10.53)$$

Thus, even though the flow outside the wake and the boundary layer is essentially irrotational, it is not accurately described by the solution for potential flow past the given body. One has to take into account the apparent change of shape of the body caused by the displacement-thickness effect of the boundary layer.

Note that the foregoing theory is not valid within a distance of order v/U from the leading edge of the plate where the thickness of the boundary layer is comparable with the distance from the leading edge.

5.11. Exercises

1. Solve by the method of inner and outer expansions

$$\varepsilon y'' + xy' - xy = 0 \Bigg\}$$
$$y(0) = a, \quad y(1) = b \Bigg]^{.}$$

2. Solve by using the method of inner and outer expansions

$$\left.\begin{array}{l} \varepsilon y'' + (1 + \alpha x)\, y' + \alpha y = 0, \quad 0 < x < 1, \quad \alpha > -1 \\ y(0) = 0, \quad y(1) = 1 \end{array}\right\}.$$

3. Solve by using the method of inner and outer expansions

$$\left.\begin{array}{l} \varepsilon\left(y'' + \dfrac{2}{x}\, y'\right) - y = 0 \\ y(0) = 0, \quad y'(1) = 0 \end{array}\right\}.$$

4. Solve by using the method of inner and outer expansions (Friedrichs, 1942)

$$\left.\begin{array}{l} \varepsilon y'' + y' = a \\ y(0) = 0, \quad y(1) = 1 \end{array}\right\}.$$

5. Solve using the method of matched asymptotic expansions –

$$\varepsilon y'' + y y' - x y = 0,$$
$$y(0) = 1, \quad y(1) = -1.$$

6. Solve by using both the method of inner/outer expansions and the method of composite expansion

$$\left.\begin{array}{l} \varepsilon y'' + (2x + 1)\, y' + 2y = 0 \\ y(0) = a, \quad y(1) = b \end{array}\right\}.$$

7. Solve by using both method of inner/outer expansion and the method of composite expansion

$$\left.\begin{array}{l} \varepsilon y'' - (2x + 1)\, y' + 2y = 0 \\ y(0) = a, \quad y(1) = b \end{array}\right\}.$$

8. Solve by using Langer's method

$$\left.\begin{array}{l} y'' + \omega^2 (\sin x)\, y = 0, \quad \omega \gg 1 \\ y(0) = y(1) = 0 \end{array}\right\}.$$

9. Consider the eigenvalue problem (Holmes, 1995)

$$\frac{d}{dx}\left[p(x)\frac{dy}{dx}\right] - \left[r(x) - \lambda^2 q(x)\right] y = 0, \quad \lambda \gg 1,$$
$$y(0) = 0, \quad y(1) = 0,$$

where $p(x)$, $q(x)$, and $r(x)$ are smooth, positive functions. By putting (Liouville, 1837),

$$y(x) = \frac{w(x)}{\sqrt{p(x)}},$$

show that the above equation becomes

$$p(x)\, w''(x) + \left[\lambda^2 q(x) - f(x)\right] w = 0$$

where

$$f(x) \equiv r(x) + \frac{1}{2}\, \sqrt{p(x)} \left(\frac{p'(x)}{\sqrt{p(x)}}\right)'.$$

Use a first-term WKBJ approximation to show that, for large λ,

$$\lambda \sim \lambda_n = n\pi \left[\int_0^1 \sqrt{\frac{q(x)}{p(x)}}\; dx\right]^{-1}.$$

10. Use the JWKB method to find an approximate solution of (Holmes, 1995)

$$\varepsilon y'' + 2y' + 2y = 0,$$
$$y(0) = 0, \quad y(1) = 1.$$

Compare the result with

*   the composite expansion obtained using the method of matched asymptotic expansions,

*   the exact solution.

5.12. Appendix 1

Initial-Value Problem for Partial Differential Equations

The fact that each equation in characteristic form involves a particular linear combination of the derivatives can be used to gain insight into the structure of solutions of the equations, such as the correct number of boundary conditions and the domain of dependence, by considering a construction of the solution at successive small time increments. In order to see this, let us consider a hyperbolic partial differential equation in the characteristic form

$$\frac{d\psi_k}{dt} + f_k(x, t, \psi) = 0 \quad \text{on} \quad \frac{dx}{dt} = c_k(x, t, \psi). \qquad \text{(A1.1)}$$

Consider the initial value problem in $x > 0$, $t > 0$, with data prescribed on the x-axis (which is transverse to the characteristics, i.e., nowhere tangent to them) at $t = 0$.[3] If P and Q_k are two neighboring points on the kth characteristic, then one obtains from equation (A1.1)

$$\left. \begin{array}{l} \psi_k(P) - \psi_k(Q_k) + f_k(Q_k)\left[t(P) - t(Q_k)\right] = 0 \\ x(P) - x(Q_k) = c_k(Q_k)\left[t(P) - t(Q_k)\right] \end{array} \right\} . \qquad \text{(A1.2)}$$

Further, the values at P will depend only on the data between P_1 and P_2 on the x-axis where PP_1 and PP_2 are the two characteristics through P (see Figure 5.18). In other words, $P_1 P_2$ is the domain of dependence of P. Thus, for the fully initial problem, with ψ_k given on $t = 0$, $-\infty < x < \infty$, the solution can be constructed in $t > 0$ and it is unique.

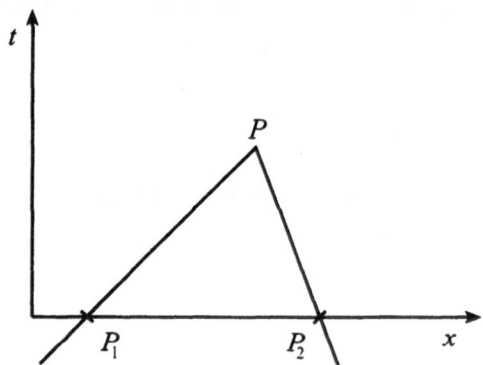

Figure 5.18. Characteristics through a point.

Therefore, it is as if the characteristics carry information from the boundaries into the region concerned. Physically, the characteristics correspond to paths of waves propagating with the velocities c_k.

---

[3] If data are prescribed on the characteristic, the differential equation does not determine the solution at any point *not* on the characteristic.

5.13. Appendix 2

Review of Nonlinear Hyperbolic Equations

Consider

$$u_t + uu_x = 0, \quad u(x,0) = \phi(x). \tag{A2.1}$$

The characteristic curves of this equation are given by

$$C : \frac{dx}{dt} = u(x,t). \tag{A2.2}$$

Observe that $u(x,t)$ is constant on C, because

$$\frac{d}{dt}\left[u(x(t),t)\right] = u_t + \frac{dx}{dt} u_x = u_t + uu_x = 0. \tag{A2.3}$$

Further, each characteristic is a straight line (see Figure 5.19).

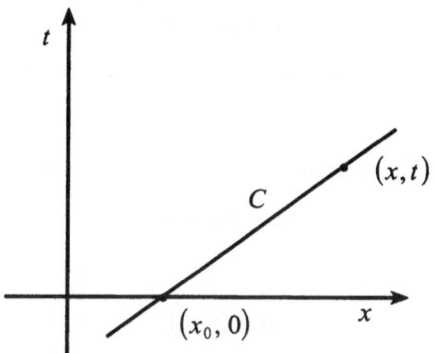

Figure 5.19. Characteristic.

The characteristic through $(x_0,0)$ and (x,t) is given by

$$C : \frac{x - x_0}{t - 0} = \frac{dx}{dt} = u(x,t) = u(x_0,0) = \phi$$

or

$$C : x = x_0 + \phi(x_0)\, t, \tag{A2.4}$$

which gives x_0 implicitly as a function of x and t.

Thus, the solution $u(x,t)$ is given in the parametric form –

$$u(x,t) = \phi\big(x_0(x,t)\big), \tag{A2.5}$$

where

$$x = x_0 + \phi(x_0)\, t. \tag{A2.4}$$

Note: We have

$$u_x = \phi'(x_0)\, x_{0_x}, \quad u_t = \phi'(x_0)\, x_{0_t} \tag{A2.6}$$

where we have from (A2.4),

$$\left.\begin{array}{l} 1 = x_{0_x} + t\,\phi'(x_0)\, x_{0_x} \\ 0 = x_{0_t} + \phi(x_0) + t\,\phi'(x_0)\, x_{0_t} \end{array}\right\}. \tag{A2.7}$$

Thus, (A2.6) becomes

$$u_x = \frac{\phi'(x_0)}{1 + t\,\phi'(x_0)}, \quad u_t = \frac{-\phi(x_0)}{1 + t\,\phi'(x_0)} \tag{A2.8}$$

which lead to

$$u_t + u u_x = 0. \tag{A2.9}$$

Example A.1: For the linear case

$$u_t + c\, u_x = 0, \tag{A2.10}$$

the solution becomes

$$u = \phi(x_0) \tag{A2.11}$$

where

$$x = x_o + ct.$$

Thus,

$$u = \phi(x - ct) \tag{A2.12}$$

as expected!

Example A.2: Consider the initial-value problem

$$u_t + u u_x = 0, \quad t > 0 \tag{A2.13}$$

$$t = 0 : u = \begin{cases} 0, & x < 0, \\ 1, & x > 0. \end{cases} \qquad (A2.14)$$

Observe that $u(x,0)$ is an increasing function of x. A continuous solution is an expansion wave –

$$u(x,t) = \begin{cases} 0, & x \le 0, \\ x/t, & 0 \le x \le t, \\ 1, & x \ge t, \end{cases} \qquad (A2.15)$$

because

$$(x/t)_t + (x/t)(x/t)_x = 0, \qquad (A2.16)$$

(see Figure 5.20).

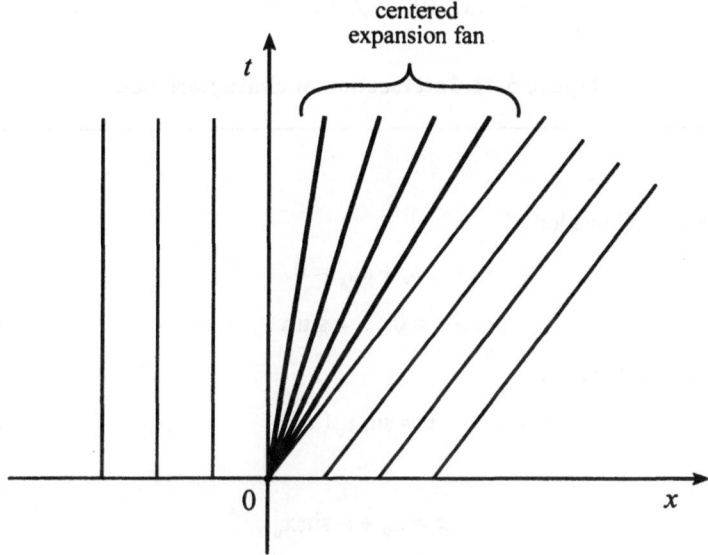

Figure 5.20. Centered expansion fan.

If two characteristics intersect – see Figure 5.21 (this happens, for instance, when $u(x,0)$ is a decreasing function of x which leads to larger-amplitude waves traveling faster than the smaller-amplitude ones), the solution $u(x,t)$ ceases to be single-valued at the point of intersection. This situation has to be resolved by inserting a jump discontinuity called a shock which is *not* along a characteristic. (By contrast, in the linear problem, discontinuities for $t > 0$ arise from discontinuities in the initial conditions and occur along characteristics.) The integral form of the partial differential equation dictates that certain compatibility conditions have to be satisfied across such a discontinuity.

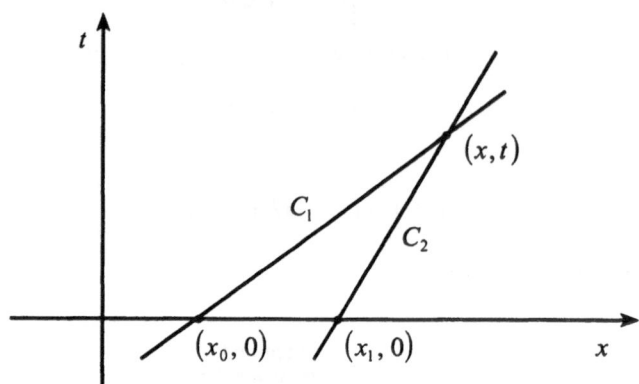

Figure 5.21. Intersection of characteristics.

Example A.3: Consider

$$u_t + uu_x = 0 \qquad \text{(A2.17)}$$

$$t = 0 : u = \sin x. \qquad \text{(A2.18)}$$

The solution is therefore

$$u = \sin x_0 (x, t) \qquad \text{(A2.19)}$$

where

$$x = x_0 + t \cdot \sin x_0.$$

Thus,

$$u = \sin(x - ut) \qquad \text{(A2.20)}$$

which shows a steepening of the initial waveform because for this problem the "crest" moves faster than the "trough". The solution eventually becomes triple-valued which is to be remedied by inserting a "shock" as shown in Figure 5.22.

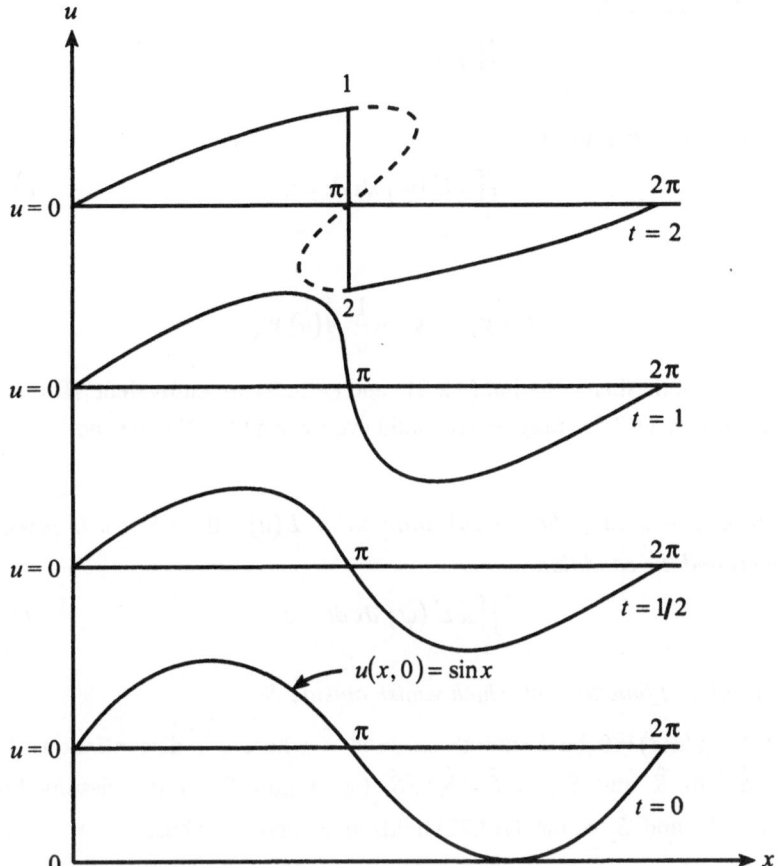

Figure 5.22. Successive solutions of $u_t + uu_x = 0$, $u(x,0) = \sin x$ at times $t = 0, \dfrac{1}{2}, 1, 2$ showing many-valuedness when $t > 1$.

In order to discuss discontinuous "shock" solutions, let us write equation (A2.1) in the conservation form –

$$L(u) \equiv u_t + \{A(u)\}_x = 0 \qquad (A2.21)$$

where $x, t \in S$ and $A(u) = u^2/2$.

For $\hat{S} \subset S$, let ψ be an arbitrarily smooth test function which vanishes outside $\hat{S}$. Then, we have

$$\iint_{\hat{S}} \psi \, L(u) \, dx \, dt = 0.$$ (A2.22)

Integration by parts leads to

$$\iint_{\hat{S}} u \, L^*(\psi) \, dx \, dt = 0$$ (A2.23)

where

$$L^*(\psi) \equiv -\psi_t - \frac{1}{u} A(u) \, \psi_x.$$

If u is smooth, then equation (A2.21) and (A2.23) are equivalent. However, if u is not smooth, (A2.23) may remain valid even when (A2.21) does not.

Definition: u is said to be a weak solution of $L(u) = 0$ in S if u is piecewise continous and the condition

$$\iint_{\hat{S}} u \, L^*(\psi) \, dx \, dt = 0$$ (A2.24)

holds for all test functions ψ which vanish outside $\hat{S}$.

Let $C : g(x,t) = 0$ be a smooth curve in $\hat{S}$ where u is discontinuous. Let C divide $\hat{S}$ into $\hat{S}_1$ and $\hat{S}_2$, so $\hat{S} = \hat{S}_1 \cup \hat{S}_2$ (see Figure 5.23). u is assumed to be smooth in $\hat{S}_1$ and $\hat{S}_2$ so that (A2.22) holds in $\hat{S}_1$ and $\hat{S}_2$. Thus,

$$\iint_{\hat{S}} u \, L^*(\psi) \, dx \, dt = \iint_{\hat{S}_1} u \, L^*(\psi) \, dx \, dt + \iint_{\hat{S}_2} u \, L^*(\psi) \, dx \, dt = 0.$$ (A2.25)

Note that,

$$\iint_{\hat{S}_1} u \, L^*(\psi) \, dx \, dt = \int_C \psi \left\{ u^+ g_t + A(u^+) \, g_x \right\} ds - \iint_{\hat{S}_1} \psi \, L(u) \, dx \, dt$$ (A2.26)

$$\iint_{\hat{S}_2} u \, L^*(\psi) \, dx \, dt = -\int_C \psi \left\{ u^- g_t + A(u^-) \, g_x \right\} ds - \iint_{\hat{S}_2} \psi \, L(u) \, dx \, dt$$ (A2.27)

where $u^\pm$ are the values taken by u on C as the limits are taken from the regions $\hat{S}_1$ and $\hat{S}_2$, respectively, and we have noted that the outward normal n for $\hat{S}_1$ is the inward normal for $\hat{S}_2$.

Using (A2.26) and (A2.27), equation (A2.25) gives

$$\int_C \psi \left\{ g_t [u] + g_x [A(u)] \right\} ds = 0 \qquad \text{(A2.28)}$$

from which

$$g_t [u] + g_x [A(u)] = 0 \qquad \text{(A2.29)}$$

where $[q]$ denotes the jump in q across C.

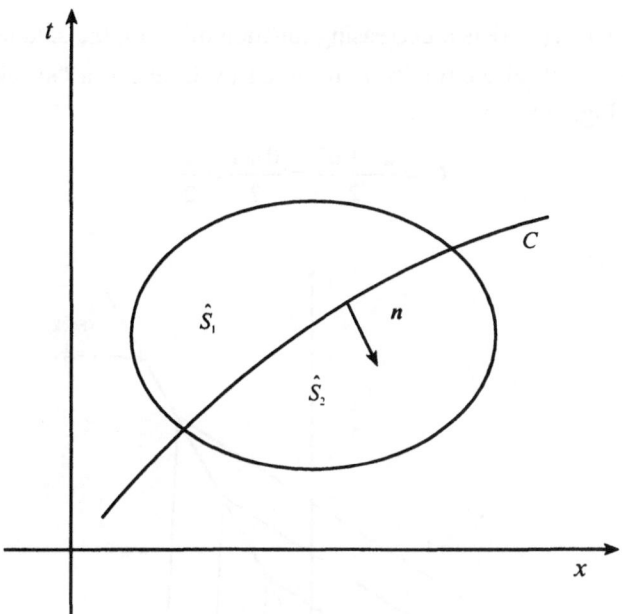

Figure 5.23. The shock curve.

If $U = -g_t/g_x$ denotes the speed of propagation of the "shock" C, we have

$$U[u] = [A(u)]. \qquad \text{(A2.30)}$$

Noting that $A(u) = u^2/2$, this yields

$$U = \frac{u^+ + u^-}{2}. \qquad \text{(A2.31)}$$

So, the speed of the shock is the average of the values of u ahead and behind the shock.

Example A.4: Consider the initial-value problem –

$$u_t + uu_x = 0, \quad t > 0 \tag{A2.32}$$

$$t = 0 : u = \begin{cases} 1, & x < 0, \\ 0, & x > 0. \end{cases} \tag{A2.33}$$

Observe that $u(x,0)$ is a decreasing function of x. So, the solution $u(x,t)$ is discontinuous. The discontinuity is resolved by inserting a "shock" given by $x = Ut$ (see Figure 5.24), where

$$U = \frac{u^+ + u^-}{2} = \frac{0+1}{2} = \frac{1}{2}. \tag{A2.34}$$

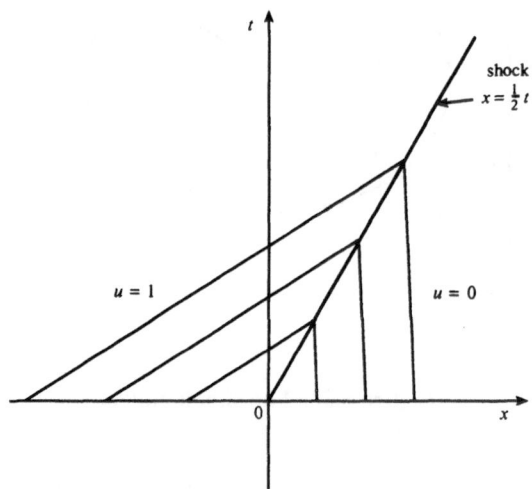

Figure 5.24. Intersection of characteristics.

Chapter 6

Method of Multiple Scales

6.1. Introduction

In the method of matched asymptotic expansions (Chapter 5), the solution is constructed in different regions that are then patched together to form a composite expansion. The method of multiple scales[1], on the other hand, starts with a generalized version of a composite expansion. This involves separate coordinates for each region, which are considered to be independent of one another. Consequently, the given equation is transformed into a partial differential equation even if it was an ordinary differential equation to begin with. On the other hand, the method of multiple scales may also be viewed as a generalization of the method of strained parameters in that the relevant scales are given implicitly rather than explicitly in terms of the original variables (Kevorkian, 1966).

6.2. Differential Equations with Constant Coefficients

In order to illustrate the method of multiple scales, consider the equation

$$\ddot{x} + x = \varepsilon x, \qquad \varepsilon \ll 1 \tag{6.2.1}$$

which has the exact solution

$$x(t,\varepsilon) = a\cos\left(\sqrt{1-\varepsilon}\, t + \alpha\right). \tag{6.2.2}$$

---

[1] This name is a bit awkward because the method of matched asymptotic expansions also uses multiple scales though each scale is effectively confined to a certain region in the latter method.

Expanding (6.2.2) in powers of ε, we have

$$x(t,\varepsilon) \sim a\cos(t+\alpha) + \frac{1}{2} \varepsilon a t \sin(t+\alpha) +$$

$$+\frac{1}{8} \varepsilon^2 a \left[t \sin(t+\alpha) - \frac{t^2}{2} \cos(t+\alpha) \right] + \cdots \qquad (6.2.3)$$

which is a nonuniform approximation. A better approximation is obtained by noting that

$$(1-\varepsilon)^{1/2} t = \left(1 - \frac{1}{2}\varepsilon - \frac{1}{8}\varepsilon^2\right) t + O(\varepsilon^3 t) \qquad (6.2.4)$$

so that (6.2.2) may be approximated by

$$x(t,\varepsilon) \sim a\cos\left[\left(1-\frac{1}{2}\varepsilon\right)t+\alpha\right] + \frac{1}{8}\varepsilon^2 a t \sin\left[\left(1-\frac{1}{2}\varepsilon\right)t+\alpha\right]$$

$$+O(\varepsilon^4 t^2) \qquad (6.2.5)$$

which shows that the problem in question is characterized by two time scales t and $\varepsilon t = \tilde{t}$, say. Thus, $x(t;\varepsilon) = x(t,\tilde{t};\varepsilon)$. We may now look for a solution of the form

$$x(t,\varepsilon) \sim x_0(t,\tilde{t}) + \varepsilon x_1(t,\tilde{t}) + \cdots. \qquad (6.2.6)$$

We then obtain to various orders in ε:

$$O(1): \frac{\partial^2 x_0}{\partial t^2} + x_0 = 0 \qquad (6.2.7)$$

$$O(\varepsilon): \frac{\partial^2 x_1}{\partial t^2} + x_1 = x_0 - 2\frac{\partial^2 x_0}{\partial t \partial \tilde{t}} \qquad (6.2.8)$$

$$O(\varepsilon^2): \frac{\partial^2 x_2}{\partial t^2} + x_2 = x_1 - 2\frac{\partial^2 x_1}{\partial t \partial \tilde{t}} - \frac{\partial^2 x_0}{\partial \tilde{t}^2} \qquad (6.2.9)$$

etc.

Solving equation (6.2.7), we obtain

$$x_0(t,\tilde{t}) = \frac{1}{2}\left[A_0(\tilde{t}) e^{it} + \overline{A}_0(\tilde{t}) e^{-it}\right] \qquad (6.2.10)$$

where the overhead bar denotes the complex conjugate.

Using (6.2.10), equation (6.2.8) becomes

$$\frac{\partial^2 x_1}{\partial t^2} + x_1 = \frac{1}{2} e^{it}\left(A_0 - 2i A_0'\right) + \text{c.c.} \qquad (6.2.11)$$

where c.c. denotes complex conjugate terms.

Removal of the secular terms in equation (6.2.11) requires

$$A_0 - 2i A_0' = 0 \tag{6.2.12}$$

from which

$$A_0 = A e^{-\frac{1}{2} i \tilde{t}}. \tag{6.2.13}$$

Using (6.2.13), (6.2.10) becomes

$$x_0 = A \cos\left(t - \frac{1}{2} \tilde{t} + \alpha \right). \tag{6.2.14}$$

Equation (6.2.11), then, has the solution

$$x_1 = \frac{1}{2} \left[A_1\left(\tilde{t}\right) e^{it} + \overline{A}_1\left(\tilde{t}\right) e^{-it} \right]. \tag{6.2.15}$$

Using (6.2.14) and (6.2.15), equation (6.2.9) becomes

$$\frac{\partial^2 x_2}{\partial t^2} + x_2 = \frac{1}{2} \left(A_1 - 2i A_1' + \frac{1}{4} A e^{-\frac{1}{2} i \tilde{t}} \right) e^{it} + \text{c.c.} \tag{6.2.16}$$

Removal of secular terms in equation (6.2.16) requires

$$A_1 - 2i A_1' + \frac{1}{4} A e^{-\frac{1}{2} i \tilde{t}} = 0 \tag{6.2.17}$$

from which

$$A_1 = -\frac{1}{8} i A \tilde{t} e^{-\frac{1}{2} i \tilde{t}}. \tag{6.2.18}$$

Using (6.2.18), (6.2.15) becomes

$$x_1 = \frac{1}{8} A \tilde{t} \sin\left(t - \frac{1}{2} \tilde{t} + \alpha \right). \tag{6.2.19}$$

Using (6.2.14) and (6.2.19), equation (6.2.16) gives the final solution,

$$x \sim A \cos\left(t - \frac{1}{2} \tilde{t} + \alpha \right) + \frac{1}{8} \varepsilon A \tilde{t} \sin\left(t - \frac{1}{2} \tilde{t} + \alpha \right) + O(\varepsilon^3) \tag{6.2.20}$$

which is same as (6.2.5)!

Example 1: Consider the initial-value problem (Nayfeh, 1973)

$$\ddot{x} + x = -2\varepsilon \dot{x}, \quad t > 0 \tag{6.2.21}$$

$$t = 0 : x = 0, \quad \dot{x} = 1. \tag{6.2.22}$$

If we take the solution of this initial-value-problem to be a regular perturbation expansion of the form –

$$x(t; \varepsilon) \sim x_0(t) + \varepsilon x_1(t) + \cdots \tag{6.2.23}$$

we then obtain

$$x(t) \sim \sin t - \varepsilon t \sin t + \cdots. \tag{6.2.24}$$

On the other hand, the exact solution of the initial-value problem (6.2.21) and (6.2.22) is

$$x(t) = \frac{1}{\sqrt{1 - \varepsilon^2}} e^{-\varepsilon t} \sin \sqrt{1 - \varepsilon^2} \, t \tag{6.2.25}$$

which may be approximated by

$$x(t) \approx \frac{1}{\sqrt{1 - \varepsilon^2}} e^{-\varepsilon t} \sin \left(t - \frac{1}{2} \varepsilon^2 t \right). \tag{6.2.26}$$

Figure 6.1 shows the comparison between the regular perturbation expansion (6.2.24) and the exact solution (6.2.25). The two curves are reasonably close to one another for small t, but differ from one another significantly for large t.

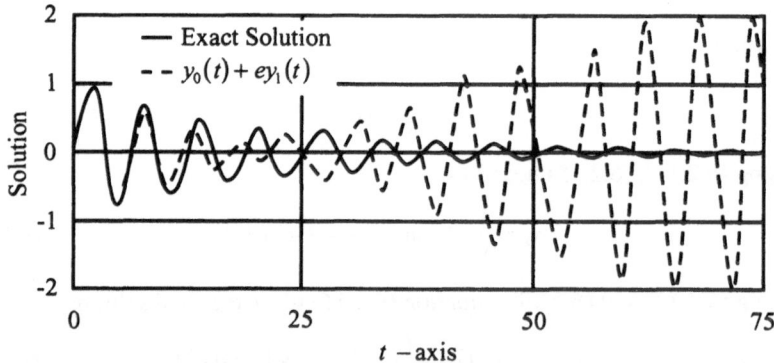

Figure 6.1. Comparison between the regular perturbation expansion in (6.2.24) and the exact solution given in (6.2.25) when $\varepsilon = 2 \times 10^{-1}$ (from Holmes, 1995).

Let us now use the method of multiple scales to construct a more accurate solution than the one given by the regular perturbation expansion (6.2.24). Let us look for a solution of the form

$$x(t;\varepsilon) \sim \sum_{m=0}^{M-1} \varepsilon^m x_m (T_0, T_1, T_2, \ldots) + O(\varepsilon^M) \qquad (6.2.27)$$

where

$$T_m = \varepsilon^m t. \qquad (6.2.28)$$

We then obtain, from equation (6.2.21), to various orders in ε:

$$O(1) : \frac{\partial^2 x_0}{\partial T_0^2} + x_0 = 0 \qquad (6.2.29)$$

$$O(\varepsilon) : \frac{\partial^2 x_1}{\partial T_0^2} + x_1 = -2 \frac{\partial x_0}{\partial T_0} - 2 \frac{\partial^2 x_0}{\partial T_0 \partial T_1} \qquad (6.2.30)$$

$$O(\varepsilon^2) : \frac{\partial^2 x_2}{\partial T_0^2} + x_2 = -2 \frac{\partial x_1}{\partial T_0} - 2 \frac{\partial^2 x_1}{\partial T_0 \partial T_1} - \frac{\partial^2 x_0}{\partial T_1^2} - 2 \frac{\partial^2 x_0}{\partial T_0 \partial T_2} - 2 \frac{\partial x_0}{\partial T_1} \qquad (6.2.31)$$

etc.

We have for equation (6.2.29), the solution

$$x_0 = A_0 (T_1, T_2) e^{iT_0} + \overline{A}_0 (T_1, T_2) e^{-iT_0}. \qquad (6.2.32)$$

Using (6.2.32), equation (6.2.30) becomes

$$\frac{\partial^2 x_1}{\partial T_0^2} + x_1 = -2i \left(A_0 + \frac{\partial A_0}{\partial T_1} \right) e^{iT_0} + \text{c.c.} \qquad (6.2.33)$$

Removal of secular terms in equation (6.2.33) requires

$$A_0 + \frac{\partial A_0}{\partial T_1} = 0 \qquad (6.2.34)$$

from which,

$$A_0 = a_0 (T_2) e^{-T_1}. \qquad (6.2.35)$$

The solution of equation (6.2.33) is then given by

$$x_1 = A_1 (T_1, T_2) e^{iT_0} + \overline{A}_1 (T_1, T_2) e^{-iT_0}. \qquad (6.2.36)$$

Using (6.2.32), (6.2.35), and (6.2.36), equation (6.2.31) becomes

$$\frac{\partial^2 x_2}{\partial T_0^2} + x_2 = -\left(2i A_1 + 2i \frac{\partial A_1}{\partial T_1} - a_0 e^{-T_1} + 2i \frac{\partial a_0}{\partial T_2} e^{-T_1} \right) e^{iT_0} + \text{c.c.} \qquad (6.2.37)$$

Removal of secular terms in equation (6.2.37) requires

$$\frac{\partial A_1}{\partial T_1} + A_1 = \frac{i}{2}\left(-a_0 + 2i\,\frac{\partial a_0}{\partial T_2}\right)e^{-T_1}. \tag{6.2.38}$$

Removal of secular terms from equation (6.2.38) (to prevent x_1 from becoming $O(x_0)$), in turn, requires

$$-a_0 + 2i\,\frac{\partial a_0}{\partial T_2} = 0 \tag{6.2.39}$$

from which

$$a_0 = \tilde{a}_0\, e^{-\frac{i}{2}T_2}. \tag{6.2.40}$$

Using equation (6.2.39), equation (6.2.38) yields

$$A_1 = a_1\left(T_2\right)e^{-T_1}. \tag{6.2.41}$$

Using (6.2.32), (6.2.35), (6.2.36), (6.2.40), and (6.2.41), (6.2.27) becomes

$$x \sim e^{-T_1}\left\{\tilde{a}_0\, e^{i\left(T_0 - \frac{T_2}{2}\right)} + \bar{\tilde{a}}_0\, e^{-i\left(T_0 - \frac{T_2}{2}\right)}\right\} + O(\varepsilon), \tag{6.2.42}$$

or

$$x \sim \frac{e^{-\varepsilon t}}{\sqrt{1-\varepsilon^2}}\,\sin\left(t - \frac{1}{2}\,\varepsilon^2 t\right) + O(\varepsilon) \tag{6.2.43}$$

in agreement with the approximate version (6.2.26) of the exact solution!

Example 2: Consider the van der Pol oscillator given by

$$\frac{d^2u}{dt^2} + u = \varepsilon\left(1 - u^2\right)\frac{du}{dt}, \tag{6.2.44}$$

$$t = 0 : u = a_0,\quad \frac{\partial u}{\partial t} = 0. \tag{6.2.45}$$

Let us look for a solution of the form

$$u \sim \sum_{n=0}^{2} \varepsilon^n u_n\left(T_0, T_1, T_2\right) + O(\varepsilon^3). \tag{6.2.46}$$

We then obtain, from equation (6.2.44), to various orders in ε:

$$O(i): \frac{\partial^2 u_0}{\partial T_0^2} + u_0 = 0 \tag{6.2.47}$$

$$O(\varepsilon): \frac{\partial^2 u_1}{\partial T_0^2} + u_1 = -2 \frac{\partial^2 u_0}{\partial T_0 \partial T_1} + \left(1 - u_0^2\right) \frac{\partial u_0}{\partial T_0} \tag{6.2.48}$$

$$O(\varepsilon^2): \frac{\partial^2 u_2}{\partial T_0^2} + u_2 = -2 \frac{\partial^2 u_1}{\partial T_0 \partial T_1} - \frac{\partial^2 u_0}{\partial T_1^2} - 2 \frac{\partial^2 u_0}{\partial T_0 \partial T_2} + \left(1 - u_0^2\right) \frac{\partial u_0}{\partial T_1} +$$

$$+ \left(1 - u_0^2\right) \frac{\partial u_1}{\partial T_0} - 2 u_0 u_1 \frac{\partial u_0}{\partial T_0} \tag{6.2.49}$$

etc.

Solving equation (6.2.47), we have

$$u_0 = A(T_1, T_2) e^{iT_0} + \overline{A}(T_1, T_2) e^{-iT_0}. \tag{6.2.50}$$

Using (6.2.50), equation (6.2.48), becomes

$$\frac{\partial^2 u_1}{\partial T_0^2} + u_1 = -i \left(2 \frac{\partial A}{\partial T_1} - A + A^2 \overline{A} \right) e^{iT_0} - i A^3 e^{3iT_0} + \text{c.c.} \tag{6.2.51}$$

Removal of secular terms in equation (6.2.51) requires

$$2 \frac{\partial A}{\partial T_1} = A - A^2 \overline{A}. \tag{6.2.52}$$

Putting

$$A = \frac{1}{2} a(T_1, T_2) e^{i\varphi(T_1, T_2)}, \tag{6.2.53}$$

we obtain from equation (6.2.52),

$$\frac{\partial \varphi}{\partial T_1} = 0, \quad \frac{\partial a}{\partial T_1} = \left(1 - \frac{a^2}{4} \right) a. \tag{6.2.54}$$

On using the initial condition (6.2.45), we obtain from (6.2.54),

$$\varphi = \varphi(T_2) \tag{6.2.55}$$

$$a^2 = \frac{4}{1 + \left(\dfrac{4}{a_0^2} - 1 \right) e^{-\varepsilon t}}. \tag{6.2.56}$$

Using (6.2.52), we obtain from equation (6.2.51),

$$u_1 = B(T_1, T_2) e^{iT_0} + \frac{1}{8} i A^3 e^{3iT_0} + \text{c.c.} \tag{6.2.57}$$

We now consider the application of the method of multiple scales to partial differential equations.

Example 3: Consider the initial-value problem (Holmes, 1995),

$$u_{xx} = u_{tt} + \varepsilon u_t, \quad -\infty < x < \infty, \quad t > 0 \tag{6.2.58}$$

$$t = 0 : u = F(x), \quad u_t = 0. \tag{6.2.59}$$

Let us take the solution of (6.2.58) and (6.2.59) to be a regular perturbation expansion of the form –

$$u \sim u_0(x,t) + \varepsilon u_1(x,t) + \cdots. \tag{6.2.60}$$

We then obtain from the initial-value problem (6.2.58) and (6.2.59),

$$O(1) : u_{0_{xx}} - u_{0_{tt}} = 0, \quad -\infty < x < \infty, \quad t > 0 \tag{6.2.61}$$

$$t = 0 : u_0 = F(x), \quad u_{0_t} = 0 \tag{6.2.62}$$

$$O(\varepsilon) : u_{1_{xx}} - u_{1_{tt}} = u_{0_t}, \quad -\infty < x < \infty, \quad t > 0 \tag{6.2.63}$$

$$t = 0 : u_1 = 0, \quad u_{1_t} = 0 \tag{6.2.64}$$

etc.

We have from the initial-value problem (6.2.61) and (6.2.62),

$$u_0 = \frac{1}{2} \left[F(\xi) + F(\eta) \right] \tag{6.2.65}$$

where

$$\xi, \eta \equiv x \mp t. \tag{6.2.66}$$

Using (6.2.65), the initial-value problem (6.2.63) and (6.2.64) becomes

$$4u_{1_{\xi\eta}} = \frac{1}{2} \left[-F'(\xi) + F'(\eta) \right] \tag{6.2.67}$$

$$t = 0 : u_1 = 0, \quad -u_{1_\xi} + u_{1_\eta} = 0 \tag{6.2.68}$$

from which,

$$u_1 = \frac{1}{8} \left[-\eta F(\xi) + \xi F(\eta) \right] + \frac{1}{8} \int_{\eta}^{\xi} \left[s F'(s) - F(s) \right] ds . \tag{6.2.69}$$

The appearance of secular terms in (6.2.69) implies that the regular perturbation expansion (6.2.60) breaks down for this problem. In order to eliminate the secular terms, let us now use the method of multiple scales and look for a solution of the form –

$$u \sim u_0\left(\xi, \eta, T_1\right) + \varepsilon u_1\left(\xi, \eta, T_1\right) + \cdots \qquad (6.2.70)$$

where

$$T_m \equiv \varepsilon^m\, t. \qquad (6.2.71)$$

We then obtain from the initial-value problem (6.2.58) and (6.2.59),

$$O(1) : 4u_{0_{\xi\eta}} = 0 \qquad (6.2.72)$$

$$t = 0 : u_0 = F(x), \quad -u_{0_\xi} + u_{0_\eta} = 0 \qquad (6.2.73)$$

$$O(\varepsilon) : 4u_{1_{\xi\eta}} = 2\left(-\frac{\partial}{\partial\xi} + \frac{\partial}{\partial\eta}\right) u_{0_{T_1}} + \left(-\frac{\partial}{\partial\xi} + \frac{\partial}{\partial\eta}\right) u_0 \qquad (6.2.74)$$

$$t = 0 : u_1 = 0, \quad -u_{1_\xi} + u_{1_\eta} = -u_{0_{T_1}} \qquad (6.2.75)$$

etc.

We have from the initial-value problem (6.2.72) and (6.2.73),

$$u_0 = f_0\left(\xi, T_1\right) + g_0\left(\eta, T_1\right) \qquad (6.2.76)$$

$$f_0\left(x, T_1\right) + g_0\left(x, T_1\right) = F(x). \qquad (6.2.77)$$

Using (6.2.76), equation (6.2.74) becomes

$$4u_{1_{\xi\eta}} = -\frac{\partial}{\partial\xi}\left(2f_{0_{T_1}} + f\right) + \frac{\partial}{\partial\eta}\left(2g_{0_{T_1}} + g_0\right). \qquad (6.2.78)$$

The removal of the secular terms in equation (6.2.78) requires

$$2f_{0_{T_1}} + f_0 = 0 \qquad (6.2.79)$$

$$2g_{0_{T_1}} + g_0 = 0 \qquad (6.2.80)$$

from which, on using (6.2.77), we obtain

$$f_0\left(\xi, T_1\right) = \frac{1}{2}\, F(\xi)\, e^{-T_1/2} \qquad (6.2.81)$$

$$g_0\left(\eta, T_1\right) = \frac{1}{2}\, F(\eta)\, e^{-T_1/2}. \qquad (6.2.82)$$

Using (6.2.81) and (6.2.82), we have from (6.2.76),

$$u(x,t) \sim \frac{1}{2} \left[F(x-t) + F(x+t) \right] e^{-\varepsilon t/2} + O(\varepsilon). \qquad (6.2.83)$$

Example 4: Consider a nonlinear dispersive wave propagation described by

$$u_{tt} - u_{xx} + u = u^3 \qquad (6.2.84)$$

$$t = 0 : u = \varepsilon \cos kx, \quad u_t = \varepsilon \omega \sin kx. \qquad (6.2.85)$$

Note that the linear problem associated with (6.2.84) and (6.2.85) gives

$$u = \varepsilon \cos(kx - \omega t), \quad \omega^2 = k^2 + 1. \qquad (6.2.86)$$

In order to determine the contributions from the nonlinear terms in equation (6.2.84), let us consider the solution to be an expansion of the form –

$$u(x,t;\varepsilon) \sim \sum_{n=1}^{\infty} \varepsilon^n u_n \left(\xi, T_1, T_2 \right) \qquad (6.2.87)$$

where

$$\xi \equiv kx - \omega t, \quad T_m \equiv \varepsilon^m t. \qquad (6.2.88)$$

We then obtain from the initial-value problem (6.2.84) and (6.2.85),

$$O(\varepsilon) : \left(\omega^2 - k^2 \right) u_{1_{\xi\xi}} + u_1 = 0 \qquad (6.2.89)$$

$$T_m = 0 : u_1 = \cos kx, \quad u_{1_{T_0}} = \omega \sin kx \qquad (6.2.90)$$

$$O(\varepsilon^2) : \left(\omega^2 - k^2 \right) u_{2_{\xi\xi}} + u_2 = -2u_{1_{T_0 T_1}} \qquad (6.2.91)$$

$$T_m = 0 : u_2 = 0, \quad u_{2_{T_0}} + u_{1_{T_1}} = 0 \qquad (6.2.92)$$

$$O(\varepsilon^3) : \left(\omega^2 - k^2 \right) u_{3_{\xi\xi}} + u_3 = -2u_{2_{T_0 T_1}} - 2u_{1_{T_0 T_2}} - u_{1_{T_1 T_1}} + u_1^3, \qquad (6.2.93)$$

$$T_m = 0 : u_3 = 0, \quad u_{3_{T_0}} + u_{2_{T_1}} + u_{0_{T_2}} = 0. \qquad (6.2.94)$$

etc.

Solving the initial-value problem (6.2.89) and (6.2.90), we obtain

$$u_1 = \cos \left[\xi + \phi_1 \left(T_1, T_2 \right) \right], \quad \omega^2 = k^2 + 1. \qquad (6.2.95)$$

Using (6.2.95), the initial-value problem (6.2.91) and (6.2.92) becomes

$$\left(\omega^2 - k^2 \right) u_{2_{\xi\xi}} + u_2 = 2\omega \phi_{1_{T_1}} \sin(\xi + \phi_1), \qquad (6.2.96)$$

$$T_m = 0 : u_2 = 0, \quad u_{2_{T_0}} + u_{1_{T_1}} = 0. \tag{6.2.97}$$

Removal of secular terms in equation (6.2.96) requires

$$\phi_{1_{T_1}} = 0 \tag{6.2.98}$$

so that

$$\phi_1 = \phi_1(T_2) \tag{6.2.99}$$

and equation (6.2.96) then has the solution

$$u_2 \equiv 0. \tag{6.2.100}$$

Using (6.2.95), (6.2.99), and (6.2.100), the initial-value problem (6.2.93) and (6.2.94) become

$$\left(\omega^2 - k^2\right) u_{3_{\xi\xi}} + u_3 = -2\left(\omega\,\phi_{1_{T_2}} - \frac{3}{8}\right)\cos(\xi + \phi_1) \tag{6.2.101}$$

$$T_m = 0 : u_3 = 0, \quad u_{3_{T_0}} = 0. \tag{6.2.102}$$

Removal of secular terms in equation (6.2.101) requires

$$\omega\,\phi_{1_{T_2}} - \frac{3}{8} = 0 \tag{6.2.103}$$

from which we obtain

$$\phi_1 = \frac{3}{8\omega}\,T_2. \tag{6.2.104}$$

Using (6.2.95), (6.2.100), and (6.2.104), we obtain

$$u \sim \varepsilon \cos\left[kx - \left(\omega - \frac{3\varepsilon^2}{8\omega}\right)t\right] + O(\varepsilon^2) \tag{6.2.105}$$

showing the amplitude-dependent frequency shift produced by the nonlinear terms.

Example 5: Consider a nonlinear dispersive wave propagation described by (Nayfeh, 1973)

$$u_{tt} - u_{xx} - u = u^3 \tag{6.2.106}$$

$$t = 0 : u = \varepsilon \cos kx, \quad u_t = 0. \tag{6.2.107}$$

Note that the linear problem associated with (6.2.106) and (6.2.107) gives

$$u = \varepsilon \cos \sigma t \cos kx, \quad \sigma^2 = k^2 - 1. \tag{6.2.108}$$

Let us determine here a solution valid for wavenumbers near the cut-off value $k = k_c = 1$ (when the linear frequency σ vanishes). For this purpose, we introduce

$$\xi = kx \tag{6.2.109}$$

so that the initial-value problem (6.2.106) and (6.2.107) becomes

$$u_{tt} - k^2 u_{\xi\xi} - u = u^3 \tag{6.2.110}$$

$$t = 0 : u = \varepsilon \cos \xi, \quad u_t = 0. \tag{6.2.111}$$

Let us take the solution of the initial-value problem (6.2.110) and (6.2.111) to be an expansion of the form

$$u(\xi, t; \varepsilon) \sim \sum_{n=1}^{\infty} \varepsilon^n u_n(\xi, T_0, T_1, T_2) \tag{6.2.112}$$

$$k = 1 + \varepsilon^2 K \tag{6.2.113}$$

where

$$T_m = \varepsilon^m t; \quad m = 0, 1, 2. \tag{6.2.114}$$

We then obtain, from the initial-value problem (6.2.110) and (6.2.111), to various orders in ε:

$$O(\varepsilon) : \frac{\partial^2 u_1}{\partial T_0^2} - \frac{\partial^2 u_1}{\partial \xi^2} - u_1 = 0 \tag{6.2.115}$$

$$T_m = 0 : u_1 = \cos \xi, \quad \frac{\partial u_1}{\partial T_0} = 0 \tag{6.2.116}$$

$$O(\varepsilon^2) : \frac{\partial^2 u_2}{\partial T_0^2} - \frac{\partial^2 u_2}{\partial \xi^2} - u_2 = -2 \frac{\partial^2 u_1}{\partial T_0 \partial T_1} \tag{6.2.117}$$

$$T_m = 0 : u_2 = 0, \quad \frac{\partial u_2}{\partial T_0} + \frac{\partial u_1}{\partial T_1} = 0 \tag{6.2.118}$$

$$O(\varepsilon^3) : \frac{\partial^2 u_3}{\partial T_0^2} - \frac{\partial^2 u_3}{\partial \xi^2} - u_3 = u_1^3 + 2K \frac{\partial^2 u_1}{\partial \xi^2} - \frac{\partial^2 u_1}{\partial T_1^2} - 2 \frac{\partial^2 u_2}{\partial T_0 T_1} - 2 \frac{\partial^2 u_1}{\partial T_0 \partial T_2} \tag{6.2.119}$$

$$T_m = 0 : u_3 = 0, \quad \frac{\partial u_3}{\partial T_0} + \frac{\partial u_2}{\partial T_1} + \frac{\partial u_1}{\partial T_2} = 0 \tag{6.2.120}$$

etc.

Solving the initial-value problem (6.2.115) and (6.2.116), we obtain

$$u_1 = a(T_1, T_2) \cos \xi, \tag{6.2.121}$$

with

$$a(0,0) = 1. \tag{6.2.122}$$

Using (6.2.121), the initial-value problem (6.2.117), and (6.2.118) becomes

$$\frac{\partial^2 u_2}{\partial T_0^2} - \frac{\partial^2 u_2}{\partial \xi^2} - u_2 = 0 \tag{6.2.123}$$

$$T_m = 0 : u_2 = 0, \quad \frac{\partial u_2}{\partial T_0} + \frac{\partial a}{\partial T_1} \cos \xi = 0. \tag{6.2.124}$$

Removal of secular terms in (6.2.124) requires

$$T_m = 0 : \frac{\partial a}{\partial T_1} = 0. \tag{6.2.125}$$

We have from equation (6.2.123),

$$u_2 = b(T_1, T_2) \cos \xi. \tag{6.2.126}$$

Using (6.2.126), (6.2.124) gives

$$b(0,0) = 0. \tag{6.2.127}$$

Using (6.2.121) and (6.2.126), equation (6.2.119) becomes

$$\frac{\partial^2 u_3}{\partial T_0^2} - \frac{\partial^2 u_3}{\partial \xi^2} - u_3 = \left(\frac{3}{4} a^3 - 2Ka - \frac{\partial^2 a}{\partial T_1^2} \right) \cos \xi + \frac{1}{4} a^3 \cos 3\xi. \tag{6.2.128}$$

Removal of secular terms in equation (6.2.128) requires

$$\frac{\partial^2 a}{\partial T_1^2} + \left(2K - \frac{3}{4} a^2 \right) a = 0. \tag{6.2.129}$$

Using the initial conditions, (6.2.122) and (6.2.125), we obtain, from equation (6.2.129),

$$\left(\frac{\partial a}{\partial T_1} \right)^2 = \frac{3}{8} (a^2 - 1)(a^2 - \beta) \tag{6.2.130}$$

where

$$\beta = \frac{16K}{3} - 1. \tag{6.2.131}$$

Since a is real, the right-hand side in (6.2.130) must be positive, hence a^2 must be outside the interval $(1, \beta)$ or $(\beta, 1)$, depending on whether $\beta > 1$ or $\beta < 1$. Since $a(0) = 1$, a^2 increases without bound if $\beta < 1$ and oscillates between 0 and 1 if $\beta > 1$ (see Figure 6.2). The case $\beta = 1$ corresponds to neutral stability, which, from (6.2.113) and (6.2.131), is therefore described by

$$k = 1 + \frac{3}{8} \varepsilon^2 + O(\varepsilon^3). \tag{6.2.132}$$

Equation (6.2.132), graphically shown in Figure 6.3 (a situation similar to the one in Section 3.8.2), indicates that the waves in question grow even at $k = k_3 = 1$, despite the cut-off predicted by the linear theory. On the other hand, the onset of instability even below the linear stability threshold when the disturbance has finite amplitude implies that this instability is a *subcritical* instability.

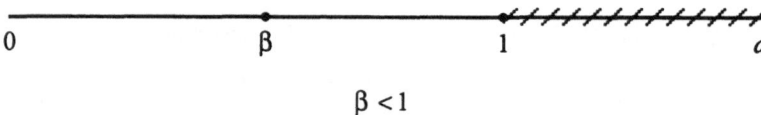

$$\beta < 1$$

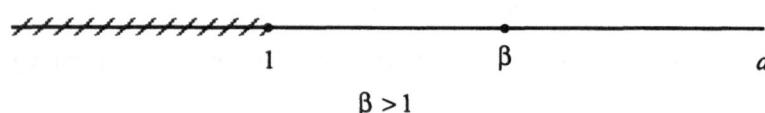

$$\beta > 1$$

Figure 6.2. Unbounded $(\beta < 1)$ and bounded $(\beta > 1)$ solutions
(shaded region is where solution exists).

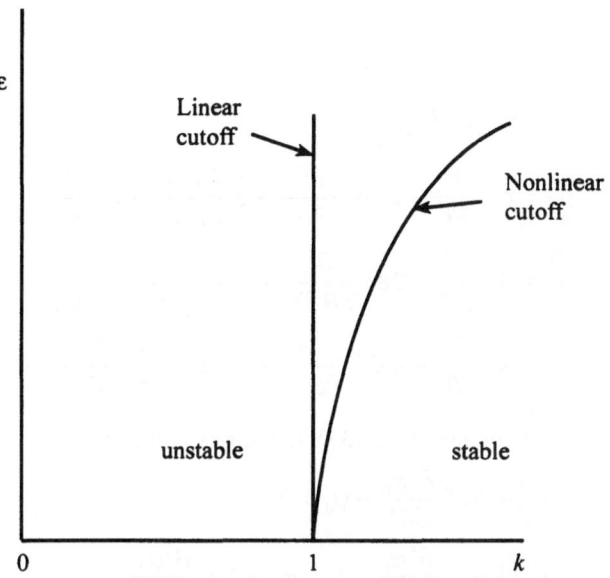

Figure 6.3. Linear and nonlinear cut-offs.

Example 6: Consider a nonlinear dispersive wave propagation described by (Bretherton, 1964),

$$\varphi_{tt} + \varphi_{xxxx} + \varphi_{xx} + \varphi = \varepsilon \varphi^3. \qquad (6.2.133)$$

The linear problem associated with equation (6.2.133) has the solution

$$\left. \begin{array}{l} \varphi = a\cos\theta, \quad \theta = kx - \omega t \\ \omega^2 = k^4 - k^2 + 1 \end{array} \right\}. \qquad (6.2.134)$$

For the nonlinear problem $(\varepsilon \neq 0)$, let us look for a solution of the form (Nayfeh and Hassan, 1971)

$$\varphi \sim \varphi_0\left(\theta, X_1, T_1\right) + \varepsilon \varphi_1\left(\theta, X_1, T_1\right) + \dots \qquad (6.2.135)$$

where

$$X_1 = \varepsilon x, \quad T_1 = \varepsilon t \qquad (6.2.136)$$

and we assume,

$$k = k\left(X_1\right), \quad \omega = \omega\left(T_1\right).$$

Noting the following rules of differentiation,

$$\frac{\partial}{\partial t} = -\omega \frac{\partial}{\partial \theta} + \varepsilon \frac{\partial}{\partial T_1}$$

$$\frac{\partial}{\partial x} = k \frac{\partial}{\partial \theta} + \varepsilon \frac{\partial}{\partial X_1}$$

$$\frac{\partial^2}{\partial t^2} = \omega^2 \frac{\partial^2}{\partial \theta^2} - 2\varepsilon\omega \frac{\partial^2}{\partial \theta \partial T_1} - \varepsilon \frac{\partial \omega}{\partial T_1} \frac{\partial}{\partial \theta} + \varepsilon^2 \frac{\partial^2}{\partial T_1^2}$$

$$\frac{\partial^2}{\partial x^2} = k^2 \frac{\partial^2}{\partial \theta^2} + 2\varepsilon k \frac{\partial^2}{\partial \theta \partial X_1} + \varepsilon \frac{\partial k}{\partial X_1} \frac{\partial}{\partial \theta} + \varepsilon^2 \frac{\partial^2}{\partial X_1^2}$$

$$\frac{\partial^4}{\partial x^4} = k^4 \frac{\partial^4}{\partial \theta^4} + 4\varepsilon k^3 \frac{\partial^4}{\partial \theta^3 \partial X_1} + 6\varepsilon k^2 \frac{\partial k}{\partial X_1} \frac{\partial^3}{\partial \theta^3} + \cdots \qquad (6.2.137)$$

we then obtain, from equation (6.2.133), to various orders in ε:

$$O(1): \left(\omega^2 + k^2\right) \frac{\partial^2 \varphi_0}{\partial \theta^2} + k^4 \frac{\partial^4 \varphi_0}{\partial \theta^4} + \varphi_0 = 0 \qquad (6.2.138)$$

$$O(\varepsilon): \left(\omega^2 + k^2\right) \frac{\partial^2 \varphi_1}{\partial \theta^2} + k^4 \frac{\partial^4 \varphi_1}{\partial \theta^4} + \varphi_1 = \varphi_0^3 + 2\omega \frac{\partial^2 \varphi_0}{\partial \theta \partial T_1}$$

$$- 2k \frac{\partial^2 \varphi_0}{\partial \theta \partial x_1} - 4k^3 \frac{\partial^4 \varphi_0}{\partial \theta^3 \partial x_1} + \left(\frac{\partial \omega}{\partial T_1} - \frac{\partial k}{\partial x_1}\right) \frac{\partial \varphi_0}{\partial \theta} - 6k^2 \frac{\partial k}{\partial x_1} \frac{\partial^3 \varphi_0}{\partial \theta^3}$$

$$(6.2.139)$$

etc.

Solving equation (6.2.138), we have

$$\varphi_0 = A(X_1, T_1) e^{i\theta} + \overline{A}(X_1, T_1) e^{-i\theta} \qquad (6.2.140)$$

where

$$\omega^2 = k^4 - k^2 + 1. \qquad (6.2.141)$$

Using (6.2.140), equation (6.2.139) becomes

$$\left(\omega^2 + k^2\right) \frac{\partial^2 \varphi_1}{\partial \theta^2} + k^4 \frac{\partial^4 \varphi_1}{\partial \theta^4} + \varphi_1 = \left[2i\omega \frac{\partial A}{\partial T_1} + 2ik\left(2k^2 - 1\right)\frac{\partial A}{\partial x_1} + i\frac{\partial \omega}{\partial T_1} A\right.$$

$$\left. + i\left(6k^2 - 1\right)\frac{\partial k}{\partial x_1} A + 3A^2 \overline{A}\right] e^{i\theta} + A^3 e^{3i\theta}$$

$$+ \text{c.c.}$$

$$(6.2.142)$$

Removal of secular terms in equation (6.2.142) leads to the modulation equation –

$$2i\omega \frac{\partial A}{\partial T_1} + 2ik\left(2k^2 - 1\right)\frac{\partial A}{\partial x_1} + i\frac{\partial \omega}{\partial T_1}A + i\left(6k^2 - 1\right)\frac{\partial k}{\partial x_1}A + 3A^2\overline{A} = 0. \quad (6.2.143)$$

Now, from (6.2.141), we have

$$\omega\omega' = k\left(2k^2 - 1\right),$$

and

$$\omega\omega''\frac{\partial k}{\partial x_1} + \omega'^2\frac{\partial k}{\partial x_1} = \left(6k^2 - 1\right)\frac{\partial k}{\partial x_1}. \quad (6.2.144)$$

Further, we have from the compatibility condition

$$\frac{\partial k}{\partial T_1} + \frac{\partial \omega}{\partial x_1} = 0 \quad (6.2.145)$$

the following result –

$$\frac{\partial k}{\partial T_1} + \omega'\frac{\partial k}{\partial x_1} = 0. \quad (6.2.146)$$

Using (6.2.146), we have

$$\frac{\partial \omega}{\partial T_1} = \omega'\frac{\partial k}{\partial T_1} = -\omega'^2\frac{\partial k}{\partial x_1} \quad (6.2.147)$$

so that we have, from (6.2.144),

$$\frac{\partial \omega}{\partial T_1} + \left(6k^2 - 1\right)\frac{\partial k}{\partial x_1} = \omega\omega''\frac{\partial k}{\partial x_1}. \quad (6.2.148)$$

Using (6.2.144)-(6.2.148), equation (6.2.143) becomes

$$2\frac{\partial A}{\partial T_1} + 2\omega'\frac{\partial A}{\partial x_1} + \omega''\frac{\partial k}{\partial x_1}A = \frac{3i}{\omega}A^2\overline{A}. \quad (6.2.149)$$

Putting

$$A = \frac{a}{2}e^{i\beta} \quad (6.2.150)$$

we obtain from equation (6.2.149)

$$\frac{\partial a^2}{\partial T_1} + \frac{\partial}{\partial x_1}\left(\omega'a^2\right) = 0 \quad (6.2.151)$$

$$\frac{\partial \beta}{\partial T_1} + \omega'\frac{\partial \beta}{\partial x_1} = \frac{3a^2}{8\omega}. \quad (6.2.152)$$

Equation (6.2.151) implies that a quantity proportional to a^2 (which may be energy in some sense) propagates with the group velocity ω', as we saw in Section 4.4.

Example 7: Consider the nonlinear-diffusion problem (Berman, 1978 and Holmes, 1995) –

$$u_t = \varepsilon u_{xx} + f(u), \quad -\infty < x < \infty, \quad t > 0 \qquad (6.2.153)$$

$$t = 0 : u = g(x), \quad -\infty < x < \infty. \qquad (6.2.154)$$

Let $0 \le u \le 1$, and we impose

$$\left.\begin{array}{l} f(0) = 0, \quad f'(0) > 0 \\ f(1) = 0, \quad f'(1) < 0 \end{array}\right\} \qquad (6.2.155)$$

and

$$\left.\begin{array}{l} 0 \le g(x) \le 1 \\ \lim_{x \Rightarrow -\infty} g(x) = 1, \quad \lim_{x \Rightarrow \infty} g(x) = 0 \end{array}\right\}. \qquad (6.2.156)$$

Good choices for $f(u)$ and $g(x)$ satisfying (6.2.155) and (6.2.156) are

$$f(u) = u(1 - u) \qquad (6.2.157)$$

$$g(x) = \frac{1}{1 + e^{\lambda x}}, \quad \lambda > 0. \qquad (6.2.158)$$

Let us look for a solution of the initial-value problem (6.2.153) and (6.2.154) of the form –

$$u \sim u_0\left(x, T_0, T_1\right) + \varepsilon u_1\left(x, T_0, T_1\right) + \cdots \qquad (6.2.159)$$

where

$$T_m = \varepsilon^m t. \qquad (6.2.160)$$

We then obtain from the initial-value problem (6.2.153) and (6.2.154),

$$O(1) : u_{0_{T_0}} = f(u_0) \qquad (6.2.161)$$

$$T_m = 0 : u_0 = g(x) \qquad (6.2.162)$$

$$O(\varepsilon) : u_{1_{T_0}} = f'(u_0) u_1 + u_{0_{xx}} - u_{0_{T_1}} \qquad (6.2.163)$$

$$T_m = 0 : u_1 = 0 \qquad (6.2.164)$$

etc.

We obtain from the initial-value problem (6.2.161) and (6.2.162),

$$\int_{1/2}^{u_0} \frac{dv}{f(v)} = T_0 + \theta(x, T_1) \tag{6.2.165}$$

with

$$\theta(x, 0) = \int_{1/2}^{g(x)} \frac{dv}{f(v)}. \tag{6.2.166}$$

Equations (6.2.165) and (6.2.166) may be formally expressed as

$$u_0 = U_0(s), \quad s \equiv T_0 + \theta(x, T_1). \tag{6.2.167}$$

Using (6.2.167), and noting the chain rules –

$$\frac{\partial}{\partial T_1} = u_{0_{T_1}} \frac{\partial}{\partial u_0}, \quad \frac{\partial}{\partial x} = u_{0_x} \frac{\partial}{\partial u_0}$$

$$\frac{\partial^2}{\partial x^2} = u_{0_{xx}} \frac{\partial}{\partial u_0} + u_{0_x}^2 \frac{\partial^2}{\partial u_0^2} \tag{6.2.168}$$

and hence, the relations –

$$\left.\begin{array}{l} \theta_{T_1} = u_{0_{T_1}} \dfrac{1}{f(u_0)}, \quad \theta_x = u_{0_x} \dfrac{1}{f(u_0)} \\[2ex] \theta_{xx} = u_{0_{xx}} \dfrac{1}{f(u_0)} - f'(u_0)\, \theta_x^2 \end{array}\right\}, \tag{6.2.169}$$

equation (6.2.163) becomes –

$$u_{1_{T_0}} - f'(u_0)\, u_1 = f(u_0)\left[\theta_{xx} - \theta_{T_1} + f'(u_0)\, \theta_x^2\right]. \tag{6.2.170}$$

Removal of secular terms in equation (6.2.170) requires

$$\theta_{xx} - \theta_{T_1} + K\, \theta_x^2 = 0 \tag{6.2.171}$$

where

$$K \equiv f'(0) > 0. \tag{6.2.172}$$

Putting

$$W(x, T_1) = e^{K\theta(x, T_1)} \tag{6.2.173}$$

equation (6.2.171) leads to

$$W_{xx} = W_{T_1}, \quad -\infty < x < \infty, \quad T_1 > 0 \tag{6.2.174}$$

from which, we have

$$W = \frac{1}{\sqrt{\pi}} \int_{-\infty}^{\infty} h\left(x + 2\sqrt{T_1}\,\xi\right) e^{-\xi^2} d\xi, \tag{6.2.175}$$

where we have from (6.2.166) and (6.2.173),

$$h(x) = W(x,0) = e^{K \int_{1/2}^{g(x)} \frac{dV}{f(V)}} \tag{6.2.176}$$

For the choice of $f(u)$ given by (6.2.157), we obtain, from (6.2.166) and (6.2.167),

$$U_0(s) = \frac{1}{1 + e^{-s}}. \tag{6.2.177}$$

Noting, from (6.2.177) that

$$U_0^{-1}(s) = \ell n \frac{s}{1-s}, \tag{6.2.178}$$

we obtain from (6.2.166),

$$\theta(x,0) = U_0^{-1}(g) = \ell n \frac{g}{1-g}. \tag{6.2.179}$$

For the choice of $g(u)$ given by (6.2.158), (6.2.179) leads to

$$\theta(x,0) = -\lambda x. \tag{6.2.180}$$

Using (6.2.166) and (6.2.180), we have from (6.2.175),

$$W(x,T_1) = \frac{1}{\sqrt{\pi}} \int_{-\infty}^{\infty} e^{-\lambda K\left(x+2\sqrt{T_1}\,\xi\right)-\xi^2} d\xi$$

or

$$W(x,T_1) = e^{-\lambda K x + \lambda^2 K^2 T_1}. \tag{6.2.181}$$

Comparing (6.2.181) with (6.2.173), we have

$$\theta(x,T_1) = -\lambda x + \lambda^2 K T_1. \tag{6.2.182}$$

Using (6.2.182), we have from (6.2.167),

$$u_0(x,T_0,T_1) = U_0\left(T_0 - \lambda x + \lambda^2 K T_1\right) \tag{6.2.183}$$

which represents a travelling wave with velocity

$$V \sim \frac{1 + \varepsilon \lambda^2 K}{\lambda}. \tag{6.2.184}$$

Observe that the speed of the wave depends on the shape of the initial profile – the steeper the initial profile (the larger λ is), the slower it moves (see Figure 6.4).

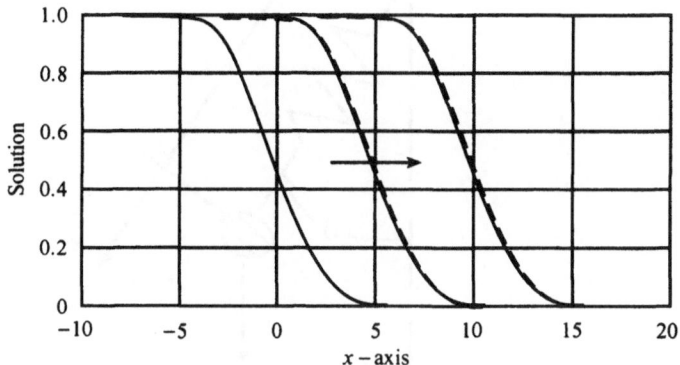

Figure 6.4. Comparison of the asymptotic traveling wave solution (6.2.183) and the numerical solution (dashed curves), of equation (6.2.153) at $t = 0$, $t = 5$, and $t = 10$. In these calculations, $\epsilon = 10^{-2}$ and $\lambda = 1$, (from Holmes, 1995).

Example 8: Consider oscillations of a spring swinging in a vertical plane (see Figure 6.5).

The Lagrangian for this system is given by (Kane and Kahn, 1968)

$$L = \frac{1}{2} m \left[\dot{x}^2 + (\ell + x)^2 \dot{\theta}^2 \right] - mg(\ell + x)(1 - \cos\theta) - \frac{1}{2} kx^2. \qquad (6.2.185)$$

The equations of motion are then

$$\ddot{x} + \frac{k}{m} x + g(1 - \cos\theta) - (\ell + x) \dot{\theta}^2 = 0 \qquad (6.2.186)$$

$$\ddot{\theta} + \frac{g}{\ell + x} \sin\theta + \frac{2}{\ell + x} \dot{x}\dot{\theta} = 0. \qquad (6.2.187)$$

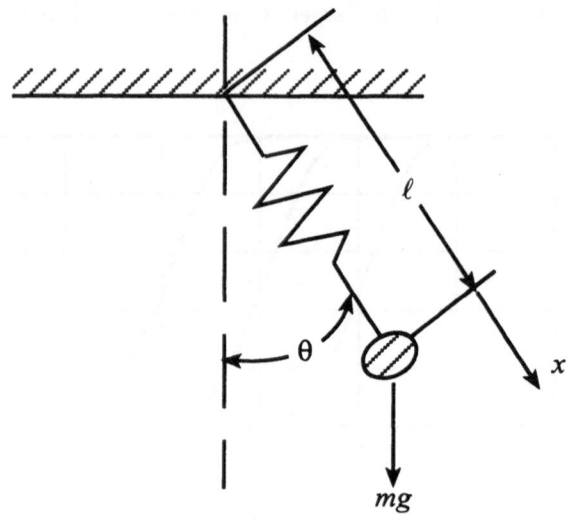

Figure 6.5. Spring-pendulum system.

Putting

$$t\,\omega_2 = \tilde{t}, \quad \frac{x}{\ell} = \varepsilon\,\tilde{x}, \quad \theta = \varepsilon\,\tilde{\theta} \tag{6.2.188}$$

and dropping the tildes, expanding the sine and the cosine functions, equations (6.2.186) and (6.2.187) become

$$\ddot{x} + \Omega^2\,x = \varepsilon\!\left(\dot{\theta}^2 - \frac{1}{2}\,\theta^2\right) \tag{6.2.189}$$

$$\ddot{\theta} + \theta = \varepsilon\!\left(x\theta - 2\dot{x}\,\dot{\theta}\right) \tag{6.2.190}$$

where

$$\Omega \equiv \frac{\omega_1}{\omega_2}, \quad \omega_1^2 \equiv \frac{k}{m}, \quad \omega_2^2 \equiv \frac{g}{\ell}. \tag{6.2.191}$$

Let us look for a solution of equations (6.2.189) and (6.2.190) of the form –

$$x(t;\varepsilon) \sim x_1(t,t_1) + \varepsilon\,x_2(t,t_1) + \cdots \tag{6.2.192}$$

$$\theta(t;\varepsilon) \sim \theta_1(t,t_1) + \varepsilon\,\theta_2(t,t_1) + \cdots \tag{6.2.193}$$

where

$$t_1 = \varepsilon t. \tag{6.2.194}$$

We then obtain from equations (6.2.189) and (6.2.190),

$$O(1): x_{1_{tt}} + \Omega^2 x_1 = 0 \tag{6.2.195}$$

$$\theta_{1_{tt}} + \theta_1 = 0 \tag{6.2.196}$$

$$O(\varepsilon): x_{2_{tt}} + \Omega^2 x_2 = -2 x_{1_{tt}} - \frac{1}{2}\theta_1^2 + \theta_{1_t}^2 \tag{6.2.197}$$

$$\theta_{2_{tt}} + \theta_2 = -2\theta_{1_{tt}} + x_1\theta_1 - 2x_{1_t}\theta_{1_t} \tag{6.2.198}$$

etc.

We have from equations (6.2.195) and (6.2.196),

$$x_1 = A(t_1) e^{i\Omega t} + \overline{A}(t_1) e^{-i\Omega t} \tag{6.2.199}$$

$$\theta_1 = B(t_1) e^{it} + \overline{B}(t_1) e^{-it}. \tag{6.2.200}$$

Using (6.2.199) and (6.2.200), equations (6.2.197) and (6.2.198) become

$$x_{2_{tt}} + \Omega^2 x_2 = -2i\Omega A_{t_1} e^{i\Omega t} - \frac{3}{2} B^2 e^{2it} + \frac{1}{2} B\overline{B} + c.c. \tag{6.2.201}$$

$$\theta_{2_{tt}} + \theta_2 = -2i B_{t_1} e^{it} + (1+2\Omega) A B e^{i(\Omega+1)t} + (1-2\Omega) A\overline{B} e^{i(\Omega-1)t} + c.c. \tag{6.2.202}$$

The removal of secular terms in equations (6.2.201) and (6.2.202), requires

$$A_{t_1} = 0, \quad B_{t_1} = 0. \tag{6.2.203}$$

We then have from equations (6.2.201) and (6.2.202),

$$x_2 = \frac{1}{2\Omega^2} B\overline{B} - \frac{3}{2} \frac{B^2}{\Omega^2 - 4} e^{2it} + c.c. \tag{6.2.204}$$

$$\theta_2 = -\frac{(1+2\Omega)}{\Omega(\Omega+2)} A B e^{i(\Omega+1)t} - \frac{(1-2\Omega)}{\Omega(\Omega-2)} A\overline{B} e^{i(\Omega-1)t} + c.c. \tag{6.2.205}$$

The solution (6.2.204) and (6.2.205) breaks down when an internal resonance sets in, i.e., when $\Omega = 2$. This is because the solution (6.2.204) and (6.2.205) misses new secular terms that arise on the right-hand sides of equations (6.2.201) and (6.2.202) when $\Omega = 2$. Remedying for this omission leads to inclusion of variations on the amplitudes of the oscillations in x and θ which become non-negligible during an internal resonance when a considerable energy exchange occurs between the x and θ motion components. In order to treat this resonance case, let us put

$$\Omega - 2 = \varepsilon\sigma. \tag{6.2.206}$$

Equations (6.2.201) and (6.2.202) then become

$$x_{2_n} + \Omega^2 x_2 = -\left(2i\Omega A_{t_1} + \frac{3}{2} B^2 e^{-i\sigma t_1}\right) e^{i\Omega t} + N.S.T. + c.c. \qquad (6.2.207)$$

$$\theta_{2_n} + \theta_2 = -\left(2i B_{t_1} - (1-2\Omega) A \overline{B} e^{i\sigma t_1}\right) e^{it} + N.S.T. + c.c. \qquad (6.2.208)$$

where *N.S.T.* denotes non-secular terms.

The removal of secular terms in equations (6.2.207) and (6.2.208) requires

$$2i\Omega A_{t_1} = -\frac{3}{2} B^2 e^{-i\sigma t_1} \qquad (6.2.209)$$

$$2i B_{t_1} = (1-2\Omega) A \overline{B} e^{i\sigma t_1}. \qquad (6.2.210)$$

Putting

$$A = -\frac{ia}{2} e^{i\Omega\alpha}, \quad B = -\frac{ib}{2} e^{i\beta}, \qquad (6.2.211)$$

equations (6.2.209) and (6.2.210) yield

$$a_{t_1} = \frac{3}{8\Omega} b^2 \cos\gamma \qquad (6.2.212)$$

$$b_{t_1} = -\frac{3}{4} a b \cos\gamma \qquad (6.2.213)$$

$$\alpha_{t_1} = -\frac{3}{8a\Omega^2} b^2 \sin\gamma \qquad (6.2.214)$$

$$\beta_{t_1} = -\frac{3}{4} a \sin\gamma \qquad (6.2.215)$$

where

$$\gamma \equiv \Omega\alpha - 2\beta + (\Omega - 2) t. \qquad (6.2.216)$$

Equations (6.2.212) and (6.2.213) give

$$2aa_{t_1} + bb_{t_1} = 0 \qquad (6.2.217)$$

from which

$$2a^2 + b^2 = \text{constant}. \qquad (6.2.218)$$

Equation (6.2.218) may be rewritten as

$$\omega_1^2 a^2 + \omega_2^2 b^2 = \text{constant}, \qquad (6.2.219)$$

which describes the redistribution of energy between the two natural modes with frequencies ω_1 and ω_2 of the system during an internal resonance.

6.3. Struble's Method

This method deals with weakly nonlinear wave systems of the form

$$\frac{d^2y}{dt^2} + \omega_0^2 y = \varepsilon f\left(y, \frac{dy}{dt}, t\right), \quad \varepsilon \ll 1 \tag{6.3.1}$$

and consists in looking for a solution of the form (Struble, 1962) –

$$y = a\cos(\omega_0 t - \varphi) + \sum_{n=1}^{\infty} \varepsilon^n y_n(t) \tag{6.3.2}$$

where a and φ are slowly varying functions of t and $y_n(t)$ describe the higher harmonics.

Example 9: Consider a system of two nonlinearly coupled oscillators (Shivamoggi and Varma, 1988) –

$$\frac{d^2x}{dt^2} + \omega_1^2 x = -\varepsilon\, 2xy \tag{6.3.3}$$

$$\frac{d^2y}{dt^2} + \omega_2^2 y = \varepsilon\left(y^2 - x^2\right). \tag{6.3.4}$$

This is the constant parameter counterpart of the system considered in Example 6 in Chapter 4.

Let us look for a solution of the form –

$$x(t;\varepsilon) \sim A_1(t_1)\,\cos\big(\varphi_1(t_1)\big) + \varepsilon\, u_1(t, t_1) + O(\varepsilon^2) \tag{6.3.5}$$

$$y(t;\varepsilon) \sim A_2(t_1)\,\cos\big(\varphi_2(t_1)\big) + \varepsilon\, v_1(t, t_1) + O(\varepsilon^2) \tag{6.3.6}$$

where $t_1 = \varepsilon t$ is the slow time scale characterizing the slow variations introduced by the weak coupling among the oscillators, and

$$\varphi_s = \omega_s t - \theta_s(t_1); \quad s = 1, 2. \tag{6.3.7}$$

Note that the solution (6.3.5)-(6.3.7) expresses the fact that the solutions of equations (6.3.3) and (6.3.4), for $\varepsilon \ll 1$, are very nearly equal to the set of harmonics represented by the first terms in the expansions (6.3.5) and (6.3.6), which they would identically be if $\varepsilon = 0$. The perturbations induced by the terms of $O(\varepsilon)$ in equations (6.3.3) and (6.3.4) may then be expected to show up as slow variations in the A_k's and the θ_k's, and as higher harmonics in the solution through the u_k's and v_k's.

Substituting the above expansions (6.3.5) and (6.3.6) in equations (6.3.3) and (6.3.4), we obtain

$$\left(2\varepsilon\omega_1 A_1\theta_{1_n}\right)\cos\varphi_1 + \left(-2\varepsilon\omega_1 A_{1_n}\right)\sin\varphi_1 + \varepsilon\left(u_{1_n} + \omega_1^2 u_1\right) + \cdots$$
$$= -\varepsilon\left(2A_1 A_2 \cos\varphi_1 \cos\varphi_2\right) + \cdots \qquad (6.3.8)$$

$$\left(2\varepsilon\omega_2 A_2\theta_{2_n}\right)\cos\varphi_2 + \left(-2\varepsilon\omega_2 A_{2_n}\right)\sin\varphi_2 + \varepsilon\left(v_{1_n} + \omega_2^2 v_1\right) + \cdots$$
$$= \varepsilon\left(A_2^2 \cos^2\varphi_2 - A_1^2 \cos^2\varphi_1\right) + \cdots . \qquad (6.3.9)$$

By equating the coefficients of $\sin\varphi_1$, $\cos\varphi_1$, $\sin\varphi_2$, $\cos\varphi_2$, and the rest to zero separately, we obtain from equations (6.3.8) and (6.3.9), to $O(\varepsilon)$:

$$\theta_{1_n} = 0 \qquad (6.3.10)$$

$$A_{1_n} = 0 \qquad (6.3.11)$$

$$u_{1_n} + \omega_1^2 u_1 = -2A_1 A_2 \cos\varphi_1 \cos\varphi_2 \qquad (6.3.12)$$

$$\theta_{2_n} = 0 \qquad (6.3.13)$$

$$A_{2_n} = 0 \qquad (6.3.14)$$

$$v_{1_n} + \omega_2^2 v_1 = A_2^2 \cos^2\varphi_2 - A_1^2 \cos^2\varphi_1 . \qquad (6.3.15)$$

On solving equations (6.3.10)-(6.3.15), we obtain

$$u_1 = \frac{A_1 A_2}{\left(\omega_1 + \omega_2\right)^2 - \omega_1^2}\cos\left(\varphi_1 + \varphi_2\right) + \frac{A_1 A_2}{\left(\omega_1 - \omega_2\right)^2 - \omega_1^2}\cos\left(\varphi_1 - \varphi_2\right) \qquad (6.3.16)$$

$$v_1 = \frac{A_2^2 - A_1^2}{2\omega_2^2} - \frac{A_2^2}{6\omega_2^2}\cos 2\varphi_2 + \frac{A_1^2/2}{\left(2\omega_1\right)^2 - \omega_2^2}\cos 2\varphi_1 \qquad (6.3.17)$$

where the A_k's and θ_k's are constants to $O(\varepsilon)$. Thus, in general, the two oscillators move as if they were effectively uncoupled, and there is no appreciable energy sharing between them. Note, however, that the solutions (6.3.16) and (6.3.17) break down when $\omega_2 = 2\omega_1$, because of small divisors, which of course, corresponds to an internal resonance in the system. This difficulty arises because the straightforward perturbation solution given by (6.3.5)-(6.3.7) does not properly recognize all the potential secular terms on the right-hand sides in equations (6.3.3) and (6.3.4), and, in particular, fails to account for the ability of such terms to cause variations in the amplitudes of the oscillators. The latter becomes nonnegligible during an internal resonance when a considerable energy sharing occurs in the system. Thus, a proper treatment of the case when an internal resonance prevails would be to sort out all the potential secular terms on the right-hand sides in equations (6.3.12) and (6.3.15).

Let us write the $O(\varepsilon)$ terms on the right in equations (6.3.12) and (6.3.15), as follows:

$$-2A_1A_2 \cos\varphi_1 \cos\varphi_2 = -A_1A_2 \cos(\varphi_2 - 2\varphi_1) \cos\varphi_1 - A_1A_2 \cos(\varphi_2 - 2\varphi_1) \cos 3\varphi_1$$
$$+ A_1A_2 \sin(\varphi_2 - 2\varphi_1) \sin\varphi_1 + A_1A_2 \sin(\varphi_2 - 2\varphi_1) \sin 3\varphi_1$$

$$-A_1^2 \cos^2\varphi_1 = -\frac{A_1^2}{2} - \frac{A_1^2}{2} \cos(\varphi_2 - 2\varphi_1) \cos\varphi_2 - \frac{A_1^2}{2} \sin(\varphi_2 - 2\varphi_1) \sin\varphi_2.$$

$$(6.3.18)$$

Using these expressions, we obtain now

$$2\omega_1 A_1 \theta_{1_n} = -A_1A_2 \cos(\varphi_2 - 2\varphi_1) \tag{6.3.19}$$

$$-2\omega_1 A_{1_n} = A_1A_2 \sin(\varphi_2 - 2\varphi_1) \tag{6.3.20}$$

$$u_{1_n} + \omega_1^2 u_1 = -A_1A_2 \cos(\varphi_2 - 2\varphi_1) \cos 3\varphi_1 + A_1A_2 \sin(\varphi_2 - 2\varphi_1) \sin 3\varphi_1 \tag{6.3.21}$$

$$2\omega_2 A_2 \theta_{2_n} = -\frac{A_1^2}{2} \cos(\varphi_2 - 2\varphi_1) \tag{6.3.22}$$

$$-2\omega_2 A_{2_n} = -\frac{A_1^2}{2} \sin(\varphi_2 - 2\varphi_1) \tag{6.3.23}$$

$$v_{1_n} + \omega_2^2 v_1 = \frac{1}{2}(A_2^2 - A_1^2) + \frac{1}{2} A_2^2 \cos 2\varphi_2. \tag{6.3.24}$$

Solving equations (6.3.21) and (6.3.24), we obtain

$$u_1 = \frac{A_1A_2}{8\omega_1^2} \left[\cos(\varphi_2 - 2\varphi_1) \cos 3\varphi_1 - \sin(\varphi_2 - 2\varphi_1) \sin 3\varphi_1 \right] \tag{6.3.25}$$

$$v_1 = \frac{A_2^2 - A_1^2}{2\omega_2^2} - \frac{A_2^2}{6\omega_2^2} \cos 2\varphi_2. \tag{6.3.26}$$

Observe that the solutions (6.3.25) and (6.3.26) no longer exhibit any small divisors.

Further, we obtain from the modulation equations (6.3.20) and (6.3.23) for amplitudes A_1 and A_2,

$$\omega_1 A_1 A_{1_{t_1}} + 2\omega_2 A_2 A_{2_{t_1}} = 0 \qquad (6.3.27)$$

from which,

$$\frac{1}{2}\omega_1 A_1^2 + \omega_2 A_2^2 = \text{constant}. \qquad (6.3.28)$$

Noting that for the present case $\omega_2 = 2\omega_1$, (6.3.28) leads to

$$\omega_1^2 A_1^2 + \omega_2^2 A_2^2 = \text{constant}, \qquad (6.3.29)$$

which indicates the exchange of energy between the two oscillators during an internal resonance (see Figure 6.6).

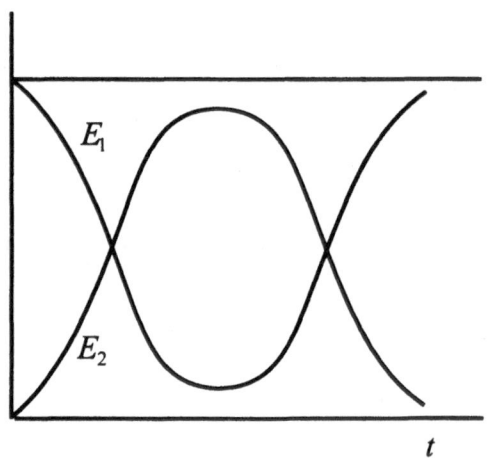

Figure 6.6. Variation of the individual actions with time during the first-order internal resonance (due to Ford and Waters 1963).

Example 10: Consider forced oscillations of the Duffing system –

$$\ddot{x} + x = \varepsilon x^3 + \varepsilon F_0 \cos \lambda t. \qquad (6.3.30)$$

Let us look for a solution of the form –

$$x(t,\varepsilon) \sim A(t_1)\cos\big(t - \theta(t_1)\big) + \varepsilon x_1(t,t_1) + \cdots \qquad (6.3.31)$$

where, $t_1 = \varepsilon t$ is the slow time scale characterizing the slow variations introduced by the weak excitation.

On substituting (6.3.31), equation (6.3.30) leads to

$$\left(2\varepsilon A\theta_{t_1}\right) \cos\left(t-\theta\right) + \left(-2\varepsilon A_{t_1}\right) \sin\left(t-\theta\right)$$
$$+ \varepsilon\left(x_{1_{tt}} + x_1\right) = \varepsilon A^3 \cos^3\left(t-\theta\right) + \varepsilon F_0 \cos\lambda t + \cdots. \tag{6.3.32}$$

By equating the coefficients of $\sin\left(t-\theta\right)$ and $\cos\left(t-\theta\right)$ and the rest to zero separately, we obtain, from equation (6.3.32), to $O(\varepsilon)$:

$$\theta_{t_1} = \frac{3}{8} A^2 \tag{6.3.33}$$

$$A_{t_1} = 0 \tag{6.3.34}$$

$$x_{1_{tt}} + x_1 = \frac{A^3}{4} \cos 3\left(t-\theta\right) + F_0 \cos\lambda t. \tag{6.3.35}$$

If $\lambda \neq 1$, equations (6.3.33)-(6.3.35) yield –

$$\theta = \theta_0 + \frac{3}{8} A^2 t_1 \tag{6.3.36}$$

$$A = \text{constant} \tag{6.3.37}$$

$$x_1 = -\left(\frac{A^3}{32}\right) \cos 3\left(t-\theta\right) + \left(\frac{F_0}{1-\lambda^2}\right) \cos\lambda t \tag{6.3.38}$$

which shows that to $O(\varepsilon)$, the amplitude of the near harmonic is not affected by the excitation.

However, if $\lambda \approx 1$, the solution (6.3.36)-(6.3.38) breaks down. In order to treat this case, we need to allow the excitation to affect the amplitude of the near harmonic especially during a resonance. For this purpose, we write

$$\varepsilon F_0 \cos\lambda t = \varepsilon F_0 \cos\left[\left(\lambda-1\right) t + \theta\right] \cos\left(t-\theta\right)$$
$$- \varepsilon F_0 \sin\left[\left(\lambda-1\right) t + \theta\right] \sin\left(t-\theta\right). \tag{6.3.39}$$

Putting

$$\lambda - 1 = \varepsilon\sigma \tag{6.3.40}$$

and using (6.3.39), we have from equation (6.3.32)

$$\theta_{t_1} = \frac{3}{8} A^2 + \frac{F_0}{2A} \cos\left[\sigma t_1 + \theta\left(t_1\right)\right] \tag{6.3.41}$$

$$A_{t_1} = \frac{F_0}{2} \sin\left[\sigma t_1 + \theta\left(t_1\right)\right] \tag{6.3.42}$$

$$x_{1_n} + x_1 = \frac{A^3}{4} \cos 3(t - \theta).$$
(6.3.43)

Equation (6.3.43) yields

$$x_1 = -\frac{A^3}{32} \cos 3(t - \theta)$$
(6.3.44)

which is well behaved at $\lambda = 1$!

Putting

$$\Phi(t_1) = \sigma t_1 + \theta(t_1),$$
(6.3.45)

equations (6.3.41) and (6.3.42) become

$$\Phi_{t_1} = \frac{3}{8} A^2 + \sigma + \frac{F_0}{2A} \cos \Phi$$
(6.3.46)

$$A_{t_1} = \frac{F_0}{2} \sin \Phi.$$
(6.3.47)

Putting further

$$a = A \cos \Phi, \quad b = A \sin \Phi,$$
(6.3.48)

we obtain from equations (6.3.46) and (6.3.47),

$$a_{t_1} = -\left[\frac{3}{8}\left(a^2 + b^2\right) + \sigma\right] b$$
(6.3.49)

$$b_{t_1} = \frac{F_0}{2} + \left[\frac{3}{8}\left(a^2 + b^2\right) + \sigma\right] a.$$
(6.3.50)

The equilibrium solutions of equations (6.3.49) and (6.3.50) correspond to

$$b = 0$$
(6.3.51)

$$\frac{F_0}{2} + \left[\frac{3}{8}\left(a^2 + b^2\right) + \sigma\right] a = 0.$$
(6.3.52)

Let us write equation (6.3.52), in the form –

$$(a + \mu)\left(a^2 - \mu a + v\right) = 0$$
(6.3.53)

where

$$v - \mu^2 = \frac{8}{3}\sigma, \quad \mu v = \frac{4}{3} F_0 > 0.$$
(6.3.54)

Since $\left(v - \mu^2\right)$ should change sign depending on whether $\sigma \lessgtr 0$, we require $v > 0$. Further, since $\mu v > 0$, this implies $\mu > 0$.

The roots of equation (6.3.53) are

$$a = -\mu, \quad \frac{1}{2}\left(\mu \pm \sqrt{\mu^2 - 4v}\right).\tag{6.3.55}$$

Thus, we have

$$4v > \mu^2 : \text{one real root } a = -\mu,$$

$$4v < \mu^2 : \text{three real roots } a = -\mu,$$

$$\alpha_1, \alpha_2 \text{ with } \alpha_1 \text{ and } \alpha_2 > 0.\tag{6.3.56}$$

Therefore, if $\sigma > 0$, there is one equilibrium point, and if σ is sufficiently negative, there will be three equilibrium points.

On the other hand, we have for the equations (6.3.49) and (6.3.50), the following integral –

$$F_0 a + \sigma\left(a^2 + b^2\right) + \frac{3}{16}\left(a^2 + b^2\right)^2 = \text{constant} = c\tag{6.3.57}$$

which shows that all solutions are bounded, since the term $\frac{3}{16}\left(a^2 + b^2\right)^2$ completely dominates all others for large a and b. The integral curves, therefore, form a family of closed paths centered about the equilibria (see Figures 6.7 and 6.8).

Thus, if $\lambda \approx 1$, for the nonlinear problem, there exist bounded, periodic responses entrained at the impressed frequency so that the resonance infinities have been removed by the nonlinearity. This is because the frequency of the natural oscillation varies with amplitude due to the nonlinearity, so that the natural oscillation does not remain in step with the excitation. The amplitude-frequency response diagram (see Figure 6.9) shows that, for a given impressed frequency, there can be more than one steady-state response. The initial conditions determine which of the possible responses actually develops. Observe the possibility of jump phenomena exhibited by the response. Note that the stability of a given branch is determined by the criterion that $|A|$ increases as F_0 increases and vice versa. Observe the possibility of jump phenomena exhibited by the response. As λ increases from below 1, $|A|$ increases along AB. At B, $|A|$ jumps abruptly to the value corresponding to E, afterwards decreasing along EC as λ increases further.

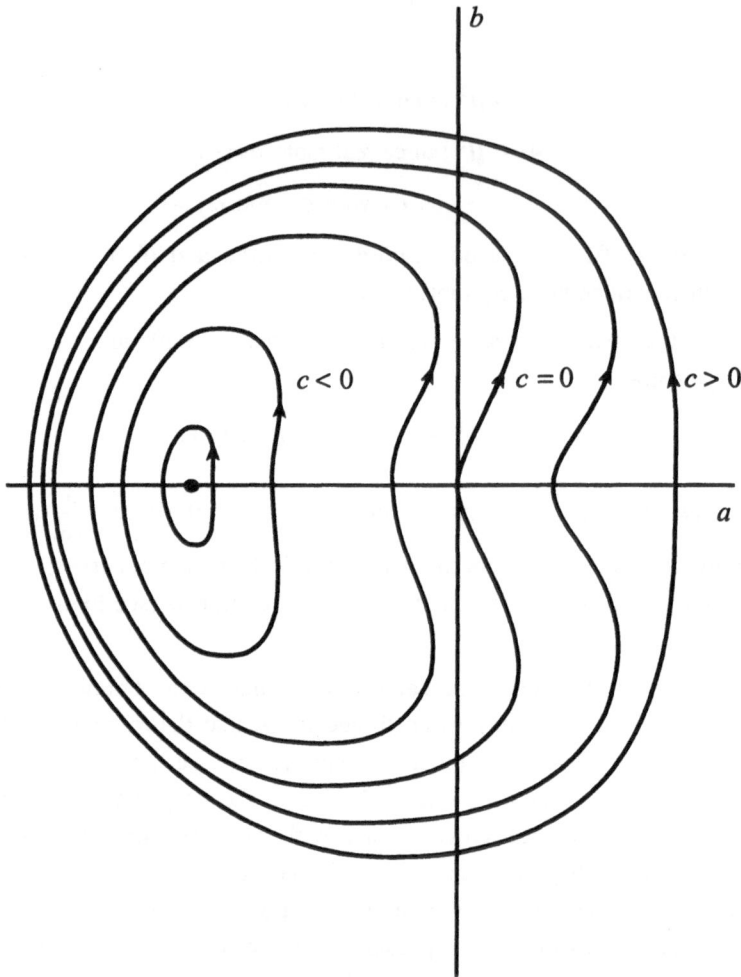

Figure 6.7. The solution curves in the *ab*-plane for the forced Duffing oscillator in the case $\sigma > 0$.

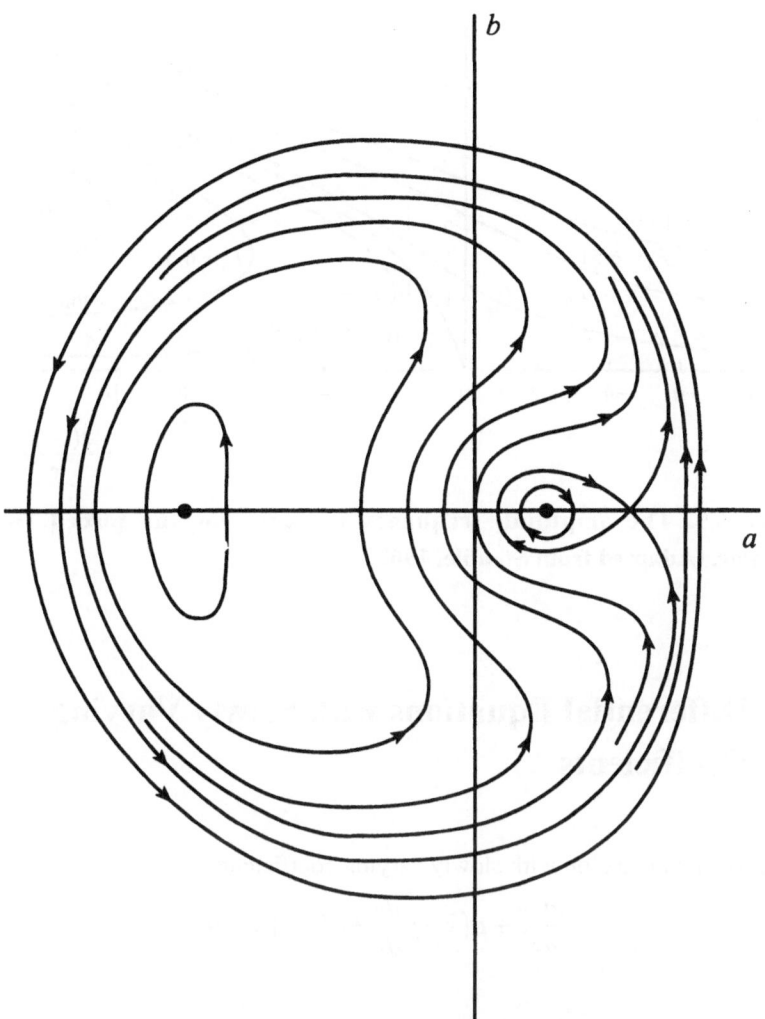

Figure 6.8. The solution curves in the *ab*-plane for the forced Duffing
oscillator in the case σ is sufficiently negative.

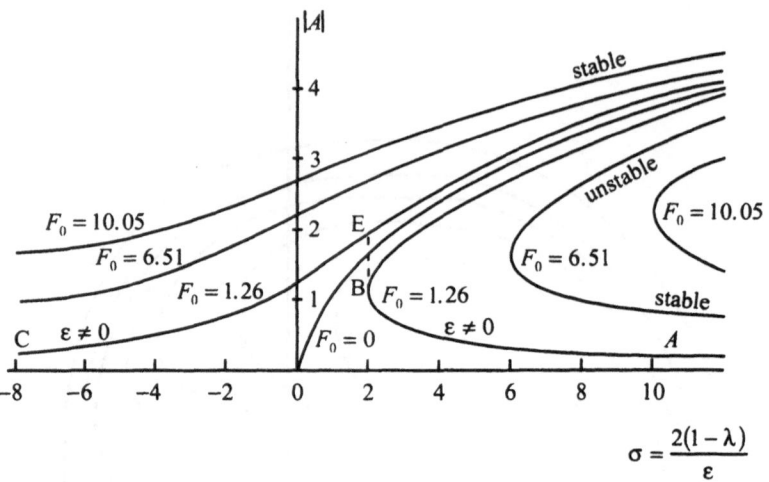

$$\sigma = \frac{2(1-\lambda)}{\varepsilon}$$

Figure 6.9. The amplitude-frequency response for the forced Duffing oscillator. (Adapted from Struble, 1962.)

6.4. Differential Equations with Slowly Varying Coefficients

Consider an equation with slowly varying coefficients,

$$\frac{d^2y}{dx^2} + p(\tilde{x};\varepsilon)\,\frac{dy}{dx} + q(\tilde{x};\varepsilon)\,y = 0 \qquad (6.4.1)$$

where

$$\tilde{x} = \varepsilon x, \quad \varepsilon \ll 1 \qquad (6.4.2)$$

and

$$p = \sum_{n=0}^{\infty} \varepsilon^n p_n(\tilde{x}), \quad q = \sum_{n=0}^{\infty} \varepsilon^n q_n(\tilde{x}). \qquad (6.4.3)$$

Let us look for a solution of the form

$$y \sim \sum_{n=0}^{\infty} \varepsilon^n A_n(\tilde{x})\,e^{\theta_1} + \sum_{n=0}^{\infty} \varepsilon^n B_n(\tilde{x})\,e^{\theta_2} \qquad (6.4.4)$$

where

$$\frac{d\theta_i}{dx} = \lambda_i(\tilde{x}) \quad i = 1, 2,$$

λ_i being the roots of

$$\lambda^2 + p_0(\tilde{x})\lambda + q_0(\tilde{x}) = 0. \tag{6.4.5}$$

Noting the following rules of differentiation,

$$\frac{d}{dx} = \lambda_1 \frac{\partial}{\partial\theta_1} + \lambda_2 \frac{\partial}{\partial\theta_2} + \varepsilon \frac{\partial}{\partial\tilde{x}}$$

$$\frac{d}{dx^2} = \lambda_1^2 \frac{\partial^2}{\partial\theta_1^2} + 2\lambda_1\lambda_2 \frac{\partial^2}{2\theta_1 2\theta_2} + \lambda_2^2 \frac{\partial^2}{\partial\theta_2^2} + \varepsilon\lambda_1 \left(\frac{\partial^2}{\partial\theta_1\partial\tilde{x}} + \frac{\partial^2}{\partial\tilde{x}\partial\theta_1} \right) \tag{6.4.6}$$

$$+ \varepsilon\lambda_2 \left(\frac{\partial^2}{\partial\theta_2\partial\tilde{x}} + \frac{\partial^2}{\partial\tilde{x}\partial\theta_2} \right) + \varepsilon\lambda_1' \frac{\partial}{\partial\theta_1} + \varepsilon\lambda_2' \frac{\partial}{\partial\theta_2} + \varepsilon^2 \frac{\partial^2}{\partial\tilde{x}^2},$$

we obtain from equation (6.4.1)

$$(2\lambda_1 + p_0) A_0' + (\lambda_1' + \lambda_1 p_1 + q_1) A_0 = 0 \tag{6.4.7}$$

$$(2\lambda_2 + p_0) B_0' + (\lambda_2' + \lambda_2 p_1 + q_1) B_0 = 0 \tag{6.4.8}$$

from which we have

$$A_0, B_0 \sim e^{-\int \frac{\lambda_i' + \lambda_i p_1 + q_1}{2\lambda_i + p_0} d\tilde{x}}, \quad i = 1, 2. \tag{6.4.9}$$

For the case $p = 0$, $q_n = 0$ for $n \geq 1$, i.e., for (Liouville, 1837 and Green, 1837)

$$\frac{d^2 y}{dx^2} + q_0(\tilde{x}) y = 0. \tag{6.4.10}$$

We therefore have from (6.4.5) and (6.4.9),

$$\left. \begin{aligned} \lambda_{1,2} &= \pm i\sqrt{q_0(\tilde{x})} \\ A_0 &= \frac{a}{\sqrt{\lambda_1}}, \quad B_0 = \frac{b}{\sqrt{\lambda_1}} \end{aligned} \right\} \tag{6.4.11}$$

where a and b are constants.

Thus,

$$y = \frac{C_1 \cos \int \sqrt{q_0(\tilde{x})} \, dx + C_2 \sin \int \sqrt{q_0(\tilde{x})} \, dx}{[q_o(\tilde{x})]^{1/4}}. \tag{6.4.12}$$

This is the JWKB approximation considered in Section 5.9.

Example 11: Consider a harmonic oscillator with a slowly-varying frequency (Shivamoggi and Muilenburg, (1991)) for which we have

$$\frac{d^2x}{dt^2} + \omega^2(\tilde{t})\, x = 0 \tag{6.4.13}$$

$$t = 0 : x = 0, \quad \frac{dx}{dt} = a \tag{6.4.14}$$

where

$$\tilde{t} = \varepsilon t, \quad \varepsilon \ll 1. \tag{6.4.15}$$

To treat this problem we shall give a systematic procedure which involves a combination of the method of strained parameters and the method of multiple scales. This is necessary since the instantaneous frequency of the basic gyration is indeed not $\omega(\tilde{t})$, but rather some average value of $\omega(\tilde{t})$ over the basic gyration (Kuzmak, 1959 and Cole, 1968).

Let us introduce a new scale,

$$\hat{t} = \int_0^t f(\varepsilon\xi)\, d\xi. \tag{6.4.16}$$

Noting the following rules of differentiation,

$$\left.\begin{aligned}
\frac{\partial}{\partial t} &= f(\tilde{t})\frac{\partial}{\partial\hat{t}} + \varepsilon\frac{\partial}{\partial\tilde{t}} \\[2mm]
\frac{\partial^2}{\partial t^2} &= f^2\frac{\partial^2}{\partial\hat{t}^2} + 2\varepsilon f\frac{\partial^2}{\partial\hat{t}\,\partial\tilde{t}} + \varepsilon f'\frac{\partial}{\partial\hat{t}} + \varepsilon^2\frac{\partial^2}{\partial\tilde{t}^2}
\end{aligned}\right\} \tag{6.4.17}$$

the initial-value problem (6.4.13) and (6.4.14) becomes

$$\frac{d^2x}{d\hat{t}^2} + \frac{\varepsilon f'(\tilde{t})}{f^2(\tilde{t})}\frac{dx}{d\hat{t}} + \frac{\omega^2(\tilde{t})}{f^2(\tilde{t})}\, x + \varepsilon\frac{2}{f}\frac{\partial^2x}{\partial\hat{t}\,\partial\tilde{t}} + \varepsilon^2\frac{1}{f^2}\frac{\partial^2x}{\partial\tilde{t}^2} = 0 \tag{6.4.18}$$

$$x(0) = 0, \quad f(0)\frac{dx(0)}{d\hat{t}} = a. \tag{6.4.19}$$

Let us look for a solution of the form

$$x(\hat{t};\varepsilon) \sim x_0(\hat{t},\tilde{t}) + \varepsilon x_1(\hat{t},\tilde{t}) + \varepsilon^2 x_2(\hat{t},\tilde{t}) + \cdots \tag{6.4.20}$$

$$f(\tilde{t}) = \omega(\tilde{t}) + \varepsilon f_1(\tilde{t}) + \varepsilon^2 f_2(\tilde{t}) + \cdots. \tag{6.4.21}$$

We then obtain from the initial-value problem (6.4.18) and (6.4.19) to various orders in ε:

$$O(1): \frac{\partial^2 x_0}{\partial \hat{t}^2} + x_0 = 0 \tag{6.4.22}$$

$$\hat{t} = 0: x_0 = 0, \quad \omega \cdot \frac{\partial x_0}{\partial \hat{t}} = a \tag{6.4.23}$$

$$O(\varepsilon): \frac{\partial^2 x_1}{\partial \hat{t}^2} + x_1 = -\frac{\omega'(\tilde{t})}{\omega^2(\tilde{t})} \frac{\partial x_0}{\partial \hat{t}} - \frac{2}{\omega(\tilde{t})} \frac{\partial^2 x_0}{\partial \hat{t} \partial \tilde{t}} + \frac{2f_1(\tilde{t})}{\omega(\tilde{t})} x_0 \tag{6.4.24}$$

$$\hat{t} = 0: x_1 = 0, \quad \frac{\partial x_1}{\partial \hat{t}} = 0 \tag{6.4.25}$$

$$O(\varepsilon^2): \frac{\partial^2 x_2}{\partial \hat{t}^2} + x_2 = -\frac{\omega'(\tilde{t})}{\omega^2(\tilde{t})} \frac{\partial x_1}{\partial \hat{t}} - \frac{2}{\omega(\tilde{t})} \frac{\partial^2 x_1}{\partial \hat{t} \partial \tilde{t}}$$

$$- \left[\frac{f_1'(\tilde{t})}{\omega^2(\tilde{t})} - \frac{2f_1(\tilde{t})\omega'(\tilde{t})}{\omega^3(\tilde{t})} \right] \frac{\partial x_0}{\partial \hat{t}} + \frac{2f_1(\tilde{t})}{\omega^2(\tilde{t})} \frac{\partial^2 x_0}{\partial \hat{t} \partial \tilde{t}}$$

$$- \frac{1}{\omega^2(\tilde{t})} \frac{\partial^2 x_0}{\partial \tilde{t}^2} + \frac{2f_1(\tilde{t})}{\omega(\tilde{t})} x_1 - \left[\frac{f_1^2(\tilde{t})}{\omega^2(\tilde{t})} - \frac{2f_2(\tilde{t})}{\omega(\tilde{t})} \right] x_0 \tag{6.4.26}$$

$$\hat{t} = 0: x_2 = 0, \quad \frac{\partial x_2}{\partial \hat{t}} = 0 \tag{6.4.27}$$

etc.

Solving equation (6.4.22), we have

$$x_0(\hat{t}, \tilde{t}) = A_0(\tilde{t}) \sin \hat{t}. \tag{6.4.28}$$

Using (6.4.28) in equation (6.4.24), the removal of secular terms in the latter then requires

$$f_1(\tilde{t}) = 0 \tag{6.4.29}$$

$$\frac{A_0(\tilde{t}) \omega'(t)}{\omega^2(\tilde{t})} + \frac{2A_0'(\tilde{t})}{\omega(\tilde{t})} = 0 \tag{6.4.30a}$$

or

$$A_0^2(\tilde{t}) \omega(\tilde{t}) = \text{constant} \equiv I \tag{6.4.30b}$$

and we may then take

$$x_1\left(\hat{t},\tilde{t}\right)\equiv 0. \qquad (6.4.31)$$

Using (6.4.28)-(6.4.31) in equation (6.4.26), the removal of secular terms in the latter then requires

$$A_0''\left(\tilde{t}\right)-2\omega\left(\tilde{t}\right)f_2\left(\tilde{t}\right)A_0\left(\tilde{t}\right)=0, \qquad (6.4.32)$$

from which, on using (6.4.30), we have

$$f_2\left(\tilde{t}\right)=\frac{3}{8}\frac{\left[\omega'\left(\tilde{t}\right)\right]^2}{\omega^3\left(\tilde{t}\right)}-\frac{1}{4}\frac{\omega''\left(\tilde{t}\right)}{\omega^2\left(\tilde{t}\right)}. \qquad (6.4.33)$$

We thus have the following solution:

$$x(t,\varepsilon)=A_0\left(\tilde{t}\right)\sin\hat{t}+O\left(\varepsilon^2\right). \qquad (6.4.34)$$

Differentiating (6.4.34), with respect to t, we obtain

$$I=\frac{1}{\omega\left(\tilde{t}\right)}\left[p+\varepsilon\frac{x\omega'\left(\tilde{t}\right)}{2\omega\left(\tilde{t}\right)}\right]^2+x^2\omega\left(\tilde{t}\right)+O\left(\varepsilon^2\right) \qquad (6.4.35a)$$

where

$$p\equiv\frac{dx}{dt}. \qquad (6.4.36)$$

Equation (6.4.35a) may be rewritten as

$$I=\frac{p^2+\omega^2\left(\tilde{t}\right)x^2}{\omega\left(\tilde{t}\right)}+\varepsilon\left[\frac{\omega'\left(\tilde{t}\right)}{\omega^2\left(\tilde{t}\right)}\,px\right]+O\left(\varepsilon^2\right) \qquad (6.4.35b)$$

which shows that the well-known adiabatic invariant $\mu=\dfrac{p^2+\omega^2x^2}{\omega}$ is the value of I (which is constant) only to the lowest order in ε! It is a pity that μ cannot look better.

Example 12: Consider a system of two nonlinearly coupled oscillators with slowly varying parameters, i.e., the linear frequencies ω_k are slowly varying functions of time (Kevorkian, 1980, Shivamoggi, 1987)

$$q_{1_{tt}}+\omega_1^2\,q_1=-\varepsilon\cdot 2q_1q_2 \qquad (6.4.37)$$

$$q_{2_{tt}} + \omega_2^2 q_2 = \varepsilon\left(q_2^2 - q_1^2\right). \tag{6.4.38}$$

Here, ε is a small parameter characterizing the weakness of the couplings between the two oscillators.

The parameters ω_1 and ω_2 may then evolve in such a way that the system passes through a state of internal resonance. Let $\omega_1 = \omega_1\left(\tilde{t}_1\right)$ and $\omega_2 = \omega_2\left(\tilde{t}_1\right)$ where $\tilde{t}_1 = \varepsilon t$, and ε is the small coupling constant that also characterizes the slow variations in ω_1 and ω_2. Let us introduce two new fast time scales

$$\hat{t}_{1,2} = \int_0^t \omega_{1,2}\left(\tilde{t}_1\right) dt \tag{6.4.39}$$

and look for solutions of the form

$$\left. \begin{array}{l} q_1\left(t;\varepsilon\right) \sim q_{10}\left(\hat{t}_1,\hat{t}_2,\tilde{t}_1,\tilde{t}_2\right) + \varepsilon q_{11}\left(\hat{t}_1,\hat{t}_2,\tilde{t}_1,\tilde{t}_2\right) + O\left(\varepsilon^2\right) \\ q_2\left(t;\varepsilon\right) \sim q_{20}\left(\hat{t}_1,\hat{t}_2,\tilde{t}_1,\tilde{t}_2\right) + \varepsilon q_{21}\left(\hat{t}_1,\hat{t}_2,\tilde{t}_1,\tilde{t}_2\right) + O\left(\varepsilon^2\right) \end{array} \right\} \tag{6.4.40}$$

where $\tilde{t}_2 = \varepsilon^2 t$ is another slow time scale introduced in the system.

Use (6.4.40), and note the following transformation rules –

$$\frac{d}{dt} = \omega_1 \frac{\partial}{\partial \hat{t}_1} + \omega_2 \frac{\partial}{\partial \hat{t}_2} + \varepsilon \frac{\partial}{\partial \tilde{t}_1} + \varepsilon^2 \frac{\partial}{\partial \tilde{t}_2},$$

$$\frac{d^2}{dt^2} = \omega_1^2 \frac{\partial^2}{\partial \hat{t}_1^2} + \omega_2^2 \frac{\partial^2}{\partial \hat{t}_2^2} + \omega_1\omega_2\left(\frac{\partial^2}{\partial \hat{t}_1 \partial \hat{t}_2} + \frac{\partial^2}{\partial \hat{t}_2 \partial \hat{t}_1}\right)$$

$$+ \varepsilon\omega_1\left(\frac{\partial^2}{\partial \hat{t}_1 \partial \tilde{t}_1} + \frac{\partial^2}{\partial \tilde{t}_1 \partial \hat{t}_1}\right) + \varepsilon\omega_2\left(\frac{\partial^2}{\partial \hat{t}_2 \partial \tilde{t}_1} + \frac{\partial^2}{\partial \tilde{t}_1 \partial \hat{t}_2}\right)$$

$$+ \varepsilon\omega_{1_{\tilde{t}_1}} \frac{\partial}{\partial \hat{t}_1} + \varepsilon\omega_{2_{\tilde{t}_1}} \frac{\partial}{\partial \hat{t}_2} + O\left(\varepsilon^2\right). \tag{6.4.41}$$

(In (6.4.41) the mixed differentiation operators are written separately because in some cases they do not yield the same result when the order is reversed. Since the present problem turns out not to be one of these cases we will not distinguish the orders of the mixed derivatives in the following.) One obtains from equations (6.4.37) and (6.4.38) –

$$O(1): \frac{\partial^2 q_{10}}{\partial \hat{t}_1^2} + \frac{\omega_2^2}{\omega_1^2} \frac{\partial^2 q_{10}}{\partial \hat{t}_2^2} + 2\frac{\omega_2}{\omega_1} \frac{\partial^2 q_{10}}{\partial \hat{t}_1 \partial \hat{t}_2} + q_{10} = 0 \tag{6.4.42}$$

$$\frac{\partial^2 q_{20}}{\partial \hat{t}_2^2} + \frac{\omega_1^2}{\omega_2^2} \frac{\partial^2 q_{20}}{\partial \hat{t}_1^2} + 2 \frac{\omega_1}{\omega_2} \frac{\partial^2 q_{20}}{\partial \hat{t}_1 \partial \hat{t}_2} + q_{20} = 0 \tag{6.4.43}$$

$$O(\varepsilon): \frac{\partial^2 q_{11}}{\partial \hat{t}_1^2} + \frac{\omega_2^2}{\omega_1^2} \frac{\partial^2 q_{11}}{\partial \hat{t}_2^2} + 2 \frac{\omega_2}{\omega_1} \frac{\partial^2 q_{11}}{\partial \hat{t}_1 \partial \hat{t}_2} + q_{11}$$

$$= -\frac{\omega_{1_{\tilde{t}_1}}}{\omega_1^2} \frac{\partial q_{10}}{\partial \hat{t}_1} - \frac{\omega_{2_{\tilde{t}_1}}}{\omega_1^2} \frac{\partial q_{10}}{\partial \hat{t}_2} - \frac{2}{\omega_1} \frac{\partial^2 q_{10}}{\partial \hat{t}_1 \partial \tilde{t}_1} - 2 \frac{\omega_2}{\omega_1^2} \frac{\partial^2 q_{10}}{\partial \hat{t}_2 \partial \tilde{t}_1} - \frac{2}{\omega_1^2} q_{10} q_{20} \tag{6.4.44}$$

$$\frac{\partial^2 q_{21}}{\partial \hat{t}_2^2} + \frac{\omega_1^2}{\omega_2^2} \frac{\partial^2 q_{21}}{\partial \hat{t}_1^2} + 2 \frac{\omega_1}{\omega_2} \frac{\partial^2 q_{21}}{\partial \hat{t}_2 \partial \hat{t}_1} + q_{21}$$

$$= -\frac{\omega_{2_{\tilde{t}_1}}}{\omega_2^2} \frac{\partial q_{20}}{\partial \hat{t}_2} - \frac{\omega_{1_{\tilde{t}_1}}}{\omega_2^2} \frac{\partial q_{20}}{\partial \hat{t}_1} - \frac{2}{\omega_2} \frac{\partial^2 q_{20}}{\partial \hat{t}_2 \partial \tilde{t}_1} - 2 \frac{\omega_1}{\omega_2^2} \frac{\partial^2 q_{20}}{\partial \hat{t}_1 \partial \tilde{t}_1} + \frac{1}{\omega_2^2} \left(q_{20}^2 - q_{10}^2 \right)$$

$$\tag{6.4.45}$$

etc.

One obtains from equations (6.4.42) and (6.4.43) –

$$\left. \begin{array}{l} q_{10} = A_1 \left(\tilde{t}_1, \tilde{t}_2 \right) \cos \left[\hat{t}_1 + \varphi_1 \left(\tilde{t}_1, \tilde{t}_2 \right) \right] \\ q_{20} = A_2 \left(\tilde{t}_1, \tilde{t}_2 \right) \cos \left[\hat{t}_2 + \varphi_2 \left(\tilde{t}_1, \tilde{t}_2 \right) \right] \end{array} \right\}. \tag{6.4.46}$$

Using (6.4.46), equations (6.4.44) and (6.4.45) become

$$\frac{\partial^2 q_{11}}{\partial \hat{t}_1^2} + \frac{\omega_2^2}{\omega_1^2} \frac{\partial^2 q_{11}}{\partial \hat{t}_2^2} + 2 \frac{\omega_2}{\omega_1} \cdot \frac{\partial^2 q_{11}}{\partial \hat{t}_1 \partial \hat{t}_2} + q_{11}$$

$$= \left(\frac{2A_{1_{\tilde{t}_1}}}{\omega_1} + \frac{A_1 \omega_{1_{\tilde{t}_1}}}{\omega_1^2} \right) \sin \left(\hat{t}_1 + \varphi_1 \right) + \frac{2A_1}{\omega_1} \varphi_{1_{\tilde{t}_1}} \cos \left(\hat{t}_1 + \varphi_1 \right)$$

$$- \frac{2A_1 A_2}{\omega_1^2} \cos \left(\hat{t}_1 + \varphi_1 \right) \cos \left(\hat{t}_2 + \varphi_2 \right) \tag{6.4.47}$$

$$\frac{\partial^2 q_{21}}{\partial \hat{t}_2^2} + \frac{\omega_1^2}{\omega_2^2} \frac{\partial^2 q_{21}}{\partial \hat{t}_1^2} + 2 \frac{\omega_1}{\omega_2} \cdot \frac{\partial^2 q_{21}}{\partial \hat{t}_1 \partial \hat{t}_2} + q_{21}$$

$$= \left(\frac{2A_{2_{\tilde{t}_1}}}{\omega_2} + \frac{A_2 \omega_{2_{\tilde{t}_1}}}{\omega_2^2} \right) \sin \left(\hat{t}_2 + \varphi_2 \right) + \frac{2A_2}{\omega_2} \varphi_{2_{\tilde{t}_1}} \cos \left(\hat{t}_2 + \varphi_2 \right)$$

$$+ \frac{1}{\omega_2^2} \left[A_2^2 \cos^2 \left(\hat{t}_2 + \varphi_2 \right) - A_1^2 \cos^2 \left(\hat{t}_1 + \varphi_1 \right) \right]. \tag{6.4.48}$$

The removal of the secularity-inducing terms in equation (6.4.47) requires –

$$\frac{2A_{1_{\tilde{t}_1}}}{\omega_1} + \frac{A_1\omega_{1_{\tilde{t}_1}}}{\omega_1^2} = 0$$

or

$$A_1^2\,\omega_1 = C_1\left(\tilde{t}_2\right) \tag{6.4.49}$$

and

$$\varphi_{1_{\tilde{t}_1}} = 0 \quad \text{or} \quad \varphi_1 = \varphi_1\left(\tilde{t}_2\right). \tag{6.4.50}$$

The solution of equation (6.4.47) is then given by

$$q_{11} = \frac{A_1A_2}{\left(\omega_1 + \omega_2\right)^2 - \omega_1^2}\,\cos\left[\left(\hat{t}_1 + \hat{t}_2\right) + \left(\varphi_1 + \varphi_2\right)\right]$$

$$+ \frac{A_1A_2}{\left(\omega_1 - \omega_2\right)^2 - \omega_1^2}\,\cos\left[\left(\hat{t}_1 - \hat{t}_2\right) + \left(\varphi_1 - \varphi_2\right)\right]. \tag{6.4.51}$$

One has to add the homogeneous solution in (6.4.51), but this can be included in q_{10} by redefining A_1.

One obtains from equation (6.4.48) similarly

$$\frac{2A_{2_{\tilde{t}_1}}}{\omega_2} + \frac{A_2\omega_{2_{\tilde{t}_1}}}{\omega_2^2} = 0$$

or

$$A_2^2\,\omega_2 = C_2\left(\tilde{t}_2\right) \tag{6.4.52}$$

and

$$\varphi_{2_{\tilde{t}_1}} = 0 \quad \text{or} \quad \varphi_2 = \varphi_2\left(\tilde{t}_2\right) \tag{6.4.53}$$

$$q_{21} = \frac{A_2^2 - A_1^2}{2\omega_2^2} - \frac{A_2^2}{6\omega_2^2}\,\cos 2\left(\hat{t}_2 + \varphi_2\right) + \frac{A_1^2/2}{\left(2\omega_1\right)^2 - \omega_2^2}\,\cos 2\left(\hat{t}_1 + \varphi_1\right). \tag{6.4.54}$$

Equations (6.4.49) and (6.4.52) show that for times t sufficiently far from T which correspond to the advent of an internal resonance $\omega_2\left(\varepsilon t\right) = 2\omega_1\left(\varepsilon t\right)$, the two oscillators individually exhibit adiabatic action invariants. However, the latter have different values in the pre- and post-resonant regimes.

For $t \approx T$, which corresponds to $\omega_2\left(\tilde{t}_1\right) \approx 2\omega_1\left(\tilde{t}_1\right)$, the results (6.4.49)–(6.4.54) break down. In order to treat this internal resonance, let us write

$$-\frac{2A_1A_2}{\omega_1^2}\cos\left(\hat{t}_1+\varphi_1\right)\cos\left(\hat{t}_2+\varphi_2\right)$$

$$=-\frac{A_1A_2}{\omega_1^2}\cos\left[\left(\hat{t}_2-2\hat{t}_1\right)+\left(\varphi_2-2\varphi_1\right)\right]\cos\left(\hat{t}_1+\varphi_1\right)$$

$$-\frac{A_1A_2}{\omega_1^2}\cos\left[\left(\hat{t}_2-2\hat{t}_1\right)+\left(\varphi_2-2\varphi_1\right)\right]\cos 3\left(\hat{t}_1+\varphi_1\right)$$

$$+\frac{A_1A_2}{\omega_1^2}\sin\left[\left(\hat{t}_2-2\hat{t}_1\right)+\left(\varphi_2-2\varphi_1\right)\right]\sin\left(\hat{t}_1+\varphi_1\right)$$

$$+\frac{A_1A_2}{\omega_1^2}\sin\left[\left(\hat{t}_2-2\hat{t}_1\right)+\left(\varphi_2-2\varphi_1\right)\right]\sin 3\left(\hat{t}_1+\varphi_1\right)$$

$$-A_1^2\cos^2\left(\hat{t}_1+\varphi_1\right)=-\frac{A_1^2}{2}-\frac{A_1^2}{2}\cos\left[\left(\hat{t}_2-2\hat{t}_1\right)+\left(\varphi_2-2\varphi_1\right)\right]\cos\left(\hat{t}_2+\varphi_2\right)$$

$$-\frac{A_1^2}{2}\sin\left[\left(\hat{t}_2-2\hat{t}_1\right)+\left(\varphi_2-2\varphi_1\right)\right]\sin\left(\hat{t}_2+\varphi_2\right).$$

$$(6.4.55)$$

Using (6.4.55), equations (6.4.47) and (6.4.48) give

$$\frac{2A_{1_{\hat{t}_1}}}{\omega_1}+\frac{A_1\omega_{1_{\hat{t}_1}}}{\omega_1^2}=-\frac{A_1A_2}{\omega_1^2}\sin\left[\left(\hat{t}_2-2\hat{t}_1\right)+\left(\varphi_2-2\varphi_1\right)\right]\qquad(6.4.56)$$

$$\frac{2A_1}{\omega_1}\varphi_{1_{\hat{t}_1}}=\frac{A_1A_2}{\omega_1^2}\cos\left[\left(\hat{t}_2-2\hat{t}_1\right)+\left(\varphi_2-2\varphi_1\right)\right]\qquad(6.4.57)$$

$$\frac{\partial^2 q_{11}}{\partial\hat{t}_1^2}+\frac{\omega_2^2}{\omega_1^2}\frac{\partial^2 q_{11}}{\partial\hat{t}_2^2}+2\frac{\omega_2}{\omega_1}\cdot\frac{\partial^2 q_{11}}{\partial\hat{t}_1\partial\hat{t}_2}+q_{11}$$

$$=-\frac{A_1A_2}{\omega_1^2}\cos\left[\left(\hat{t}_2-2\hat{t}_1\right)+\left(\varphi_2-2\varphi_1\right)\right]\cos 3\left(\hat{t}_1+\varphi_1\right)$$

$$+\frac{A_1A_2}{\omega_1^2}\sin\left[\left(\hat{t}_2-2\hat{t}_1\right)+\left(\varphi_2-2\varphi_1\right)\right]\sin 3\left(\hat{t}_1+\varphi_1\right)\quad(6.4.58)$$

$$\frac{2A_{2_{\hat{t}_1}}}{\omega_2}+\frac{A_2\omega_{2_{\hat{t}_1}}}{\omega_2^2}=\frac{A_1^2}{2\omega_2^2}\sin\left[\left(\hat{t}_2-2\hat{t}_1\right)+\left(\varphi_2-2\varphi_1\right)\right]\qquad(6.4.59)$$

$$\frac{2A_2}{\omega_2}\varphi_{2_{\hat{t}_1}}=\frac{A_1^2}{2\omega_2^2}\cos\left[\left(\hat{t}_2-2\hat{t}_1\right)+\left(\varphi_2-2\varphi_1\right)\right]\qquad(6.4.60)$$

$$\frac{\partial^2 q_{21}}{\partial \hat{t}_2^2} + \frac{\omega_1^2}{\omega_2^2} \frac{\partial^2 q_{21}}{\partial \hat{t}_1^2} + 2\frac{\omega_1}{\omega_2} \cdot \frac{\partial^2 q_{21}}{\partial \hat{t}_1 \partial \hat{t}_2} + q_{21}$$

$$= \frac{1}{2\omega_2^2}\left(A_2^2 - A_1^2\right) + \frac{1}{2\omega_2^2} A_2^2 \cos 2\left(\hat{t}_2 + \varphi_2\right). \qquad (6.4.61)$$

It might appear that, in equations (6.4.56), (6.4.57), (6.4.59), and (6.4.60), there is some inconsistency because the A's and φ's depend on $\tilde{t}_1$ and $\tilde{t}_2$ whereas the right-hand sides involve $\hat{t}_1$ and $\hat{t}_2$. This can be resolved by noting that the function $\left(\hat{t}_2 - 2\hat{t}_1\right)$ is nearly constant because its derivative $\left(\omega_2 - 2\omega_1\right)$ is very small near an internal resonance, and we assume that $\left(\hat{t}_2 - 2\hat{t}_1\right)$ is constant enough to remove the apparent inconsistency.

On solving equations (6.4.58) and (6.4.61), one obtains

$$q_{11} = \frac{A_1 A_2}{8\omega_1^2}\left[\cos\left\{\left(\hat{t}_2 - 2\hat{t}_1\right) + \left(\varphi_2 - 2\varphi_1\right)\right\} \cos 3\left(\hat{t}_1 + \varphi_1\right)\right.$$

$$\left. - \sin\left\{\left(\hat{t}_2 - 2\hat{t}_1\right) + \left(\varphi_2 - 2\varphi_1\right)\right\} \sin 3\left(\hat{t}_1 + \varphi_1\right)\right],$$

$$q_{21} = \frac{A_2^2 - A_1^2}{2\omega_2^2} - \frac{A_2^2}{6\omega_2^2}\cos 2\left(\hat{t}_2 + \varphi_2\right). \qquad (6.4.62)$$

Observe that (6.4.62) no longer exhibits any small divisors. Also, one obtains from equations (6.4.56) and (6.4.59)

$$\left(A_1^2 \omega_1\right)_{\tilde{t}_1} + 2\left(A_2^2 \omega_2\right)_{\tilde{t}_1} = 0$$

or

$$A_1^2 \omega_1 + A_2^2 \omega_2 = C\left(\tilde{t}_2\right). \qquad (6.4.63)$$

If one introduces the action $J_k \equiv A_k^2 \omega_k$, then (6.4.63) gives

$$J_1 + 2J_2 = C\left(\tilde{t}_2\right). \qquad (6.4.64)$$

Equation (6.4.64) exhibits the fact that the energies of the two oscillators are redistributed as the system evolves through the internal resonance.

If the system parameters ω_1 and ω_2 are constant, equation (6.4.64) becomes (on noting that $\omega_2 \approx 2\omega_1$)

$$J_1\omega_1 + J_2\omega_2 = C\left(\tilde{t}_2\right) \qquad (6.4.65)$$

as deduced previously in Example 9 in Chapter 6.

6.5. Generalized Multiple-Scale Method

We illustrate a generalized method of multiple scales (Nayfeh, 1964) by the following example.

Example 13: Consider the following boundary-value problem:

$$\varepsilon \frac{d^2 y}{dx^2} + a(x) \frac{dy}{dx} + b(x) y = 0, \quad 0 < x < 1 \tag{6.5.1}$$

$$y(0) = \alpha, \quad y(1) = \beta \tag{6.5.2}$$

where $a(x)$ is continuously differentiable and positive in $(0,1)$.

Let us introduce new independent variables

$$\xi = x, \quad \eta = \frac{g(x)}{\varepsilon} \quad \text{with} \quad g(0) = 0, \tag{6.5.3}$$

and note the following rules of differentiation –

$$\frac{d}{dx} = \frac{d\eta}{dx} \frac{\partial}{\partial \eta} + \frac{\partial}{\partial \xi},$$

$$\frac{d^2}{dx^2} = \left(\frac{d\eta}{dx}\right)^2 \frac{\partial^2}{\partial \eta^2} + \frac{d^2\eta}{dx^2} \frac{\partial}{\partial \eta} + 2 \frac{d\eta}{dx} \frac{\partial^2}{\partial \eta \partial \xi} + \frac{\partial^2}{\partial \xi^2}. \tag{6.5.4}$$

Looking for a solution of the form

$$y \sim \sum_{n=0}^{N-1} \varepsilon^n y_n (\xi, \eta) + O(\varepsilon^N) \tag{6.5.5}$$

we then obtain, from the boundary-value problem (6.5.1) and (6.5.2) to various orders in ε:

$$0(1): \left[g' \frac{\partial^2 y_0}{\partial \eta^2} + a(\xi) \frac{\partial y_0}{\partial \eta} \right] g' = 0 \tag{6.5.6}$$

$$y_0 (0,0) = \alpha, \quad y_0 \left(1, \frac{g(1)}{\varepsilon} \right) = \beta, \tag{6.5.7}$$

$$0(\varepsilon): \left[g' \frac{\partial^2 y_1}{\partial \eta^2} + a(\xi) \frac{\partial y_1}{\partial \eta} \right] g' = -2g' \frac{\partial^2 y_0}{\partial \xi \partial \eta} - g'' \frac{\partial y_0}{\partial \eta} - a(\xi) \frac{\partial y_0}{\partial \xi} - b(\xi) y_0$$

$$\tag{6.5.8}$$

$$y_1(0,0) = 0, \quad y_1\left(1, \frac{g(1)}{\varepsilon}\right) = 0 \tag{6.5.9}$$

etc.

Let us choose for convenience

$$g(x) = \int_0^x a(t)\, dt. \tag{6.5.10}$$

The boundary-value problem (6.5.6) and (6.5.7) then becomes

$$\frac{\partial^2 y_0}{\partial \eta^2} + \frac{\partial y_0}{\partial \eta} = 0 \tag{6.5.11}$$

$$y_0(0,0) = \alpha, \quad y_0(1,\infty) = \beta. \tag{6.5.12}$$

Equation (6.5.11) yields

$$y_0 = A_0(\xi) + B_0(\xi)\, e^{-\eta}. \tag{6.5.13}$$

Imposing the boundary conditions (6.5.12), we obtain

$$A_0(0) + B_0(0) = \alpha \tag{6.5.14}$$

$$A_0(1) = \beta. \tag{6.5.15}$$

Using (6.5.13)-(6.5.15), the boundary-value problem (6.5.8), and (6.5.9) becomes

$$\left(\frac{\partial^2 y_1}{\partial \eta^2} + \frac{\partial y_1}{\partial \eta}\right) = \left[\frac{B_0'}{a} + \frac{(a'-b)}{a^2}\, B_0\right] e^{-\eta} - \left(\frac{A_0'}{a} + \frac{b\, A_0}{a^2}\right) \tag{6.5.16}$$

$$y_1(0,0) = 0, \quad y_1(1,\infty) = 0. \tag{6.5.17}$$

Removal of secular terms in equation (6.5.16) requires

$$a\, A_0' + b\, A_0 = 0 \tag{6.5.18}$$

$$a\, B_0' + (a'-b)\, B_0 = 0 \tag{6.5.19}$$

from which we have

$$A_0(\xi) = \beta \exp\left[\int_x^1 \frac{b(t)}{a(t)}\, dt\right] \tag{6.5.20}$$

$$B_0(\xi) = \frac{C}{a(x)} \exp\left[\int_0^x \frac{b(t)}{a(t)}\, dt\right] \tag{6.5.21}$$

where C is an arbitrary constant.

Using these results, the boundary conditions (6.5.14) then yield –

$$\beta \cdot \exp\left[\int_0^1 \frac{b(t)}{a(t)} \, dt\right] + \frac{C}{a(0)} = \alpha.$$ (6.5.22)

Using (6.5.22), (6.5.21) becomes

$$B_0(\xi) = \frac{\alpha\, a(0) - \beta\, a(0)\, e^{\int_0^1 \frac{b(t)}{a(t)} dt}}{a(x)} \; e^{\int_0^x \frac{b(t)}{a(t)} dt}.$$ (6.5.23)

Example 14: Consider the boundary-value problem

$$\varepsilon\, y'' + y' + y = 0$$ (6.5.24)

$$y(0) = a, \quad y(1) = b$$ (6.5.25)

which was solved in Section 5.3 by using the method of matched asymptotic expansions.

Let us introduce new independent variables

$$\xi = x, \quad \eta = \frac{x}{\varepsilon}$$ (6.5.26)

and note the following rules of differentiation –

$$\frac{d}{dx} = \frac{\partial}{\partial \xi} + \frac{1}{\varepsilon} \frac{\partial}{\partial \eta},$$

$$\frac{d^2}{dx^2} = \frac{\partial^2}{\partial \xi^2} + \frac{2}{\varepsilon} \frac{\partial^2}{\partial \xi \partial \eta} + \frac{1}{\varepsilon^2} \frac{\partial^2}{\partial \eta^2}.$$ (6.5.27)

Looking for a solution of the form

$$y \sim y_0(\xi,\eta) + \varepsilon\, y_1(\xi,\eta) + \cdots$$ (6.5.28)

we then obtain, from equation (6.5.24), to various orders in ε:

$$0(1): \frac{\partial^2 y_0}{\partial \eta^2} + \frac{\partial y_0}{\partial \eta} = 0$$ (6.5.29)

$$0(\varepsilon): \frac{\partial^2 y_1}{\partial \eta^2} + \frac{\partial y_1}{\partial \eta} = -2\frac{\partial^2 y_0}{\partial \xi \partial \eta} - \frac{\partial y_0}{\partial \xi} - y_0$$ (6.5.30)

etc.

Solving equation (6.5.29), we have

$$y_0 = A(\xi) + B(\xi)\, e^{-\eta}.$$ (6.5.31)

Using (6.5.31), equation (6.5.30) becomes

$$\frac{\partial^2 y_1}{\partial \eta^2} + \frac{\partial y_1}{\partial \eta} = -(A' + A) + (B' - B)\, e^{-\eta}.$$ (6.5.32)

The removal of secular terms in equation (6.5.32) requires

$$A' + A = 0$$ (6.5.33)

$$B' - B = 0$$ (6.5.34)

from which we have

$$A = \alpha e^{-\xi}$$ (6.5.35)

$$B = \beta e^{\xi}.$$ (6.5.36)

Using (6.5.35) and (6.5.36), (6.5.31) becomes

$$y = \alpha e^{-x} + \beta e^{x - \frac{x}{\varepsilon}} + O(\varepsilon).$$ (6.5.37)

Imposing the boundary conditions (6.5.25), we finally obtain

$$y = b e^{1-x} + (a - be)\, e^{x - x/\varepsilon} + O(\varepsilon)$$ (6.5.38)

in agreement with the results in Example 1 in Section 5.3.

6.6. Applications to Solid Mechanics: Dynamic Buckling of a Thin Elastic Plate

The classical study of buckling of elastic systems (see Section 3.7) is based on the assumption that the external loads are applied statically. This means that their magnitude increases with sufficient slowness that any inertial forces may be neglected. Such an approach is not valid when one adds a periodic load to the stationary one, and the former has a rapidly varying magnitude. Then the forced vibrations of the elastic body take the place of the static deflections. The study of the evolution of the response of an elastic body under time-dependent loading is the subject of dynamic stability.

The classical linear theory (see Bolotin, 1964), that considers infinitesimal deflections and establishes the conditions under which a thin plate under the action of a uniaxial dynamic compressive load can take up equilibrium configurations other than the one corresponding to a uniform compression, reveals that the forced vibrations of the elastic plate show a parametric resonance, so that periodic forces acting in the middle plane of the plate can excite intense transverse vibrations for certain relationships between the forcing frequency and the natural frequency of transverse vibrations of the plate. A distinguishing feature of the parametric resonance (see Section 3.2) is that whereas the ordinary resonance occurs when the frequency of the exciting force equals the natural frequency of the system, in the parametric resonance the exciting frequencies may be multiples of the natural frequency. There is also a possibility of the occurrence of excited vibrations for the frequencies lower than the principal resonance frequency. On the other hand, the forced vibrations are bounded if the frequency and amplitude of the applied load are properly chosen. In particular, if the applied load $P(t)$ consists of a stationary part and a periodic part,

$$P(t) = P_0 + P_1 \cos \Omega t,$$

and if P_0 were a compressive load greater than the static critical load P_{cr} for the onset of buckling, the plate in the undeflected configuration is unstable. But, if P_1 and Ω are chosen properly, the small motions in the neighborhood of the undeflected configuration can be stabilized (this is possible even if the time average of $P(t)$ over a cycle can be much larger than the static critical load for the onset of buckling).

Here we consider the dynamic buckling of a rectangular thin elastic plate under a uniaxial harmonically varying load (with a compressive stationary part). The linear theory is known not to lead to a valid description of the dynamic post-buckling behavior of the plate because it predicts vibration-amplitudes that increase unboundedly with time. It is found that experimentally observed amplitudes do not increase indefinitely with time under these circumstances, but level off to a stationary value signifying the advent of a supercritical state of equilibrium. In order to determine whether the linearly unstable vibrations become stationary, and the magnitude of the stationary amplitudes, one has to consider the nonlinear problem (Shivamoggi, 1977b).

Nonlinear theory of a vibrating plate results by abandoning one or more assumptions made in developing the linear theory. Two types of nonlinearities arise:

* nonlinear inertia,

* nonlinear elasticity.

Nonlinear inertia constitutes additional inertia forces which arise during coplanar displacements u,v; the latter are coupled nonlinearly with the transverse deflection of the plate, w. Nonlinear elasticity arises in considering the dynamic response of thin plates loaded beyond their buckling limits so that one needs to retain in the expressions for strain components, the squares and products of the deflections and their derivatives. Further, the linear theory involves an assumption that the middle plane of the plate is free from stress, so that there are no resultant forces, called the membrane forces, acting in the middle plane of the plate. However, the latter arise if

(i) there are external loads acting in the middle plane of the plate;

(ii) the plate is bent into a nondevelopable surface that results in strains in the middle plane of the plate.

Under these circumstances, it is necessary to take into consideration the effect of bending vibrations of the plate of the stresses acting in the middle plane of the plate. Thus, in the case of a plate, the elastic nonlinearity is not only geometrical but also physical in origin. The membrane forces produce an additional resultant transverse load,

$$N_x \frac{\partial^2 w}{\partial x^2} - 2N_{xy} \frac{\partial^2 w}{\partial x \partial y} + N_y \frac{\partial^2 w}{\partial y^2},$$

where the membrane forces N_x, N_{xy}, N_y are measured per unit length.

We consider only the inertia effect due to the lateral motion of the plate. Then w will satisfy a nonlinear partial differential equation,

$$\nabla^4 w + \frac{\rho h}{D} \frac{\partial^2 w}{\partial t^2} = \frac{1}{D} \left(N_x \frac{\partial^2 w}{\partial x^2} - 2N_{xy} \frac{\partial^2 w}{\partial x \partial y} + N_y \frac{\partial^2 w}{\partial y^2} \right), \qquad (6.6.1)$$

where,

$$D = Eh^3 / 12 \left(1 - \upsilon^2 \right),$$

ρ denotes the mass density, E being the modulus of elasticity, and υ the Poisson's ratio. Equation (6.6.1) in the static case reduces to von Kármán's celebrated nonlinear differential equation for the bending of thin plates corresponding to large deflections (von Kármán, 1910).

Consider the dynamic post-buckling response of a rectangular thin elastic plate of constant thickness h, simply supported along the edges and subjected to a compressive distributed load $P(t)$ per unit length at the edges $x = 0, a$. We confine our attention to the middle plane (taken to be the x,y plane) so that we may ignore the displacement components u, v. When a uniaxial applied periodic load

$$P(t) = P_0 + P_1 \cos \Omega t$$

per unit length at edges $x = 0, a$, where P_0 is stationary, and P_1 is of small magnitude, acts in the middle plane of the plate (see Figure 6.10), upon using (see Timoshenko and Woinowsky-Krieger, 1959 for the derivation of strain terms),

$$N_x = \frac{Eh}{1 - v^2} \left[\frac{1}{2} \left(\frac{\partial w}{\partial x} \right)^2 + \frac{1}{2} v \left(\frac{\partial w}{\partial y} \right)^2 \right] - (P_0 + P_1 \cos \Omega t),$$

$$N_{xy} = \frac{Eh}{1 + v} \left(\frac{\partial w}{\partial x} \frac{\partial w}{\partial y} \right),$$

$$N_y = \frac{Eh}{1 - v^2} \left[\frac{1}{2} \left(\frac{\partial w}{\partial y} \right)^2 + \frac{1}{2} v \left(\frac{\partial w}{\partial x} \right)^2 \right], \qquad (6.6.2)$$

equation (6.6.1) becomes

$$\mu \nabla^4 w = \frac{\partial^2 w}{\partial x^2} \left[-\frac{(1 - v^2)}{Eh} (P_0 + P_1 \cos \Omega t) + \frac{1}{2} \left(\frac{\partial w}{\partial x} \right)^2 + \frac{1}{2} v \left(\frac{\partial w}{\partial y} \right)^2 \right]$$

$$+ \frac{\partial^2 w}{\partial y^2} \left[\frac{1}{2} \left(\frac{\partial w}{\partial y} \right)^2 + \frac{1}{2} v \left(\frac{\partial w}{\partial x} \right)^2 \right]$$

$$+ (1 - v) \frac{\partial^2 w}{\partial x \partial y} \left(\frac{\partial w}{\partial x} \frac{\partial w}{\partial y} \right) - \frac{\rho h (1 - v^2)}{Eh^2} \frac{\partial^2 w}{\partial t^2}, \qquad (6.6.3)$$

where $\mu \equiv h/2$, with the boundary conditions

$$\left. \begin{array}{l} x = 0, a : w = 0, \quad \partial^2 w / \partial x^2 = 0 \\ y = 0, b : w = 0, \quad \partial^2 w / \partial y^2 = 0 \end{array} \right\}. \qquad (6.6.4)$$

In deriving equation (6.6.3), we have, in effect, assumed that the forces $P(t)$ applied at the edges of the plate are transmitted throughout the plate to a sufficiently close approximation without any change in their magnitude along the x-direction.

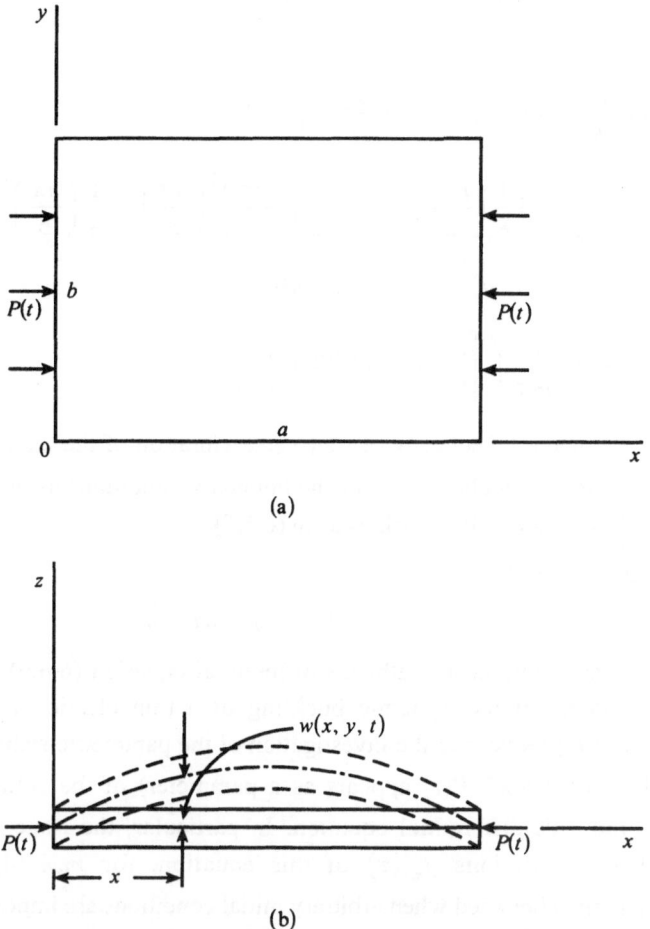

Figure 6.10. Geometry of the loaded plate: (a) plan, (b) section.

Consider cases wherein the amplitude P_1 of the periodic load is much smaller than the static critical load P_{cr} for the onset of buckling. Seek solutions of the form

$$w(x,y,t;\varepsilon) = \varepsilon \sum_{m=1}^{\infty} \sum_{n=1}^{\infty} f_{mn}(t;\varepsilon) \sin(m\pi x/a) \sin(n\pi y/b) \qquad (6.6.5)$$

where ε is a small quantity, which we shall identify below, and then equation (6.6.3) gives, upon neglecting the nonlinear modal coupling,

$$\frac{d^2 f_{mn}}{d\tau^2} + (\alpha + \varepsilon \cos \tau)\, f_{mn} + \varepsilon^2 c f_{mn}^3 = 0 \qquad (6.6.6)$$

where

$$\alpha = \frac{\omega_{mn}^2}{\Omega^2}\,(1 - p_0), \quad \varepsilon = -\frac{\omega_{mn}^2}{\Omega^2}\, p_1 \ll 1,$$

$$c = \frac{D}{\rho h \mu \Omega^2}\left[\frac{1}{8}\left(\frac{m\pi}{a}\right)^4 - \left(1 - \frac{v}{2}\right)\left(\frac{m\pi}{a}\right)^2\left(\frac{n\pi}{b}\right)^2 + \frac{1}{8}\left(\frac{n\pi}{b}\right)^4\right],$$

$$p_0 = P_0/P_{cr}, \quad p_1 = P_1/P_{cr}, \quad \tau = \Omega t,$$

$$P_{cr} = D\left(\frac{a}{m\pi}\right)^2\left[\left(\frac{m\pi}{a}\right)^2 + \left(\frac{n\pi}{b}\right)^2\right],$$

and ω_{mn} is the natural frequency of transverse vibration of the plate. It may be noted that it is the particular choice of the boundary conditions as in (6.6.4) that made possible separating the variables as in (6.6.5).

Let the initial conditions be

$$\tau = 0; \quad f_{mn} = A_{mn}, \quad df_{mn}/d\tau = 0. \qquad (6.6.7)$$

One thus has a nonlinear Mathieu's differential equation (6.6.6) so that the nonlinear problem of the dynamic buckling of a thin elastic plate under a periodic load $P(t)$ reduces to the investigation of the parametric stability (where the time-dependent load $P(t)$ appears as a parameter) of the solution of the nonlinear Mathieu's differential equation. In particular, the issue of stability requires that all solutions $f_{mn}(\tau)$ of this equation, for $m, n = 1, 2, \ldots$, and $0 \le \tau < \infty$, remain bounded when arbitrary initial conditions are imposed.

The general theory of linear differential equations with periodic coefficients (viz., Floquet theory; see Section 3.13 which becomes useful here since we are constructing nonlinear solutions in the neighborhood of the linear solution) shows that the α, ε-plane is divided into regions of stability and instability separated by transition curves. Further, along the transition curves, periodic solutions of period 2π or 4π exist (as well as linearly increasing solutions except at the so-called critical points). In the unstable regions the growth takes place exponentially. The transition curves intersect the $\varepsilon = 0$ line at the critical points:

$$\alpha_c = n^2/4; \quad n = 0, 1, 2, \ldots.$$

In order to find an approximation to the transition curves for $\varepsilon \ll 1$, one looks for $\alpha(\varepsilon)$ such that periodic solutions of period 2π or 4π result. Consider the cases for which the stationary part P_0 of the applied load $P(t)$ is slightly greater than the static critical load P_{cr} for the onset of buckling so that, corresponding to the neighborhood of the critical point $\alpha_c = 0$, one writes

$$\alpha(\varepsilon) = -\varepsilon\alpha_1 - \varepsilon^2\alpha_2 + O(\varepsilon^3), \tag{6.6.8}$$

and seeks solutions of the form

$$f_{mn}(\tau;\varepsilon) \sim f_{mno}(\tau,\tilde{\tau}) + \varepsilon f_{mn1}(\tau,\tilde{\tau}) + \varepsilon^2 f_{mn2}(\tau,\tilde{\tau}) + O(\varepsilon^3), \tag{6.6.9}$$

where

$$\tilde{\tau} = \varepsilon\tau.$$

Here, τ characterizes the time scale of simple harmonic oscillations at the critical points, and $\tilde{\tau}$ characterizes the time scale of the effective spring forces $\varepsilon\cos\tau$ and $\varepsilon^2 c f_{mn}^2$, that are expected to produce a cumulative effect of long times on the solution.

Using (6.6.8) and (6.6.9), equation (6.6.6) gives

$$O(1): \frac{\partial^2 f_{mno}}{\partial\tau^2} = 0 \tag{6.6.10}$$

$$O(\varepsilon): \frac{\partial^2 f_{mn1}}{\partial\tau^2} = -2\frac{\partial^2 f_{mno}}{\partial\tau\partial\tilde{\tau}} + (\alpha_1 - \cos\tau) f_{mno} \tag{6.6.11}$$

$$O(\varepsilon^2): \frac{\partial^2 f_{mn2}}{\partial\tau^2} = -2\frac{\partial^2 f_{mn1}}{\partial\tau\partial\tilde{\tau}} + (\alpha_1 - \cos\tau) f_{mn1} - \frac{\partial^2 f_{mno}}{\partial\tilde{\tau}^2} + \alpha_2 f_{mno} - c f_{mno}^3 \tag{6.6.12}$$

etc.

In order to have bounded solutions, one has, from equation (6.6.10),

$$f_{mno}(\tau,\tilde{\tau}) = B_0(\tilde{\tau}). \tag{6.6.13}$$

Using (6.6.13), in order to have bounded solutions for equation (6.6.11), one requires

$$\alpha_1 = 0. \tag{6.6.14}$$

The solution of equation (6.6.11) is then given by

$$f_{mn1}(\tau,\tilde{\tau}) = B_1(\tilde{\tau}) + B_0(\tilde{\tau}) \cos\tau. \tag{6.6.15}$$

Using (6.6.13), (6.6.15), equation (6.6.12) becomes

$$\frac{\partial^2 f_{mn2}}{\partial \tau^2} = 2 \frac{dB_0}{d\tilde{\tau}} \sin\tau - \left[B_1\left(\tilde{\tau}\right) + B_0\left(\tilde{\tau}\right) \cos\tau \right] \cos\tau$$

$$- \frac{d^2 B_0}{d\tilde{\tau}^2} + \alpha_2 B_0\left(\tilde{\tau}\right) - c\left[B_0\left(\tilde{\tau}\right) \right]^3. \tag{6.6.16}$$

The removal of the secular terms in equation (6.6.16) requires

$$\frac{d^2 B_0}{d\tilde{\tau}^2} + \left(-\alpha_2 + \frac{1}{2} \right) B_0 + cB_0^3 = 0. \tag{6.6.17}$$

Using equation (6.6.7), one obtains from equation (6.6.17),

$$\left(\frac{dB_0}{d\tilde{\tau}} \right)^2 = -\frac{c}{2}\left(A_{mn}^2 - B_0^2 \right)\left(\tilde{c} - B_0^2 \right) \tag{6.6.18}$$

where

$$\tilde{c} = \frac{2\left[-\alpha_2 + (1/2) \right]}{-c} - A_{mn}^2.$$

Different cases arise depending on the signs and magnitudes of c and $\tilde{c}$. If $c < 0$, since $B_0(0) = A_{mn}$, B_0^2 cannot exceed A_{mn}^2 if $\tilde{c} > A_{mn}^2$, and cannot be smaller than A_{mn}^2 if $\tilde{c} < A_{mn}^2$; otherwise, $dB_0/d\tilde{\tau}$ will be imaginary. Thus, if $\tilde{c} > A_{mn}^2$, B_0 is bounded and oscillates between A_{mn} and $-A_{mn}$, and if $\tilde{c} < A_{mn}^2$, B_0 is unbounded. The special case $\tilde{c} = A_{mn}^2$ separates the stable oscillations from unstable oscillations, and hence, the transition curve (there is only one in this case) corresponds to

$$\alpha_2 = \frac{1}{2} + c\, A_{mn}^2. \tag{6.6.19}$$

On the other hand, if $c > 0$, B_0^2 is bounded and oscillates between A_{mn} and $\tilde{c}$ if $\tilde{c} > 0$, and oscillates between 0 and A_{mn} if $\tilde{c} < 0$. In all cases, the solution of equation (6.6.18a) is given by Jacobian elliptic functions.

Using (6.6.19), one obtains from (6.6.8),

$$\alpha = -\varepsilon^2 \left(\frac{1}{2} + c\, A_{mn}^2 \right) + O\left(\varepsilon^3 \right), \tag{6.6.20}$$

which determines the amplitudes of a periodic vibration corresponding to a given load (see Figure 6.11). Note the branching of a nonlinear solution with respect to the undeflected configuration at

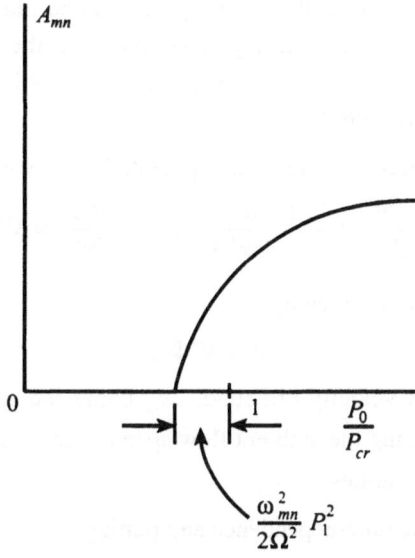

Figure 6.11. Plate vibration amplitude vs. load (from Shivamoggi, 1977b).

$$P_0 = 1 - \frac{\omega_{mn}^2}{2\Omega^2} \, P_1^2$$

which signifies that the load is a monotonically increasing function of the deformation. It is also important to note the backward shift of the branching point caused by the periodic load P_1 which signifies that the state of equilibrium corresponding to $P_0 \approx P_{cr}$ can always be destabilized by a periodic load irrespective of its frequency. However, once the branching occurs, the amplitude A_{mn} of the vibration depends only on the stationary part P_0 of the applied load $P(t)$.

6.7. Applications to Fluid Dynamics

6.7.1 The problem of aerodynamically generated sound

As we saw in Section 3.8.1, the problem of sound-wave propagation is one of singular perturbation type. Here, we consider sound radiation from low Mach-number flows. The method of multiple scales provided an elegant approach to this problem by linking the mathematical requirement of uniform validity of the solution directly to the physical possibility of a sound radiation in the far field (Shivamoggi, 1981a). (Alternatively, one may use the method of matched asymptotic expansions, wherein a wavefield is matched to an incompressible source flow field (Sears, 1969).)

For linearized, unsteady, compressible potential flow, we have

$$\frac{1}{c_\infty'^2}\frac{\partial^2 \Phi'}{\partial t'^2} + \frac{2M_\infty}{c_\infty'}\frac{\partial^2 \Phi'}{\partial x' \partial t'} + M_\infty^2 \frac{\partial^2 \Phi'}{\partial x'^2} = \nabla'^2 \Phi' \qquad (6.7.1)$$

where the fluid velocity is given by

$$v' \equiv \nabla' \Phi',$$

c_∞' being the ambient velocity of sound, M_∞' being the ambient Mach number ($M_\infty' \equiv U_\infty'/c_\infty'$, U_∞' being the ambient flow speed in the x-direction). The primes denote dimensional quantities.

Assuming harmonic time-dependence and putting

$$\Phi'(x', y', z', t') = \phi'(x', y', z') e^{-i\omega' t'}, \qquad (6.7.2)$$

equation (6.7.1) becomes

$$\phi'_{y'y'} + \phi'_{z'z'} + \beta^2 \phi'_{x'x'} + 2iM_\infty^2 \frac{\omega'}{c_\infty^2} \phi'_{x'} + \frac{\omega'^2}{c_\infty'^2} \phi' = 0, \qquad (6.7.3)$$

where

$$\beta^2 = 1 - M_\infty^2.$$

Looking for solutions of the form

$$\phi'(x', y', z') = \psi'(X', y', z') e^{i\frac{M_\infty x' \omega'}{c_\infty' \beta}}, \qquad (6.7.4)$$

where

$$X' \equiv \frac{x'}{\beta},$$

equation (6.7.3) becomes

$$\psi'_{x'x'} + \psi'_{y'y'} + \psi'_{z'z'} + \chi^2 \psi' = 0, \qquad (6.7.5)$$

where

$$\chi' \equiv \frac{\omega'}{c_\infty' \beta}.$$

Non-dimensionalizing according to

$$x = \frac{X'}{b'}, \quad y = \frac{y'}{b'}, \quad z = \frac{z'}{b'}, \quad \psi = \frac{\psi'}{2b'^2\omega'}, \tag{6.7.6}$$

$$\varepsilon \equiv \frac{\omega'b'}{c_\infty' \beta} \ll 1,$$

b' being a length scale characterizing the source field, and seeking solutions of the form

$$\psi(x,y,z,\varepsilon) = \psi_0\left(x,\tilde{x},y,\tilde{y},z,\tilde{z}\right) + \varepsilon^2\psi_1\left(x,\tilde{x},y,\tilde{y},z,\tilde{z}\right) + O\left(\varepsilon^4\right), \tag{6.7.7}$$

where

$$\tilde{x} = \varepsilon y, \quad \tilde{y} = \varepsilon y, \quad \tilde{z} = \varepsilon z,$$

we obtain from equation (6.7.5),

$$O(1): \ \psi_{0xx} + \psi_{0yy} + \psi_{0zz} = 0 \tag{6.7.8}$$

$$O\left(\varepsilon^2\right): \ \psi_{1xx} + \psi_{1yy} + \psi_{1zz} = -\left(\psi_{0\tilde{x}\tilde{x}} + \psi_{0\tilde{y}\tilde{y}} + \psi_{0\tilde{z}\tilde{z}} + \psi_0\right) \tag{6.7.9}$$

etc.

According to equation (6.7.9), ψ_1 would behave asymptotically as

$$\psi_1 \sim \frac{1}{4\pi r} \ \tilde{\Box}^2 \int_{\upsilon} \psi_0\left(x,\tilde{x},y,\tilde{y},z,\tilde{z}\right) dV(x,y,z), \tag{6.7.10}$$

where υ is the volume of the source field, and

$$r = \left(x^2 + y^2 + z^2\right)^{1/2}, \quad \tilde{\Box}^2 \equiv \left(\frac{\partial^2}{\partial\tilde{x}^2} + \frac{\partial^2}{\partial\tilde{y}^2} + \frac{\partial^2}{\partial\tilde{z}^2} + 1\right),$$

so that ψ_1 will at least be of the same order as ψ_0 in the far field. This means that (6.7.7) will not provide a uniformly valid approximation to the properties of the flow. In order to preclude this, one requires

$$\tilde{\Box}^2\psi_0 = 0 \tag{6.7.11}$$

or, if

$$\Psi_0 \equiv \psi_0 c^{-i\tilde{t}}, \quad \tilde{t} = \omega' t', \tag{6.7.12}$$

equation (6.7.11) implies

$$\left(\tilde{\nabla}^2 - \frac{\partial^2}{\partial\tilde{t}^2}\right)\Psi_0 = 0, \tag{6.7.13}$$

where

$$\tilde{\nabla}^2 = \frac{\partial^2}{\partial \tilde{x}^2} + \frac{\partial^2}{\partial \tilde{y}^2} + \frac{\partial^2}{\partial \tilde{z}^2}.$$

Equation (6.7.13) simply implies the existence of a sound radiation in the far field.

One then has the following program: ψ_0 can be represented in terms of transformed incompressible singularities. In order to make ψ_0 uniformly valid, one replaces the latter by the corresponding acoustic singularities.

Thus, for a source in motion, one obtains

$$\Phi' = \frac{1}{r'} f'\left(t' - \frac{r'}{c'_\infty \beta^2} + \frac{M_\infty x'}{c'_\infty \beta^2}\right), \tag{6.7.14}$$

where

$$r' = \left[(x' + U't')^2 + \beta^2 (y'^2 + z'^2)\right]^{1/2},$$

U' being the velocity of the source.

6.7.2 Mathematical formalization of Lighthill's theory of aerodynamically generated sound

According to Lighthill's (1952) theory, which was proposed to serve as a model for treating the sound radiation field by relatively small regions of distributed sources embedded in an infinite homogeneous fluid at rest with sound speed c_∞ (and Curle's (1955) extension to incorporate the effect of the presence of solid boundaries on the sound field), one has

$$\rho - \rho_\infty = \frac{1}{4\pi c_\infty^2} \frac{\partial^2}{\partial x_i \partial x_j} \int_v \frac{T_{ij}\left[y, t - (|x - y|/c_\infty)\right]}{|x - y|} dV(y)$$

$$+ \frac{1}{4\pi c_\infty^2} \frac{\partial}{\partial x_i} \int_S \frac{\ell_j P_{ij}\left[y, t - (|x - y|/c_\infty)\right]}{|x - y|} dS(y) \tag{6.7.15}$$

where ρ is the mass density of the fluid,

$$T_{ij} = \rho v_i v_j + P_{ij} - c_\infty^2 \rho \delta_{ij},$$

$$P_{ij} = p\delta_{ij} - \mu\left[\left(\frac{\partial v_i}{\partial x_j} + \frac{\partial v_j}{\partial x_i}\right) - \frac{2}{3}\left(\frac{\partial v_k}{\partial x_k}\right)\delta_{ij}\right].$$

v_i is the fluid velocity, p is the pressure, μ is the coefficient of viscosity which is taken to be constant, ℓ_i is the direction of cosines of the outward normal to the solid surface S from the fluid, and the solid surface S is supposed to be either fixed or moving in its own plane. The physical situation under consideration is a flow occupying a volume υ in the presence of solid boundaries S, and the subscript ∞ denotes the uniform conditions outside υ.

Lighthill adopted a force-oscillation view for this aerodynamically generated sound on the grounds that the latter is so weak that it does not react significantly upon the flow producing it. Thus, sound is conjectured to be generated in the quiescent atmosphere in the same manner as that produced in a classical stationary medium by a continuous distribution of acoustic dipoles of instantaneous strength $l_j P_{ij}$ per unit area over the solid surface S. Further assumptions in the Lighthill theory are that:

* the influence of viscosity is negligible on the ground that the term accounting for viscous stresses contains an additional spatial derivative and this leads to a source of essentially higher order and lower efficiency than those due to the momentum fluxes;

* the existence of some fictitious sources is then postulated in order to simulate the phenomena of convection, refraction, and dissipation of the sound in a real medium;

* if the temperature is nearly uniform in the flow field, then the quantity

$$\frac{\partial^2}{\partial x_i \partial x_j} \left[\left(p - c_\infty^2 \rho \right) \delta_{ij} \right]$$

is small, so that the flow-noise generation is assumed to be insensitive to deviations of the pressure from its adiabatic approximation;

* the speed of sound is constant and the sources of sound are allowed to move, whereas the medium into which they radiate is assumed to be stationary.

Given these limitations, it therefore appears that Lighthill's theory is satisfactory for either low-convection Mach number or low frequencies with wavelengths much larger than the linear dimension of the region of the distributed sources. One may then expect to formalize the approximations otherwise made intuitively in Lighthill's theory by developing a small Mach-number expansion of the Navier-Stokes equations (Shivamoggi, 1977c).

At low speeds, a valid approach to study the compressibility effects is through a formal perturbation of the basic equations for incompressible flow, using the ambient Mach number M_∞ as the perturbation parameter, as seen in Section 6.7.1.

The Navier-Stokes equations for a fluid are

$$\frac{\partial \tilde{\rho}}{\partial t} + \tilde{\nabla} \cdot \left(\tilde{\rho} \tilde{v} \right) = 0 \tag{6.7.16}$$

$$\frac{\partial}{\partial \tilde{t}} \left(\tilde{\rho} \tilde{v} \right) + \tilde{\nabla} \cdot \left(\tilde{\rho} \tilde{v} \tilde{v} \right) + \nabla \tilde{p} = \tilde{\mu} \left[\tilde{\nabla}^2 \tilde{v} + \frac{1}{3} \tilde{\nabla} \left(\tilde{\nabla} \cdot \tilde{v} \right) \right]. \tag{6.7.17}$$

One has for an ideal gas, the equation of state

$$\tilde{p} = \tilde{\rho} \tilde{R} \tilde{T}, \tag{6.7.18}$$

where $\tilde{R}$ is the gas constant.

Consider an isothermal situation for analytical convenience and nondimensionalize as follows:

$$x = \frac{\tilde{U}\tilde{x}}{\tilde{v}_\infty}, \quad t = \frac{\tilde{U}^2 \tilde{t}}{\tilde{v}_\infty}, \quad \tilde{v} = \frac{\tilde{\mu}}{\tilde{\rho}_\infty}, \quad v = \frac{\tilde{v}}{\tilde{U}},$$

$$p = \frac{\tilde{p}}{\tilde{p}_\infty} = \rho = \frac{\tilde{\rho}}{\tilde{\rho}_\infty}, \quad M_\infty = \frac{\tilde{U}}{\tilde{c}_\infty} = \frac{\tilde{U}}{\sqrt{\tilde{p}_\infty / \tilde{\rho}_\infty}}, \tag{6.7.19}$$

where the adiabatic exponent γ (i.e., $p \sim \rho^\gamma$) is taken to be unity accordingly. However, it may be readily verified that the results given in the following are valid for any value of γ, namely, an adiabatic case.

One then obtains

$$\frac{\partial \rho}{\partial t} + \nabla \cdot (\rho v) = 0 \tag{6.7.20}$$

$$\frac{\partial \rho v}{\partial t} + \nabla \cdot (\rho v v) - \left[\nabla^2 v + \frac{1}{3} \nabla (\nabla \cdot v) \right] = -\frac{1}{M_\infty^2} \nabla \rho. \tag{6.7.21}$$

It is a simple matter to derive from equations (6.7.20) and (6.7.21),

$$\nabla^2 \rho - M_\infty^2 \left[\frac{\partial^2 \rho}{\partial t^2} - \nabla \nabla : (\rho v v) + \nabla^2 (\nabla \cdot v) + \frac{1}{3} \nabla^2 (\nabla \cdot v) \right] = 0. \tag{6.7.22}$$

Consider the cases for which $M_\infty \ll 1$, and seek the solutions of the form

$$\left.\begin{array}{l} \rho(x,t;M_\infty) \sim 1 + M_\infty^2\, \rho_1(\xi,\eta,t) + M_\infty^4\, \rho_2(\xi,\eta,t) + O(M_\infty^6) \\[2mm] v(x,t;M_\infty) \sim v_0(\xi,\eta,t) + M_\infty^2\, v_1(\xi,\eta,t) + O(M_\infty^4) \end{array}\right\}, \quad (6.7.23)$$

where

$$\xi = x, \quad \eta \equiv M_\infty x.$$

ξ and η, respectively, characterize the source field and the sound field.

Note that

$$\nabla = \nabla_\xi + M_\infty \nabla_\eta. \tag{6.7.24}$$

Using (6.7.23) in equation (6.7.22), and noting (6.7.24), one obtains

$$O(M_\infty^2): \quad \nabla_\xi^2 \rho_1 = -\nabla_\xi \nabla_\xi : v_0 v_0 \tag{6.7.25}$$

$$O(M_\infty^4): \quad \nabla_\xi^2 \rho_2 = \left(\frac{\partial^2 \rho_1}{\partial t^2} - \nabla_\eta^2 \rho_1\right) + \nabla_\xi^2\left(\frac{4}{3}\,\nabla_\xi \cdot v_1\right)$$

$$- \nabla_\xi \nabla_\xi : \left(v_0 v_1 + v_1 v_0 + \rho_1 v_0 v_0\right) \tag{6.7.26}$$

etc.

The general solution of equation (6.7.25) is

$$\rho_1(\xi,\eta,t) = \frac{1}{4\pi} \int_V \frac{\nabla_\zeta \nabla_\zeta : v_0 v_0(\zeta,\eta,t)}{|\xi - \zeta|}\, dV(\zeta)$$

$$- \frac{1}{4\pi} \int_S \rho_1(\zeta,\eta,t)\, \nabla_\zeta\left(\frac{1}{|\xi - \zeta|}\right) \cdot dS(\zeta)$$

$$+ \frac{1}{4\pi} \int_S \frac{1}{|\xi - \zeta|}\, \nabla_\zeta \rho_1(\zeta,\eta,t) \cdot dS(\zeta), \tag{6.7.27}$$

where the solid surface S has been assumed to be stationary and impervious.

Using the divergence theorem and noting that the fluid velocity normal to the surface S is zero, one finds

$$\rho_1(\xi,\eta,t) = \frac{1}{4\pi} \nabla_\xi \nabla_\xi : \int_V \frac{v_0 v_0(\zeta,\eta,t)}{|\xi - \zeta|}\, dV(\zeta)$$

$$+ \frac{1}{4\pi} \nabla_\xi \int_S \frac{\rho_1(\zeta,\eta,t)}{|\xi - \zeta|} \cdot dS(\zeta). \tag{6.7.28}$$

Noting that

$$\frac{4}{3} \nabla_\xi \cdot \mathbf{v}_1 (\xi, \eta, t) = -\frac{4}{3} \frac{\partial}{\partial t} \nabla_\xi \nabla_\xi : \int_V \frac{\mathbf{v}_0 \mathbf{v}_0 (\zeta, \eta, t)}{|\xi - \eta|} dV(\zeta), \quad (6.7.29)$$

it is seen that the first term in equation (6.7.25) introduces a monopole contribution to ρ_2, whereas ρ_1 has only a quadruple contribution. To elaborate, note that in the far field, ρ_1 behaves asymptotically as

$$\rho_1 (\xi, \eta, t) \sim \frac{1}{4\pi} \frac{\hat{\xi}\hat{\xi}}{|\xi|^3} : \int_V \mathbf{v}_0 \mathbf{v}_0 (\zeta, \eta, t) \, dV(\zeta),$$

where

$$\hat{\xi} = \frac{\xi}{|\xi|},$$

while ρ_2 would behave asymptotically as

$$\rho_2 (\xi, \eta, t) \sim -\frac{1}{4\pi |\xi|} \left(\frac{\partial^2}{\partial t^2} - \nabla_\eta^2 \right) \int_V \rho_1 (\zeta, \eta, t) \, dV(\zeta),$$

so that ρ_2 would dominate ρ_1 in the far field. This means that a perturbation expansion in terms of a small Mach number will not provide a uniformly valid approximation to the properties of the flow. In order to preclude this, one requires

$$\frac{\partial^2 \rho_1}{\partial t^2} - \nabla_\eta^2 \rho_1 = 0, \quad (6.7.30)$$

which physically involves admission of the possibility of sound radiation in the far field. This further determines the functional form of the solution (6.7.28) with regard to η, and one has

$$\rho_1 (\xi, \eta, t) = \frac{1}{4\pi} \nabla_\xi \nabla_\xi : \int_V \frac{\mathbf{v}_0 \mathbf{v}_0 (\zeta, t - |\eta|)}{|\xi - \zeta|} dV(\zeta)$$

$$+ \frac{1}{4\pi} \nabla_\xi \int_S \frac{\rho_1 (\zeta, t - |\eta|)}{|\xi - \zeta|} \cdot dS(\zeta), \quad (6.7.31)$$

or

$$\rho_1(\boldsymbol{\xi},\boldsymbol{\eta},t) = \frac{1}{4\pi\,\tilde{c}_\infty^2}\,\tilde{\nabla}_{\tilde{x}}\tilde{\nabla}_{\tilde{x}} : \int_V \frac{\tilde{v}_0\tilde{v}_0\left\{\zeta,\left(\tilde{U}^2/\tilde{v}_\infty\right)\left[\tilde{t} - \left(|\tilde{x}|/\tilde{c}_\infty\right)\right]\right\}}{|x-\zeta|}\cdot dV(\zeta)$$

$$\frac{1}{4\pi\,\tilde{c}_\infty^2}\,\tilde{\nabla}_{\tilde{x}} \int_S \frac{(\tilde{\rho}-\tilde{\rho}_\infty)\big/\tilde{\rho}_\infty\left\{\zeta,\left(\tilde{U}^2/\tilde{v}_\infty\right)\left[\tilde{t} - \left(|\tilde{x}|/\tilde{c}_\infty\right)\right]\right\}}{|\tilde{x}-\zeta|}\cdot dS(\zeta).$$

$$(6.7.32)$$

Corresponding to the far field and cases of weak sound-radiation, if one interprets this as the first term in a Taylor series expansion in the retarded time $\tilde{t} - |\tilde{x}|/\tilde{c}_\infty$, the subsequent terms in the expansion can be incorporated by writing this as

$$\frac{\tilde{\rho}-\tilde{\rho}_\infty}{\tilde{\rho}_\infty} = \frac{1}{4\pi\,\tilde{c}_\infty^2}\,\tilde{\nabla}_x\tilde{\nabla}_x : \int_V \frac{\tilde{v}_0\tilde{v}_0\left\{\zeta,\left(\tilde{U}^2/\tilde{v}_\infty\right)\left[\tilde{t} - \left(|\tilde{x}-\zeta|/\tilde{c}_\infty\right)\right]\right\}}{|\tilde{x}-\zeta|}\,dV(\zeta)$$

$$+\frac{1}{4\pi\,\tilde{c}_\infty^2}\,\tilde{\nabla}_{\tilde{x}} \int \frac{(\tilde{\rho}-\tilde{\rho}_\infty)\big/\tilde{\rho}_\infty\left\{\zeta,\left(\tilde{U}^2/\tilde{v}_\infty\right)\left[\tilde{t} - \left(|\tilde{x}-\zeta|/\tilde{c}_\infty\right)\right]\right\}}{|\tilde{x}-\zeta|}\cdot dS(\zeta).$$

$$(6.7.33)$$

Incidentally, (6.7.33) shows that:

* the so-called "Reynolds stresses" are the principal generators of aerodynamic sound;

* upon a comparison with Curle's (1955) result, that the quantity

$$\frac{\partial^2}{\partial\tilde{x}_i\,\partial\tilde{x}_j}\left[\left(\tilde{\rho}-\tilde{c}_\infty^2\,\tilde{p}\right)\delta_{ij}\right],$$

is small under nearly isothermal situations;

both in conformity with Lighthill's remarkable conjectures made intuitively.

6.7.3 Nonlinear Shallow Water Waves

There are two types of wave motions on a water surface. Shallow-water waves arise when the wavelength of the waves is much greater than the depth of water. Surface waves correspond to disturbances that do not extend far below the surface. The wavelength is much less than the depth of water.

The features that make an analysis of surface wave motions difficult are

*   the presence of nonlinearities;

*   the free surface being unknown a priori, besides being variant with time.

In order to make progress with the theory of surface wave motions it is, in general, necessary to simplify the model by making special hypotheses of one kind or another which suggest themselves on the basis of general physical circumstances contemplated in a given class of problems. Thus, two approximate theories result when

*   the amplitude of the surface waves is considered small, or

*   the depth of the water is considered small with respect to wavelength.

The first hypothesis leads to a linear theory and to boundary-value problems of nearly classical type; while the second hypotheses leads to a nonlinear theory for initial-value problems, which, in the lowest order is of the type corresponding to sound wave propagation in compressible fluids.

6.7.3(a) Governing Equations

The equilibrium configuration of a liquid in a container of finite size is one of rest with a plane surface. One may produce a wavemotion on the surface which is due to the action of gravity that acts in the direction of restoring the undisturbed state of rest. If the wavemotion is assumed to have started from rest relative to the undisturbed state of flow, which is itself assumed irrotational, then, the wavemotions will be irrotational. The wave propagation is taken to occur along the x-direction, and the gravity acts opposite to the z-direction. We will assume that the fluid flow is two-dimensional and there is no dependence on the transverse horizontal coordinate.

The velocity potential Φ defined according to

$$v = \nabla \Phi, \tag{6.7.34}$$

$v = (u, v, w)$ being the fluid velocity, satisfies

$$\frac{\partial^2 \Phi}{\partial x^2} + \frac{\partial^2 \Phi}{\partial z^2} = 0. \tag{6.7.35}$$

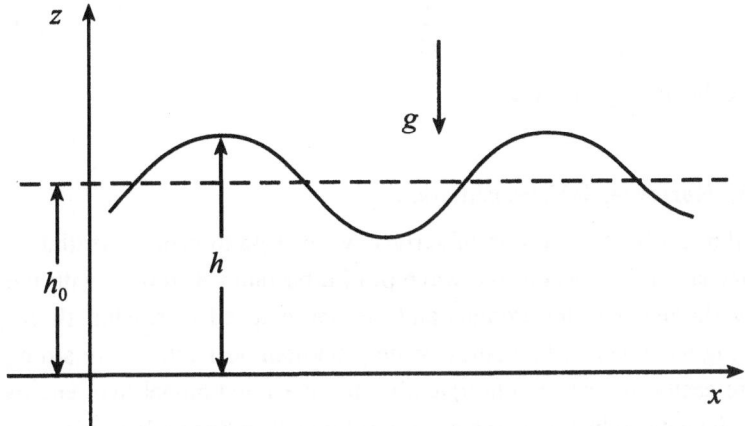

Figure 6.12. Surface waves on water of mean depth h_0.

At a rigid stationary boundary, we have the fluid impenetrability condition –

$$\frac{\partial \Phi}{\partial n} = 0. \tag{6.7.36}$$

If $z = h(x,t)$ denotes the displacement of the free surface, since a fluid particle on that surface will remain there, we have the following kinematic condition expressing the fact that the free surface remains a material surface:

$$z = h : \frac{D}{Dt}(z - h) = 0 \tag{6.7.37}$$

or

$$z = h : \frac{\partial \Phi}{\partial z} = \frac{\partial h}{\partial t} + \nabla \Phi \cdot \nabla h. \tag{6.7.38}$$

The dynamic condition at the free surface is,

$$z = h : p = T\left(\frac{1}{R_1}\right) \tag{6.7.39}$$

where R_1 is the radius of curvature of the free surface, counted positive when the center of curvature is above the surface,

$$\frac{1}{R_1} = \frac{\partial^2 h}{\partial x^2}\left[1 + \left(\frac{\partial h}{\partial x}\right)^2\right]^{-3/2}. \tag{6.7.40}$$

T is the surface tension, and the pressure p is given by the Bernoulli integral –

$$\frac{p}{\rho} = -\frac{\partial \Phi}{\partial t} - gz - \frac{u^2 + w^2}{2}, \tag{6.7.41}$$

ρ being the density of water.

6.7.3(b) Korteweg-deVries Equation

Nonlinear effects in a train of surface waves lead to many essential observed phenomena. In deep water, the wave profile becomes distorted, with the crests slightly sharper than the troughs and the phase speed increasing slightly with increasing wave slope. In shallow water, the nonlinear effects are stronger and so more easily observed; consequently, the linearized model here entails much more stringent conditions on the wave amplitude than that in deep water.

Consider small (but finite)-amplitude long waves of amplitude a and wavelength λ in shallow water of depth h so that

$$\alpha \equiv \frac{a}{h} << 1 \quad \text{and} \quad \beta \equiv \frac{h}{\lambda} << 1. \tag{6.7.42}$$

Two limiting cases arise for such waves depending on whether the parameter $S \equiv \dfrac{\alpha}{\beta^2} = \dfrac{a\lambda^2}{h^3}$ is much greater than one (the *Stokes* approximation) or is near one (the *Boussinesq* approximation). The first case leads to nonlinear periodic waves while the second case affords a balance between nonlinearity and dispersion and leads to solitary waves.

Consider a homogeneous, incompressible fluid whose undisturbed depth above a rigid horizontal boundary is h_0. Let the disturbed free surface be at a height $h(x,t)$ (see Figure 6.12).

One, then, has the following boundary-value problem for the velocity potential Φ:

$$\frac{\partial^2 \Phi}{\partial x^2} + \frac{\partial^2 \Phi}{\partial z^2} = 0 \tag{6.7.43}$$

$$z = 0 : \frac{\partial \Phi}{\partial z} = 0 \tag{6.7.44}$$

$$z = h : \frac{\partial \Phi}{\partial z} = \frac{\partial h}{\partial t} + \frac{\partial \Phi}{\partial x}\frac{\partial h}{\partial x} \tag{6.7.45}$$

$$\frac{\partial \Phi}{\partial t} + \frac{1}{2}\left[\left(\frac{\partial \Phi}{\partial x}\right)^2 + \left(\frac{\partial \Phi}{\partial z}\right)^2\right] + g(h - h_0) = 0. \tag{6.7.46}$$

Consider propagation of waves moving only to the right, and so introduce

$$\left.\begin{aligned}\xi = \sqrt{\varepsilon}\,(x - c_0 t), \quad \tau = \varepsilon^{3/2} t, \quad \psi = \varepsilon^{1/2}\Phi \\ c_0 = \sqrt{gh_0}, \quad \varepsilon \ll 1 \end{aligned}\right\} \tag{6.7.47}$$

so that the boundary-value problem (6.7.43)-(6.7.46) becomes

$$\varepsilon \frac{\partial^2 \psi}{\partial \xi^2} + \frac{\partial^2 \psi}{\partial z^2} = 0 \tag{6.7.48}$$

$$z = 0 : \frac{\partial \psi}{\partial z} = 0 \tag{6.7.49}$$

$$z = h : \frac{\partial \psi}{\partial z} = \varepsilon^2 \frac{\partial h}{\partial \tau} + \varepsilon\left(\frac{\partial \psi}{\partial \xi} - c_0\right)\frac{\partial h}{\partial \xi} \tag{6.7.50}$$

$$\varepsilon^2 \frac{\partial \psi}{\partial \tau} - \varepsilon c_0 \frac{\partial \psi}{\partial \xi} + \frac{1}{2}\left[\varepsilon\left(\frac{\partial \psi}{\partial \xi}\right)^2 + \left(\frac{\partial \psi}{\partial z}\right)^2\right] + \varepsilon g(h - h_0) = 0. \tag{6.7.51}$$

Look for solutions of the form –

$$\left.\begin{aligned}h \sim h_0 + \varepsilon h_1 + \varepsilon^2 h_2 + \cdots \\ \psi \sim \varepsilon \psi_1 + \varepsilon^2 \psi_2 + \cdots \end{aligned}\right\}, \tag{6.7.52}$$

then equation (6.7.48) leads to

$$O(\varepsilon) : \frac{\partial^2 \psi_1}{\partial z^2} = 0 \tag{6.7.53}$$

$$O(\varepsilon^n) : \frac{\partial^2 \psi_n}{\partial z^2} + \frac{\partial^2 \psi_{n-1}}{\partial \xi^2} = 0, \quad n > 1. \tag{6.7.54}$$

One obtains, from equations (6.7.53) and (6.7.54), on using the boundary condition (6.7.49),

$$O(\varepsilon) : \psi_1 = \psi_1(\xi, \tau),$$

$$O(\varepsilon^2) : \psi_2 = -\frac{1}{2} z^2 \frac{\partial^2 \psi_1}{\partial \xi^2},$$

$$O(\varepsilon^3) : \psi_3 = \frac{1}{24} z^4 \frac{\partial^4 \psi_1}{\partial \xi^4},$$

etc. $\tag{6.7.55}$

The boundary conditions (6.7.50) and (6.7.51) are imposed at a boundary which is unknown without the solution and varies slightly with the perturbation parameter ε. If the solution is analytic in spatial coordinates, the transfer of the boundary conditions to the basic configuration of the boundary corresponding to $\varepsilon = 0$ is effected by expanding the solution in a Taylor series about the values at the basic configuration, as in Section 3.8.2. Thus, (6.7.50) and (6.7.51) give,

$$z = h_0 : O\left(\varepsilon^2\right): \quad \frac{\partial \psi_2}{\partial z} = -c_0 \frac{\partial h_1}{\partial \xi},$$

$$O\left(\varepsilon^3\right): \quad \frac{\partial \psi_3}{\partial z} + h_1 \frac{\partial^2 \psi_2}{\partial z^2} = \frac{\partial h_1}{\partial \tau} - c_0 \frac{\partial h_2}{\partial \xi} + \frac{\partial \psi_1}{\partial \xi} \frac{\partial h_1}{\partial \xi},$$

etc.   (6.7.56)

and

$$z = h_0 : O\left(\varepsilon^2\right): \quad -c_0 \frac{\partial \psi_1}{\partial \xi} + g h_1 = 0,$$

$$O\left(\varepsilon^3\right): \quad \frac{\partial \psi_1}{\partial \tau} - c_0 \frac{\partial \psi_2}{\partial \xi} + \frac{1}{2}\left(\frac{\partial \psi_1}{\partial \xi}\right)^2 + g h_2 = 0,$$

etc.   (6.7.57)

From (6.7.55)-(6.7.57), one derives (Jeffrey and Kakutani, 1972)

$$\frac{\partial h_1}{\partial \tau} + \frac{3c_0}{2h_0} h_1 \frac{\partial h_1}{\partial \xi} = -\frac{c_0 h_0^2}{6} \frac{\partial^3 h_1}{\partial \xi^3} \qquad (6.7.58)$$

which is the Korteweg-deVries equation.

Thus, in the shallow-water approximation, the elliptic Laplace equation (6.7.43) is replaced by the nonlinear hyperbolic equation (6.7.58).

6.7.3(c) Solitary Waves

Write the Korteweg-deVries equation (6.7.58) in the form

$$\phi_t + \kappa \phi \phi_x + \phi_{xxx} = 0. \qquad (6.7.59)$$

Looking for steady, progressing, and localized wave solutions of the form

$$\phi(x,t) = \phi(\xi), \quad \xi = x - Ut, \qquad (6.7.60)$$

equation (6.7.59) gives

$$\phi_\xi \left(\kappa\phi - U \right) + \phi_{\xi\xi\xi} = 0,$$

$$|\xi| \to \infty : \phi, \phi_\xi, \phi_{\xi\xi} \to 0. \qquad (6.7.61)$$

Upon integrating, equation (6.7.61) gives

$$\phi_{\xi\xi} = U\phi - \frac{\kappa}{2}\phi^2 \qquad (6.7.62)$$

and again, equation (6.7.62) gives

$$\frac{1}{2}\phi_\xi^2 = \frac{U}{2}\phi^2 - \frac{\kappa}{6}\phi^3 \qquad (6.7.63)$$

from which

$$\int_{\phi_{max}}^{\phi} \frac{d\phi}{\sqrt{\frac{U}{2}\phi^2 - \frac{\kappa}{6}\phi^3}} = \sqrt{2}\,\xi. \qquad (6.7.64)$$

The integral in (6.7.64) is of the form

$$I = \int \frac{d\phi}{\phi\sqrt{1 - \sigma\phi}}. \qquad (6.7.65)$$

Putting

$$\psi = \sqrt{1 - \sigma\phi}, \quad \sigma \equiv \frac{\kappa}{3U}, \qquad (6.7.66)$$

(6.7.65) becomes

$$I = -2\int \frac{d\psi}{1 - \psi^2} = \ln \frac{1 - \psi}{1 + \psi}. \qquad (6.7.67)$$

Using (6.7.67), (6.7.64) leads to

$$\frac{1 - \psi}{1 + \psi} = e^{\sqrt{U}\,\xi}, \qquad (6.7.68)$$

from which

$$\psi = \frac{1 - e^{\sqrt{U}\,\xi}}{1 + e^{\sqrt{U}\,\xi}}. \qquad (6.7.69)$$

Noting, from (6.7.66), that

$$\phi = \frac{1 - \psi^2}{\sigma} \qquad (6.7.70)$$

one obtains, on using (6.7.69),

$$\phi = \frac{1}{\sigma}\frac{4e^{\sqrt{U}\xi}}{\left(1+e^{\sqrt{U}\xi}\right)^2} = \frac{1}{\sigma}\,\text{sech}^2\,\frac{\sqrt{U}\,\xi}{2} = \frac{3U}{\kappa}\,\text{sech}^2\left[\frac{\sqrt{U}}{2}(x-Ut)\right] \qquad (6.7.71)$$

which represents a unidirectional solitary wave. Equation (6.7.71) shows that

*   this solitary wave moves with a velocity proportional to the square root of its amplitude (this is a consequence of the nonlinearity of the wave);

*   the width of the solitary wave is inversely proportional to the square root of its amplitude (this is a consequence of the spreading of the wave due to dispersion); this result, of course, confirms the condition $\alpha \sim \beta^2$ discussed before (see (6.7.42))!

Solitary waves are localized waves propagating without change of shape and velocity[2]. The essential quality of a solitary wave is the balance between nonlinearity, which tends to steepen the wavefront in consequence of the increase of the wavespeed with amplitude, and dispersion, which tends to spread the wavefront. Solitary waves are found to preserve their identity and to be stable in processes of mutual collisions. Solitary waves are strictly nonlinear phenomena with no counterparts in linear theory (see Drazin and Johnson, 1989, for further details).

6.7.3(d) Stokes Waves

Irrotational, steady, progressive waves were considered first by Stokes and are called Stokes waves. The Korteweg-deVries equation (6.7.58) can be written as

$$\frac{\partial h_1}{\partial t} + c_0\left(1+\frac{3}{2}\frac{h_1}{h_0}\right)\frac{\partial h_1}{\partial x} + \gamma\frac{\partial^3 h_1}{\partial x^3} = 0 \qquad (6.7.72)$$

where

$$\gamma \equiv c_0\,h_0^2/6.$$

---

[2] Solitary waves were first observed by J. Scott Russell on the Edinburgh-Glasgow canal in 1834. Russell also performed laboratory experiments on solitary waves and empirically deduced that the speed u of the solitary wave is given by

$$u^2 = g(h_0 + a).$$

Let us find the next approximation to the linear periodic wavetrain using the method of strained parameters. Thus let

$$\left. \begin{array}{l} \dfrac{h_1}{h_0} \sim \varepsilon h_1^{(1)}(\theta) + \varepsilon^2 h_1^{(2)}(\theta) + \varepsilon^3 h_1^{(3)}(\theta) + \cdots \\[2mm] \omega = \omega_0(k) + \varepsilon \omega_1(k) + \varepsilon^2 \omega_2(k) + \cdots \end{array} \right\}$$ (6.7.73)

where

$$\theta \equiv kx - \omega t.$$

Using (6.7.73), equation (6.7.72) leads to the following hierarchy of equations –

$$O(\varepsilon): \ (\omega - c_0 k) \, h^{(1)'} - \gamma k^3 h^{(1)'''} = 0 \tag{6.7.74}$$

$$O(\varepsilon^2): \ (\omega - c_0 k) \, h^{(2)'} - \gamma k^3 h^{(2)'''} = \frac{3}{2} c_0 k h_1^{(1)} h^{(1)'} - \omega_1 h^{(1)'} \tag{6.7.75}$$

$$O(\varepsilon^3): \ (\omega - c_0 k) \, h^{(3)'} - \gamma k^3 h^{(3)'''} = \frac{3}{2} c_0 k \left(h_1^{(1)} h_1^{(2)} \right)' - \omega_2 h^{(1)'} \tag{6.7.76}$$

etc.

Equation (6.7.74) gives the linear result –

$$h_1^{(1)} = \cos\theta, \quad \omega_0(k) = c_0 k - \gamma k^3. \tag{6.7.77}$$

Using (6.7.77), the removal of secular terms on the right-hand side in equation (6.7.75) requires

$$\omega_1 = 0. \tag{6.7.78}$$

Using (6.7.77) and (6.7.78), equation (6.7.75) gives

$$h_1^{(2)} = \frac{c_0}{8\gamma k^2} \cos 2\theta. \tag{6.7.79}$$

Using (6.7.77)-(6.7.79), the removal of secular terms on the right-hand side in equation (6.7.76) requires

$$\omega_2 = \frac{3c_0^2}{32\gamma k}. \tag{6.7.80}$$

Using (6.7.77)-(6.7.80), (6.7.73) becomes

$$\left.\begin{array}{l} \dfrac{h_1}{h_0} \sim \varepsilon \cos\theta + \dfrac{3\varepsilon^2}{4k^2 h_0^2} \cos 2\theta + \cdots \\[3mm] \dfrac{\omega}{c_0 k} = 1 - \dfrac{1}{6} k^2 h_0^2 + \dfrac{9\varepsilon^2}{16 k^2 h_0^2} + \cdots \end{array}\right\} \qquad (6.7.81)$$

Equation (6.7.81) exhibits two essential effects of nonlinearities –

* periodic solutions of the form $e^{i(kx - \omega t)}$ may exist, but they are no longer sinusoidal;

* the amplitude appears in the dispersion relation.

6.7.3(e) Perturbed Solitary Wave Propagation

Consider a perturbed Korteweg-deVries equation describing nonlinear wave propagation in a slowly-varying medium or a channel of slowly-varying cross section (Smyth, 1984)

$$u_t + 6uu_x + u_{xxx} = \varepsilon u \qquad (6.7.82)$$

where ε is a small parameter. The above perturbation adds energy to the wave in question.

We will analyze the behavior of the perturbed solitary wave solution of equation (6.4.66) using the principle of energy conservation. The perturbed solitary wave turns out not to conserve "mass". So, following Johnson (1973) and Grimshaw (1979), we introduce a "tail" behind the solitary wave to restore the satisfaction of "mass" conservation. Further, following Knickerbocker and Newell (1980) and Smyth (1984), the "tail" is divided into two distinct "near" and "far" tail regions.

Let us assume that the solution of equation (6.7.82) comprises mainly a solitary wave with slowly-varying parameters given by the expansion –

$$u \sim u_0(\xi, \tau) + \varepsilon u_1(\xi, \tau) + \cdots \qquad (6.7.83)$$

where

$$\xi = x - \int_0^t \omega(\tau)\, dt,$$

τ being the slow-time scale $\tau = \varepsilon t$. This ansatz is valid if the solitary-wave width is small compared with the scale length of the variations in the medium.

Substituting (6.7.83) into equation (6.7.82), we obtain to $O(1)$,

$$-\omega u_{0\xi} + 6u_0\, u_{0\xi} + u_{0\xi\xi\xi} = 0\,. \tag{6.7.84}$$

Equation (6.7.84) has the solitary-wave solution

$$u_0 = \frac{\omega}{2}\, \text{sech}^2\, \frac{\sqrt{\omega}}{2}\, \xi\,. \tag{6.7.85}$$

In order to determine $\omega_0(\tau)$, consider the energy-conservation law for equation (6.7.82) –

$$\frac{d}{dt} \int_{-\infty}^{\infty} \frac{1}{2}\, u^2 dx = \varepsilon \int_{-\infty}^{\infty} u^2 dx\,. \tag{6.7.86}$$

Substituting (6.7.85) into (6.7.86), we obtain

$$\omega' = \frac{4}{3}\, \omega \tag{6.7.87}$$

from which

$$\omega = \omega_0\, e^{\frac{4}{3}\tau} \tag{6.7.88}$$

where ω_0 is the initial value of ω.

The solitary-wave solution (6.7.85) does not satisfy the "mass" conservation law –

$$\frac{d}{dt} \int_{-\infty}^{\infty} u\, dx = \varepsilon \int_{-\infty}^{\infty} u\, dx \tag{6.7.89}$$

because, substituting (6.7.85), the left-hand side of equation (6.7.89) gives $\frac{4}{3}\, \varepsilon \sqrt{\omega}$ while the right-hand side gives $2\varepsilon \sqrt{\omega}$.

So, one needs to introduce a "tail" behind the solitary wave to remedy this "mass" defect. Further, it turns out that the expansion (6.7.83) is not uniformly valid as $x \to -\infty$, so one needs to introduce a new expansion to describe the "tail" region. In order to see this, first note that, substituting (6.7.83) into equation (6.7.82), we obtain to $O(\varepsilon)$,

$$-\omega u_{1\xi} + 6u_0 u_{1\xi} + 6u_{0\xi} u_1 + u_{1\xi\xi\xi} = u_0 - u_{0\tau}\,. \tag{6.7.90}$$

Let v be a solution of the adjoint equation associated with the left-hand side of equation (6.7.90), i.e.,

$$-\omega v_{\xi} + 6u_0 v_{\xi} + v_{\xi\xi\xi} = 0\,. \tag{6.7.91}$$

Multiplying equation (6.7.90) by v, equation (6.7.91) by u, and adding, we obtain

$$-\omega\left(vu_1\right)+\left(vu_{1\xi\xi}\right)_\xi -\left(v_\xi u_{1\xi}\right)_\xi +\left(v_{\xi\xi}u_1\right)_\xi =\left(u_0 -u_{0\tau}\right)v. \qquad (6.7.92)$$

Integrating (6.7.92) with respect to ξ, over $(-\infty,\infty)$, we obtain

$$\int_{-\infty}^{\infty}\left(u_0 -u_{0\tau}\right)vd\xi =\left[-\omega\left(vu_1\right)+vu_{1\xi\xi} -v_\xi u_{1\xi} +v_{\xi\xi}u_1\right]_{-\infty}^{\infty}. \qquad (6.7.93)$$

Now, the solitary wave in question will have no effect on the region ahead of it so that we have the boundary condition –

$$\xi \to \infty : u_1 \to 0. \qquad (6.7.94)$$

Further,

$$\xi \to -\infty : u_1 \text{ bounded}. \qquad (6.7.95)$$

Equation (6.7.93) has the following bounded solutions –

$$v = u_0 \qquad (6.7.96)$$

$$v = 1. \qquad (6.7.97)$$

Using (6.7.94) and (6.7.95), and making the choice (6.7.96), equation (6.7.93) gives

$$\int_{-\infty}^{\infty}\left(u_0 -u_{0\tau}\right)u_0\, d\xi = 0 \qquad (6.7.98)$$

which is simply the energy-conservation law (6.7.86)!

Next, making the choice (6.7.97), and using (6.7.94), equation (6.7.93) gives

$$u_1\left(-\infty\right) \approx \frac{1}{6\sqrt{\omega}}. \qquad (6.7.99)$$

Comparison of (6.7.99) with (6.7.85) shows that the expansion (6.7.83) is not uniformly valid as $\xi \to -\infty$, so one needs to introduce a new expansion to describe the "tail" region.

As we saw above, the nonuniformity in u, as $\xi \to -\infty$, is $O(\varepsilon)$ so that the region behind the solitary wave may be expected to be of $O(\varepsilon)$ in height and a slowly-varying function of x and t. Therefore, in the near-tail region, consider an expansion of the form –

$$u \sim \varepsilon U_1\left(X,\tau\right)+\varepsilon^2 U_2\left(X,\tau\right)+\cdots \qquad (6.7.100)$$

where X is the slow space scale $X = \varepsilon x$.

Substituting (6.7.100) into equation (6.7.82), we obtain to $O(\varepsilon)$,

$$U_{1\tau} = U_1 \tag{6.7.101}$$

from which

$$U_1 = f(X) \, e^\tau. \tag{6.7.102}$$

The function $f(X)$ is to be determined by matching the near-tail solution (6.7.100) with the solitary-wave solution (6.7.85). This requires knowledge of the solitary-wave position $X = X_s(\tau)$, which is obtained by considering the solitary-wave speed

$$\frac{dX_s}{d\tau} = \omega(\tau) \tag{6.7.103}$$

from which, on assuming

$$\tau = 0 : X_s = 0, \tag{6.7.104}$$

we obtain,

$$X_s = \frac{3\omega_0}{4} \left(e^{\frac{4}{3}\tau} - 1 \right). \tag{6.7.105}$$

Matching the near-tail solution (6.7.100) with the solitary-wave solution (6.7.83), and using (6.7.85), (6.7.99), and (6.7.102), we obtain

$$f(X) \, e^\tau = \frac{1}{6\sqrt{\omega}},$$

and using (6.7.88) and (6.7.105), this leads to

$$f(X) = \frac{2}{3\sqrt{\omega_0} \left(1 + \dfrac{4X}{3\omega_0} \right)^{5/4}}. \tag{6.7.106}$$

The near-tail solution (6.7.100) remedies the "mass" defect associated with the solitary-wave solution (6.7.85). In order to see this, consider the "mass" defect

$$\Delta m = \frac{d}{dt} \int_{-\infty}^{\infty} u\, dx - \varepsilon \int_{-\infty}^{\infty} u\, dx$$

$$= \varepsilon \frac{d}{d\tau} \int_{-\infty}^{\infty} u_0\, dx + \varepsilon \frac{d}{d\tau} \int_{-\infty}^{X_s} U_1\, dX$$

$$-\varepsilon \int_{-\infty}^{\infty} u_0\, dx - \dot{\varepsilon} \int_{-\infty}^{X_s} U_1\, dX + O(\varepsilon^2)$$

and using (6.7.85), (6.7.102), (6.7.105), and (6.7.106), this becomes

$$\Delta m = O(\varepsilon^2) \tag{6.7.107}$$

so that the near-tail plus the solitary-wave solution conserve "mass" to $O(\varepsilon)$.

Next, observe that the near-tail expansion does not remain bounded for large times and is therefore not valid in this regime, so, one needs to introduce another new expansion to describe the far-tail region trailing the near-tail region (see Figure 6.13).

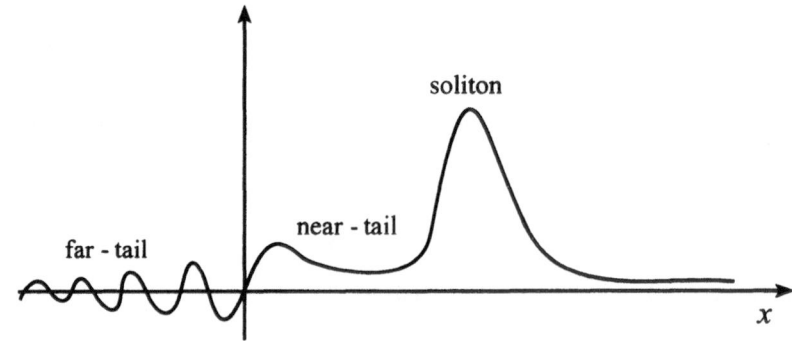

Figure 6.13. Schematic diagram of the perturbed

Korteweg-deVries solitary wave.

In the far-tail region, consider an expansion of the form

$$u \sim \varepsilon U_1(x,t,\tau) + \varepsilon^2 U_2(x,t,\tau) + \cdots. \tag{6.7.108}$$

Substituting (6.7.108) into equation (6.7.82), we obtain to $O(\varepsilon)$,

$$U_{1t} + U_{1xxx} = 0 \tag{6.7.109}$$

which has the similarity solution –

$$U_1 = B(\tau) \int\limits_{-\infty}^{x/(3t)^{1/3}} Ai(s)\, ds \qquad (6.7.110)$$

where $Ai(x)$ is the Airy function and $B(\tau)$ is a constant of integration to be determined by matching the far-tail solution (6.7.108) with the near-tail solution (6.7.100) at $x = 0$. Using (6.7.102) and (6.7.106), this leads to

$$B(\tau) = \frac{2}{3\sqrt{\omega_0}}\, e^{\tau}. \qquad (6.7.111)$$

Figure 6.13 gives a schematic diagram of the perturbed Kortweg-deVries solitary wave, which shows a decaying oscillatory far-tail.

6.7.3(f) Wave Modulation and Nonlinear Schrödinger Equation

Let us superimpose a slowly-varying weak modulation on a stationary weakly nonlinear wave, and study the evolution of such a modulation. To make the formulation simple, let us assume that, consequent to the superimposition of the modulation, the wave is still periodic, but with the amplitude and phase slowly-varying in x and t.

Consider an initially quiescent water subjected to a gravitational field g' (Figure 6.14). Here the prime denotes dimensional quantities. Let us suppose that at time $t' = 0$, a progressive wave is established such that the elevation of the free surface is raised to $y' = \eta'$ where

$$t' = 0 : \eta' = a'(\varepsilon x')\, e^{2\pi i x'/\lambda'} + c.c., \quad \varepsilon = \frac{2\pi a'}{\lambda'} \ll 1, \qquad (6.7.112)$$

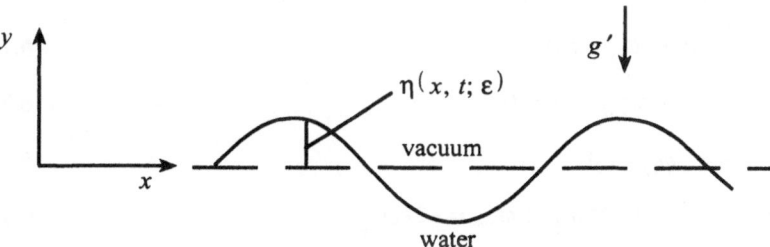

Figure 6.14. Deformed surface of water subjected to gravity.

Equation (6.7.112) represents a sinusoidal form with slowly-varying amplitude. If we nondimensionalize all the physical quantities using a reference length $\lambda'/2\pi$, and a time $(\lambda'/2\pi g')$, the wave motions for subsequent times at the disturbed free surface described by $y = \eta(x,t;\varepsilon)$ are governed by the following boundary-value problem:

$$y < \eta : \varphi_{xx} + \varphi_{yy} = 0 \tag{6.7.113}$$

$$y = \eta : \varphi_y = \eta_t + \varphi_x \eta_x \tag{6.7.114}$$

$$\varphi_t + \frac{1}{2}\left(\varphi_x^2 + \varphi_y^2\right) + \eta = 0 \tag{6.7.115}$$

$$y \to -\infty : \varphi_y \to 0, \tag{6.7.116}$$

where $\nabla\varphi$ denotes the perturbation in the velocity potential of the liquid.

Since the disturbance is assumed to be a progressive wave, let us introduce the following independent variables –

$$\xi = kx - \omega t, \quad \zeta = \varepsilon(x - Ct), \quad \tau = \varepsilon^2 t, \tag{6.7.117}$$

where C is the group velocity of the primary progressive wave, and look for solutions of the form

$$\left. \begin{aligned} \varphi(x,y,t;\varepsilon) &= \sum_{n=1}^{\infty} \varepsilon^n \varphi_n(\xi,y,\zeta,\tau) \\ \eta(x,t;\varepsilon) &= \sum_{n=1}^{\infty} \varepsilon^n \eta_n(\xi,\zeta,\tau) \end{aligned} \right\}. \tag{6.7.118}$$

This leads to a hierarchy of problems of various orders in ε:

$$O(\varepsilon): \quad y < 0 : k^2 \varphi_{1\xi\xi} + \varphi_{1yy} = 0 \tag{6.7.119}$$

$$y = 0 : \varphi_{1y} + \omega \eta_{1\xi} = 0 \tag{6.7.120}$$

$$-\omega \varphi_{1\xi} + \eta_1 = 0 \tag{6.7.121}$$

$$y \to -\infty : \varphi_{1y} \to 0 \tag{6.7.122}$$

$$O(\varepsilon^2): \quad y < 0 : k^2 \varphi_{2\xi\xi} + \varphi_{2yy} = -2k\varphi_{1\xi\zeta} \tag{6.7.123}$$

$$y = 0 : \varphi_{2y} + \omega \eta_{2\xi} = k^2 \varphi_{1\xi} \eta_{1\xi} - C\eta_{1\zeta} - \varphi_{1yy} \eta_1 \tag{6.7.124}$$

$$-\omega \varphi_{2\xi} + \eta_2 = -\frac{1}{2}\left(k^2 \varphi_{1\xi}^2 + \varphi_{1y}^2\right) + C\varphi_{1\zeta} + \omega \varphi_{1\xi y} \eta_1 \tag{6.7.125}$$

$$y \to -\infty : \varphi_{2y} \to 0 \tag{6.7.126}$$

$$O(\varepsilon^3) : \quad y < 0 : k^2 \varphi_{3\xi\xi} + \varphi_{3yy} = -2k\varphi_{2\xi\zeta} - \varphi_{1\zeta\zeta} \tag{6.7.127}$$

$$y = 0 : \quad \varphi_{3y} + \omega\eta_{3\xi} = k^2 \varphi_{2\xi}\eta_{1\xi} + k^2 \varphi_{1\xi}\eta_{2\xi} +$$

$$+ k^2 \varphi_{1\xi y}\eta_1\eta_{1\xi} + k\varphi_{1\xi}\eta_{1\zeta} + k\varphi_{1\zeta}\eta_{1\xi} - C\eta_{2\zeta} + \eta_{1\tau}$$

$$- \varphi_{2yy}\eta_1 - \frac{1}{2}\,\varphi_{1yyy}\eta_1^2 - \varphi_{1yy}\eta_2 \tag{6.7.128}$$

$$-\omega\varphi_{3\xi} + \eta_3 = -\left(k^2 \varphi_{1\xi}\varphi_{2\xi} + k^2 \varphi_{1\xi}\varphi_{1\xi y}\eta_1\right.$$

$$+ \varphi_{1y}\varphi_{2y} + \varphi_{1y}\varphi_{1yy}\eta_1 \Big) + C\varphi_{2\zeta} - k\varphi_{1\xi}\varphi_{1\zeta} - \varphi_{1\tau}$$

$$+ C\varphi_{1\zeta y}\eta_1 + \omega\varphi_{1\xi y}\eta_2 + \omega\varphi_{2\xi y}\eta_1 + \frac{1}{2}\,\omega\varphi_{1\xi yy}\eta_1^2,$$

$$\tag{6.7.129}$$

$$y \to -\infty : \varphi_{3y} \to 0. \tag{6.7.130}$$

From (6.7.119) and (6.7.122), we obtain the linear results

$$\left.\begin{aligned}
\eta_1 &= A_1(\zeta,\tau)\, e^{i\xi} + c.c. \\
\varphi_1 &= -\frac{i\omega}{k}\, A_1(\zeta,\tau)\, e^{i\xi + ky} + c.c.
\end{aligned}\right\} \tag{6.7.131}$$

and

$$\left.\begin{aligned}
\omega^2 &= k \\
C &\equiv \frac{d\omega}{dk} = \frac{1}{2\omega}
\end{aligned}\right\}. \tag{6.7.132}$$

Using (6.7.131) and (6.7.132), (6.7.123)-(6.7.126) give

$$\left.\begin{aligned}
\eta_2 &= A_1^2\, e^{2i\xi} + c.c. \\
\varphi_2 &= \frac{A_{1\zeta}}{k}\left(\frac{1}{2\omega} - \omega y\right) e^{i\xi + ky} + c.c.
\end{aligned}\right\}. \tag{6.7.133}$$

Using (6.7.131)-(6.7.133), we obtain from (6.7.127), (6.7.128), and (6.7.130):

$$\varphi_3 = \frac{i\omega}{2k}\, A_{1\zeta\zeta}\left(y^2 - \frac{y}{k}\right) e^{i\xi + ky}$$

$$+ \frac{1}{k}\left[A_{1\tau} + \frac{i}{2\omega k}\, A_{1\zeta\zeta} + i\omega k^2\left(\frac{5}{2} - 6M^2 k\right)|A_1|^2 A_1\right] e^{i\xi + ky}$$

$$+ \text{ higher harmonics } + c.c. \tag{6.7.134}$$

Using (6.7.131)-(6.7.134), elimination of the secular terms in (6.7.129) gives the nonlinear Schrödinger equation (Hasimoto and Ono, 1972) –

$$i A_{1\tau} - \frac{1}{8\omega^2} A_{1\zeta\zeta} - \frac{2k^2}{\omega} |A_1|^2 A_1 = 0. \tag{6.7.135}$$

Let us now construct a localized stationary solution of equation (6.7.135).

Let us first write equation (6.7.135) in the form

$$i \frac{\partial A_1}{\partial \tau} + \frac{1}{2} \frac{\partial^2 A_1}{\partial \zeta^2} + \kappa |A_1|^2 A_1 = 0. \tag{6.7.136}$$

Putting

$$A_1 = v(\zeta - U\tau) \, e^{i(\chi - s\tau)}, \tag{6.7.137}$$

equation (6.7.136) gives

$$\frac{1}{2} v'' + \frac{i}{2} (2\gamma - U) v' + \left(s - \frac{\gamma^2}{2}\right) v + \kappa |v|^2 v = 0. \tag{6.7.138}$$

Here primes denote differentiation with respect to the argument.

Letting

$$\gamma = \frac{U}{2}, \quad s = \frac{U^2}{2} - \frac{\alpha}{2}, \tag{6.7.139}$$

equation (6.7.138) becomes

$$v'' - \alpha v + 2\kappa v^3 = 0. \tag{6.7.140}$$

Upon integrating once, equation (6.7.139) gives

$$v'^2 = \alpha v^2 - \kappa v^4 \tag{6.7.141}$$

from which

$$\int \frac{dv}{v\sqrt{\alpha - \kappa v^2}} = (\xi - U\tau). \tag{6.7.142}$$

Putting

$$w = \sqrt{1 - \frac{\kappa v^2}{\alpha}}, \tag{6.7.143}$$

(6.7.142) leads to

$$\frac{1}{\sqrt{\alpha}} \int \frac{dw}{1 - w^2} = (\xi - U\tau), \tag{6.7.144}$$

from which

$$\frac{1}{2} \ln \frac{1-w}{1+w} = \sqrt{\alpha} \, (\xi - U\tau).$$ (6.7.145)

If $\kappa > 0$, $\alpha > 0$, one obtains, from (6.7.145),

$$v = \sqrt{\frac{\alpha}{\kappa}} \, \text{sech} \sqrt{\alpha} \, (\xi - U\tau)$$ (6.7.146)

which represents an envelope soliton (Figure 6.15) that propagates unchanged in shape and with constant velocity. The latter result arises because the nonlinearity and dispersion exactly balance each other – a result which turns out to be unique to one-dimensional solutions.

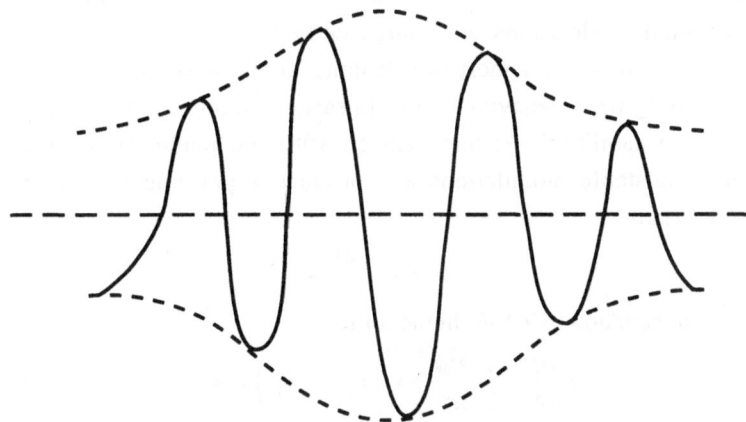

Figure 6.15. An envelope soliton.

6.7.3(g) Long-Time Evolution of the Modulation

It is found that the long-time evolution of the wave modulation exhibits a non-ergodic behavior. The numerical solution of the nonlinear Schrödinger equation (6.7.136) with periodic boundary conditions and with a modulationally unstable initial condition shows that (Lake et al., 1977) a state of maximum modulation is reached by the unstable wave system. After that, the solution demodulates and eventually returns to an unmodulated state. This process is repeated in time. Thus, the end state is neither random (no thermalization) nor steady, but consists of a time-periodic spreading and regrouping of wave energy initially confined to carrier wavenumber and its linearly unstable harmonics and sidebands (Figure 6.16). This is due to the fact that the actively participating modes in this long-time evolution are few and clearly identifiable (with those which are modulationally unstable,) and these linearly unstable modes nonlinearly evolve into a superperiodic state.

Figure 6.16. Recurring modulation and demodulation of the wave envelope (from Lake et al., 1977).

It turns out (see below) that a finite-amplitude uniform wavetrain is unstable to infinitesimal modulations with sufficiently long wavelengths, while it is stable to modulations with short wavelengths so that a threshold for instability exists. The long-time behavior of the linearly unstable modulation near this threshold for instability shows that (Janssen, 1981) the nonlinear effects stabilize the linearly unstable modulations and produce a periodic motion. For this purpose, let us put

$$A_1 = e^{-i\kappa|\phi_0|^2 t}\phi \tag{6.7.147}$$

and first write equation (6.7.136) in the form

$$i\frac{\partial\phi}{\partial t} + \frac{1}{2}\frac{\partial^2\phi}{\partial\xi^2} + \kappa\left(|\phi|^2 - |\phi_0|^2\right)\phi = 0, \tag{6.7.148}$$

where ϕ_0 corresponds to the basic wave.

In order to investigate the modulational instability of the wavetrain whose evolution is governed by equation (6.7.148), one puts

$$\phi = \rho^{1/2}\,e^{i\sigma} \tag{6.7.149}$$

so that equation (6.7.148) gives

$$\rho_t + \left(\rho\sigma_x\right)_x = 0 \tag{6.7.150}$$

$$\sigma_t + \frac{1}{2}\sigma_x^2 + \frac{1}{8\rho^2}\rho_t^2 - \frac{1}{4\rho}\rho_{xx} - \kappa(\rho-\rho_0) = 0. \tag{6.7.151}$$

In order to perform the linear stability analysis, one next puts

$$\begin{pmatrix}\rho \\ \sigma\end{pmatrix} = \begin{pmatrix}\rho_0 \\ 0\end{pmatrix} + \begin{pmatrix}\rho_1 \\ \sigma_1\end{pmatrix} e^{i(Kx-\Omega t)} \tag{6.7.152}$$

where K is the wavenumber and Ω is the frequency of the modulation. Assuming that $|\rho_1| << |\rho_0|$, and keeping only the terms linear in ρ_1 and σ_1, we obtain, from equations (6.7.150) and (6.7.151),

$$\Omega^2 = \frac{1}{4} K^2 \left(K^2 - 4\kappa\rho_0 \right). \tag{6.7.153}$$

Thus, if $\kappa > 0$, Ω^2 is negative for $|K| < \sqrt{4\kappa\rho_0}$.

We will now consider the nonlinear development of the initially linearly unstable modulation. For this purpose, we consider the initial-value problem for modulations with wavenumbers near the threshold for instability given by $K^2 = 4\kappa\rho_0$.

We look for a solution of the following form (Shivamoggi, 1990):

$$\left. \begin{aligned} \rho(x,t) &\sim \rho_0 + \varepsilon\rho_1(x,\tau) + \varepsilon^2\rho_2(x,\tau) + \cdots \\ \sigma(x,t) &\sim \varepsilon\sigma_1(x,\tau) + \varepsilon^2\sigma_2(x,\tau) + \cdots \\ K^2 &= 4\rho_0\kappa + \varepsilon^3\chi + \cdots \end{aligned} \right\} \tag{6.7.154}$$

where ε is a small parameter that characterizes the departure of K^2 from the linear stability threshold value $4\rho_0\kappa$, and $\tau = \varepsilon t$ is a slow time scale characterizing slow time evolutions near the stability threshold. We have introduced an explicit detuning parameter χ in (6.7.154).

Substituting (6.7.154) into equations (6.7.150) and (6.7.151), we obtain the following systems of equations to various orders in ε:

$$O\left(\varepsilon^n\right): L\begin{pmatrix} \rho_n \\ \sigma_n \end{pmatrix} = S_n\left(\rho_0,\rho_1,\cdots,\rho_{n-1}; \sigma_1,\cdots,\sigma_{n-1}\right), \quad n=1,2,\ldots, \tag{6.7.155}$$

where

$$L \equiv \begin{bmatrix} 0 & \rho_0 \dfrac{\partial^2}{\partial x^2} \\ \dfrac{1}{4}\dfrac{\partial^2}{\partial x^2} + \kappa\rho_0 & 0 \end{bmatrix}$$

and the function S_n depends on the solutions up to $O\left(\varepsilon^{n-1}\right)$.

We obtain, from equation (6.7.155), to $O(\varepsilon)$:

$$L\begin{pmatrix} \rho_1 \\ \sigma_1 \end{pmatrix} = 0, \tag{6.7.156}$$

the solution for which corresponds to the neutrally-stable case of the linear problem:

$$
\left.\begin{aligned}
\rho_1 &= A(\tau)\, e^{iKx} + c.c. \\
\sigma_1 &= \alpha(\tau) \\
K^2 &= 4\rho_0\kappa
\end{aligned}\right\}
\tag{6.7.157}
$$

where c.c. means complex conjugate.

Using (6.7.157), we obtain, from equation (6.7.155), to $O(\varepsilon^2)$:

$$
L\begin{pmatrix}\rho_2 \\ \sigma_2\end{pmatrix} = \left[\begin{aligned}
& -\frac{dA}{d\tau}\, e^{iKx} + c.c. \\
& -\kappa|A|^2 + \rho_0\frac{d\alpha}{d\tau} + \left\{-\frac{3}{2}\,\kappa A^2 e^{2iKx} + c.c.\right\}
\end{aligned}\right]
\tag{6.7.158}
$$

from which,

$$
\left.\begin{aligned}
\rho_2 &= \frac{1}{\kappa}\frac{d\alpha}{d\tau} - |A|^2 + \frac{A^2}{2\rho_0}\, e^{2iKx} + c.c. \\
\sigma_2 &= \frac{1}{4\rho_0^2\kappa}\frac{dA}{d\tau}\, e^{iKx} + c.c.
\end{aligned}\right\}.
\tag{6.7.159}
$$

Using (6.7.157) and (6.7.159), we obtain from equation (6.7.155), to $O(\varepsilon^3)$:

$$
L\begin{pmatrix}\rho_3 \\ \sigma_3\end{pmatrix} = \left[\begin{aligned}
& -\frac{1}{\kappa}\frac{d^2\alpha}{d\tau^2} + \frac{d|A|^2}{d\tau} \\
& \left\{\frac{1}{4\rho_0\kappa}\frac{d^2 A}{d\tau^2} + \frac{1}{4}\,\chi A + \frac{\kappa}{2\rho_0}|A|^2 A\right. \\
& \left. \quad - 2\kappa A^2\left(\frac{1}{\kappa}\frac{d\alpha}{d\tau} - |A|^2\right)\right\} e^{iKx} + c.c.
\end{aligned}\right] + N.S.T.
\tag{6.7.160}
$$

Removal of the secular terms in the first member of equation (6.7.160) requires

$$
\frac{1}{\kappa}\frac{d\alpha}{d\tau} - |A|^2 = \text{const.}
\tag{6.7.161}
$$

Let us take the constant above to be zero. Removal of the secular terms in the second member of equation (6.7.160), then, requires

$$
\frac{d^2 A}{d\tau^2} + \left(\rho_0\kappa\chi + 2\kappa^2|A|^2\right) A = 0.
\tag{6.7.162}
$$

If we impose the following initial conditions,

$$\tau = 0 : A = \hat{A}, \quad \frac{dA}{d\tau} = 0 \qquad (6.7.163)$$

and take A to be real, we obtain, from equation (6.7.162),

$$\left(\frac{dA}{d\tau}\right)^2 = \kappa^2 \left(\hat{A}^2 - A^2\right)\left(A^2 - \beta\right) \qquad (6.7.164)$$

where

$$\beta \equiv -\frac{\rho_0 \chi}{\kappa} - \hat{A}^2.$$

Equation (6.7.164) shows that A is bounded and oscillates between $\hat{A}$ and $\sqrt{\beta}$ if $\beta > 0$ and oscillates between 0 and $\hat{A}$ if $\beta < 0$. This demonstrates the nonlinear saturation of the linearly unstable modulation $(\chi < 0)$ near the linear-instability threshold.

6.7.3(h) Second-Harmonic Resonance

Suppose a typical surface disturbance is characterized by a sinusoidal travelling wave with amplitude a' and wavelength λ', then, let us nondimensionalize all physical quantities in the following with respect to a reference length $(\lambda'/2\pi)$, a time $(\lambda'/2\pi g')^{1/2}$ where g' denotes the acceleration due gravity, the primes here denote the dimensional quantities. The potential function of the motion of the water is taken to be $g'^{1/2} (\lambda'/2\pi)^{3/2} \phi$. If $y = \eta$ denotes the disturbed shape of the surface (whose mean level is given by $y = 0$), one has the following boundary-value problem:

$$y < \eta : \phi_{xx} + \phi_{yy} = 0 \qquad (6.7.165)$$

$$y = \eta : \phi_y = \eta_t + \eta_x \phi_x \qquad (6.7.166)$$

$$\phi_t + \frac{1}{2}\left(\phi_x^2 + \phi_y^2\right) + \eta = k^2 \eta_{xx}\left(1 + \eta_x^2\right)^{-3/2} \qquad (6.7.167)$$

$$y \to -\infty : \phi_y \to 0, \qquad (6.7.168)$$

where

$$k^2 \equiv \left(\frac{2\pi}{\lambda'}\right)^2 \frac{T'}{\rho' g'},$$

T' being the surface tension, ρ' being the density of water.

Let us look for a travelling wave, and introduce a new independent variable –

$$\xi = x - ct \tag{6.7.169}$$

so that the boundary-value problem (6.7.165)-(6.7.168) becomes

$$y < \eta : \phi_{\xi\xi} + \phi_{yy} = 0 \tag{6.7.170}$$

$$y = \eta : \phi_y = \left(\phi_\xi - c\right)\eta_\xi \tag{6.7.171}$$

$$\eta = c\phi_\xi - \frac{1}{2}\left(\phi_\xi^2 + \phi_y^2\right) + k^2\eta_{\xi\xi}\left(1 + \eta_\xi^2\right)^{-3/2} \tag{6.7.172}$$

$$y \to -\infty : \phi_y \to 0. \tag{6.7.173}$$

Seek solutions to (6.7.170)-(6.7.173) of the form –

$$\left. \begin{aligned} \phi(\xi, y; \varepsilon) &\sim \sum_{n=1}^{\infty} \varepsilon^n \phi_n(\xi, y) \\[2mm] \eta(\xi; \varepsilon) &\sim \sum_{n=1}^{\infty} \varepsilon^n \eta_n(\xi) \\[2mm] c(k; \varepsilon) &\sim \sum_{n=0}^{\infty} \varepsilon^n c_n(k) \\[2mm] k(\varepsilon) &= \sum_{n=0}^{\infty} \varepsilon^n k_n \end{aligned} \right\} \tag{6.7.174}$$

where

$$\varepsilon \equiv a' \cdot \frac{2\pi}{\lambda'} \ll 1.$$

Using (6.7.174), one obtains from the boundary-value problem (6.7.170)-(6.7.173), the following hierarchy of boundary-value problems:

$$0(\varepsilon) : y < 0 : \phi_{1\xi\xi} + \phi_{1yy} = 0 \tag{6.7.175}$$

$$y = 0 : \phi_{1y} = -c_0\eta_{1\xi} \tag{6.7.176}$$

$$\eta_1 = c_1\phi_{1\xi} + k_0^2\,\eta_{1\xi\xi} \tag{6.7.177}$$

$$y \to -\infty : \phi_{1y} \to 0 \tag{6.7.178}$$

$$0(\varepsilon^2) : y < 0 : \phi_{2\xi\xi} + \phi_{2yy} = 0 \tag{6.7.179}$$

$$y = 0 : \phi_{2y} + \phi_{1yy}\eta_1 = -c_0\eta_{2\xi} + \left(\phi_{1\xi} - c_1\right)\eta_{1\xi} \tag{6.7.180}$$

$$\eta_2 = c_0 \left(\phi_{2\xi} + \phi_{1y\xi} \eta_1 \right) - \frac{1}{2} \left(\phi_{1\xi}^2 + \phi_{1y}^2 \right) + c_1 \phi_{1\xi} + k_0^2 \, \eta_{2\xi\xi} + 2 k_0 k_1 \eta_{1\xi\xi}$$

(6.7.181)

$$y \to -\infty : \phi_{2y} \to 0$$

(6.7.182)

$$0 \left(\varepsilon^3 \right) : y < 0 : \phi_{3\xi\xi} + \phi_{3yy} = 0$$

(6.7.183)

$$y = 0 : \phi_{3y} + \phi_{2yy} \eta_1 + \phi_{1yy} \eta_2 + \frac{1}{2} \, \phi_{1yyy} \eta_1^2$$

$$= -c_0 \eta_{3\xi} + \left(\phi_{1\xi} - c_1 \right) \eta_{2\xi} + \left(\phi_{2\xi} + \phi_{1\xi y} \eta_1 - c_2 \right) \eta_{1\xi}$$

(6.7.184)

$$\eta_3 = c_0 \left(\phi_{3\xi} + \phi_{2\xi y} \eta_1 + \phi_{1\xi y} \eta_2 + \frac{1}{2} \, \phi_{1\xi yy} \eta_1^2 \right)$$

$$+ c_1 \left(\phi_{2\xi} + \phi_{1\xi y} \eta_1 \right) + c_2 \phi_{1\xi} - \phi_{1\xi} \left(\phi_{2\xi} + \phi_{1\xi y} \eta_1 \right) - \phi_{1y} \left(\phi_{2y} + \phi_{1yy} \eta_1 \right)$$

$$+ k_0^2 \, \eta_{3\xi\xi} + 2 k_0 k_1 \eta_{2\xi\xi} + 2 k_0 k_2 \eta_{1\xi\xi} - \frac{3}{2} \, k_0^2 \, \eta_{1\xi\xi} \eta_{1\xi}^2$$

(6.7.185)

$$y \to -\infty : \phi_{3y} \to 0 .$$

(6.7.186)

Let

$$\eta_1 = A \cos \xi .$$

(6.7.187)

Then, from (6.7.175)-(6.7.178), one obtains

$$\phi_1 \left(\xi, y \right) = A c_0 e^y \sin \xi$$

(6.7.188)

$$c_0^2 = 1 + k_0^2 .$$

(6.7.189)

Next, letting

$$\eta_2 = B \cos 2\xi$$

(6.7.190)

and using (6.7.187)-(6.7.189), one then obtains, from (6.7.179), (6.7.180), and (6.7.182),

$$\phi_2 \left(\xi, y \right) = c_0 \left(B - \frac{A^2}{2} \right) e^{2y} \sin 2\xi + c_1 A e^y \sin \xi .$$

(6.7.191)

Using (6.7.187)-(6.7.191), one finds that the removal of secular terms in (6.7.181) requires

$$k_1 = 0, \quad c_1 = 0$$

(6.7.192)

and then

$$B = \frac{c_0^2}{2\left(1 - 2k_0^2\right)} A^2 . \tag{6.7.193}$$

The case $k_0 = \pm\sqrt{1/2}$, where the above solution breaks down, corresponds to the second-harmonic resonance (Wilton, 1915 and McGoldrick, 1970), which we shall treat shortly.

Using (6.7.187)-(6.7.193), one obtains, from (6.7.183), (6.7.184), and (6.7.186),

$$\phi_3\left(\xi, y\right) = A\left[-c_0 A^2\left(\frac{c_0^2/4}{1 - 2k_0^2} + \frac{3}{8}\right) + c_2\right] e^y \sin\xi + \text{ higher harmonics.} \tag{6.7.194}$$

Using (6.7.187)-(6.7.194), one finds that the removal of secular terms in (6.7.185) requires

$$c_2 = \frac{c_0}{2}\left(\frac{c_0^2/2}{1 - 2k_0^2} + \frac{1}{2} - \frac{3}{8}\frac{k_0^2}{c_0^2}\right) A^2, \quad k_2 = 0 \tag{6.7.195}$$

which is again not valid for $k_0 = \pm\sqrt{1/2}$.

In order to treat the case of second-harmonic resonance corresponding to $k = \pm\sqrt{1/2}$, first note that for this case, the fundamental component

$$\left.\begin{array}{l} \eta_1^{(1)} = A\cos\xi \\[2mm] \phi_1^{(1)} = A c_0 e^y \sin\xi \end{array}\right\}$$

and its second harmonic

$$\left.\begin{array}{l} \eta_1^{(2)} = \hat{B}\cos 2\xi \\[2mm] \phi_1^{(2)} = \hat{B}c_0 e^{2y} \sin 2\xi \end{array}\right\}$$

have the same linear wave velocity c_0, so that the two can interact resonantly with each other. In order to treat this nonlinear resonant interaction, put

$$\eta_1 = A\cos\xi + \hat{B}\cos 2\xi \tag{6.7.196}$$

$$\phi_1 = c_0\left(Ae^y \sin\xi + \hat{B}e^{2y} \sin 2\xi\right). \tag{6.7.197}$$

Using (6.7.196) and (6.7.197), one obtains, from (6.7.179), (6.7.180), and (6.7.182),

$$\phi_2(\xi, y) = \left(-\frac{3A\hat{B}}{2} c_0 + Ac_1\right) e^y \sin \xi +$$

$$+\left(-A^2 c_0 + 2\hat{B}c_1\right) \frac{1}{2} e^{2y} \sin 2\xi + \text{ higher harmonics.} \qquad (6.7.198)$$

Using (6.7.196)-(6.7.198), one finds that the removal of secular terms in (6.7.181) requires

$$-c_0^2 \hat{B} + 2c_0 c_1 - 2k_0 k_1 = 0 \qquad (6.7.199)$$

$$-\frac{1}{2} c_0^2 A + 4c_0 \hat{B}c_1 - 8k_0 k_1 \hat{B} = 0, \qquad (6.7.200)$$

from which

$$\hat{B} = \frac{k_0 k_1}{c_0^2} \pm \sqrt{\frac{A^2}{4} - c_0^2 k_1^2} \qquad (6.7.201)$$

$$c_1 = \frac{k_1 \left(3k_0 + c_0^2/k_0^2\right)}{2c_0} \pm \frac{c_0}{2} \sqrt{\frac{A^2}{4} - c_0^2 k_1^2} \qquad (6.7.202)$$

which show that purely phase-modulated waves are possible for wavenumbers near the second-harmonic resonant values.

6.8. Applications to Plasma Physics

6.8.1 Nonlinear Longitudinal Waves in a Hot Electron-Plasma

Consider nonlinear longitudinal waves propagating in a hot electron plasma (Shivamoggi, 1982b). One has the usual governing equations:

(i) conservation of mass –

$$\frac{\partial n}{\partial t} + \frac{\partial}{\partial x}(nv) = 0, \qquad (6.8.1)$$

(ii) conservation of momentum –

$$\frac{\partial v}{\partial t} + v \frac{\partial v}{\partial x} + \frac{1}{m_e n} \frac{\partial p}{\partial x} = -\frac{e}{m_e} E, \qquad (6.8.2)$$

(iii) conservation of energy –

$$\frac{\partial p}{\partial t} + v \frac{\partial p}{\partial x} + 3p \frac{\partial v}{\partial x} = 0, \qquad (6.8.3)$$

(iv) Ampere's law –

$$\frac{\partial E}{\partial t} = 4\pi \, env,$$ (6.8.4)

where n is the electron number density, v the mass velocity, p the pressure, and E the electric field. In equation (6.8.3), note that the adiabatic exponent $\gamma = 3$, since we have considered a one-dimensional motion.

Seek solutions to equations (6.8.1)-(6.8.4), representing slowly varying wavetrains of the form

$$\left. \begin{aligned} n &\sim n_0 + \varepsilon n_1\left(\xi,\tau,\theta\right) + \varepsilon^2 n_2\left(\xi,\tau,\theta\right) + O\!\left(\varepsilon^3\right) \\ v &\sim \varepsilon v_1\left(\xi,\tau,\theta\right) + \varepsilon^2 v_2\left(\xi,\tau,\theta\right) + O\!\left(\varepsilon^3\right) \\ p &\sim p_0 + \varepsilon p_1\left(\xi,\tau,\theta\right) + \varepsilon^2 p_2\left(\xi,\tau,\theta\right) + O\!\left(\varepsilon^3\right) \\ E &\sim \varepsilon E_1\left(\xi,\tau,\theta\right) + \varepsilon^2 E_2\left(\xi,\tau,\theta\right) + O\!\left(\varepsilon^3\right), \end{aligned} \right\}$$ (6.8.5)

where

$$\xi = \varepsilon x, \quad \tau = \varepsilon t, \quad \varepsilon \ll 1,$$

$$\theta = k\left(\xi,\tau\right) x - \omega\left(\xi,\tau\right) t.$$

Using (6.8.5), one obtains from (6.8.1)-(6.8.4), to $O(\varepsilon)$,

$$\frac{\partial n_1}{\partial t} + n_0 \frac{\partial v_1}{\partial x} = 0$$ (6.8.6)

$$\frac{\partial v_1}{\partial t} + \frac{1}{m_e n_0} \frac{\partial p_1}{\partial x} + \frac{e}{m_e} E_1 = 0$$ (6.8.7)

$$\frac{\partial p_1}{\partial t} + 3 p_0 \frac{\partial v_1}{\partial x} = 0$$ (6.8.8)

$$\frac{\partial E_1}{\partial t} - 4\pi e n_0 v_1 = 0.$$ (6.8.9)

Upon solving equations (6.8.6)-(6.8.9), one obtains

$$\left. \begin{aligned} E_1 &= A\left(\xi,\tau\right) e^{i\theta} + \bar{A}\left(\xi,\tau\right) e^{-i\theta} \\ v_1 &= \frac{-i\omega}{4\pi e n_0}\left[A\left(\xi,\tau\right) e^{i\theta} - \bar{A}\left(\xi,\tau\right) e^{-i\theta}\right] \\ n_1 &= \frac{-ik}{4\pi e}\left[A\left(\xi,\tau\right) e^{i\theta} - \bar{A}\left(\xi,\tau\right) e^{-i\theta}\right] \\ p_1 &= -\frac{ik V_{Te}^2 m_e}{4\pi e}\left[A\left(\xi,\tau\right) e^{i\theta} - \bar{A}\left(\xi,\tau\right) e^{-i\theta}\right] \end{aligned} \right\}$$ (6.8.10)

where the dispersion relation is given by

$$\omega^2 = \omega_{pe}^2 + k^2 V_{Te}^2, \tag{6.8.11}$$

ω_{pe} is the plasma frequency, and V_{Te} is the electron thermal speed –

$$\omega_{pe}^2 \equiv 4\pi n_0 e^2/m_e, \quad V_{Te}^2 = 3p_0/m_e n_0.$$

Using (6.8.5), one obtains from (6.8.1)-(6.8.4), to $O(\varepsilon^2)$,

$$\frac{\partial n_2}{\partial t} + n_0 \frac{\partial v_2}{\partial x} = -\frac{\partial}{\partial x}(n_1 v_1) - \frac{\partial n_1}{\partial \tau} - n_0 \frac{\partial v_1}{\partial \xi} \tag{6.8.12}$$

$$\frac{\partial v_2}{\partial t} + \frac{1}{m_e n_0} \frac{\partial p_2}{\partial x} + \frac{e}{m} E_2 = -v_1 \frac{\partial v_1}{\partial x} + \frac{n_1}{m_e n_0^2} \frac{\partial p_1}{\partial x} - \frac{\partial v_1}{\partial \tau} - \frac{1}{m_e n_0} \frac{\partial p_1}{\partial \xi} \tag{6.8.13}$$

$$\frac{\partial p_2}{\partial t} + 3p_0 \frac{\partial v_2}{\partial x} = -v_1 \frac{\partial p_1}{\partial x} - 3p_1 \frac{\partial v_1}{\partial x} - \frac{\partial p_1}{\partial \tau} - 3p_0 \frac{\partial v_1}{\partial \xi} \tag{6.8.14}$$

$$\frac{\partial E_2}{\partial t} - 4\pi e n_0 v_2 = 4\pi e n_1 v_1 - \frac{\partial E_1}{\partial \tau}. \tag{6.8.15}$$

One deduces, from equations (6.8.12)-(6.8.15),

$$\left(\frac{\partial^2}{\partial t^2} - V_{Te}^2 \frac{\partial^2}{\partial x^2} + \omega_{pe}^2\right)\left(\frac{\partial E_2}{\partial t}\right) = \left(-\frac{\partial^2}{\partial t^2} + V_{Te}^2 \frac{\partial^2}{\partial x^2}\right)\left(\frac{\partial E_1}{\partial \tau} - 4\pi e n_1 v_1\right)$$

$$+ \frac{4\pi e}{m_e} \frac{\partial}{\partial x}\left(v_1 \frac{\partial p_1}{\partial x} + 3p_1 \frac{\partial v_1}{\partial x} + \frac{\partial p_1}{\partial \tau} + 3p_0 \frac{\partial v_1}{\partial \xi}\right)$$

$$- 4\pi n_0 e \frac{\partial}{\partial t}\left(v_1 \frac{\partial v_1}{\partial x} - \frac{n_1}{m_e n_0^2} \frac{\partial p_1}{\partial x}\right)$$

$$- 4\pi n_0 e \frac{\partial}{\partial t}\left(\frac{\partial v_1}{\partial \tau} + \frac{1}{m_e n_0} \frac{\partial p_1}{\partial \xi}\right). \tag{6.8.16}$$

Using (6.8.10) and (6.8.11), equation (6.8.16) becomes

$$\left(\frac{\partial^2}{\partial t^2} - V_{Te}^2 \frac{\partial^2}{\partial x^2} + \omega_{pe}^2\right)\left(\frac{\partial E_2}{\partial t}\right) = \left[2\omega^2 \frac{\partial A}{\partial \tau} + 2\omega k V_{Te}^2 \frac{\partial A}{\partial \xi} + \omega V_{Te}^2 A \frac{\partial k}{\partial \xi}\right] e^{i\theta}$$

$$+ \left[2\omega^2 \frac{\partial \overline{A}}{\partial \tau} + 2\omega k V_{Te}^2 \frac{\partial \overline{A}}{\partial \xi} + \omega V_{Te}^2 \overline{A} \frac{\partial k}{\partial \xi}\right] e^{-i\theta}$$

$$+ 4\pi e \left(\frac{\partial^2}{\partial t^2} - V_{Te}^2 \frac{\partial^2}{\partial x^2}\right)(n_1 v_1)$$

$$+ \frac{4\pi e}{m_e} \frac{\partial}{\partial x}\left(v_1 \frac{\partial p_1}{\partial x} + 3 p_1 \frac{\partial v_1}{\partial x}\right)$$

$$- 4\pi n_0 e \frac{\partial}{\partial t}\left(v_1 \frac{\partial v_1}{\partial x} - \frac{n_1}{m_e n_0^2} \frac{\partial p_1}{\partial x}\right). \qquad (6.8.17)$$

In order to render the assumed form of the solution (6.8.5) uniformly valid, one has to remove the secular terms in equation (6.8.17). This requires

$$\left.\begin{array}{l} 2\omega^2 \dfrac{\partial A}{\partial \tau} + 2\omega k V_{Te}^2 \dfrac{\partial A}{\partial \xi} + \omega V_{Te}^2 A \dfrac{\partial k}{\partial \xi} = 0 \\[3mm] 2\omega^2 \dfrac{\partial \overline{A}}{\partial \tau} + 2\omega k V_{Te}^2 \dfrac{\partial \overline{A}}{\partial \xi} + \omega V_{Te}^2 \overline{A} \dfrac{\partial k}{\partial \xi} = 0 \end{array}\right\}. \qquad (6.8.18)$$

Using (6.8.11), (6.8.18) becomes

$$\frac{\partial}{\partial \tau}\left[|A|^2\right] + \frac{\partial}{\partial \xi}\left[C(k)|A|^2\right] = 0, \qquad (6.8.19)$$

where $C(k)$ is the group velocity

$$C(k) = \frac{d\omega}{dk} = \frac{kV_{Te}^2}{\omega}. \qquad (6.8.20)$$

Equation (6.8.19) implies that a quantity proportional to $|A|^2$ (which may be energy in some sense) propagates with the group velocity $C(k)$, as we saw in Example 6.

6.8.2 Ion-Acoustic Solitary Waves in an Inhomogeneous Plasma

Nonlinear waves get modified in an essential way in an inhomogeneous plasma. For instance, the amplitude, width, and propagation velocity of an ion-acoustic solitary wave will change as the latter moves in an inhomogeneous plasma (Nishikawa and Kaw, 1975, Shivamoggi, 1981b, 1988b, Kuehl and Imen, 1985).

Theoretical treatments describing the properties of an ion-acoustic solitary wave in an inhomogeneous plasma have been based on the Korteweg-deVries equation, modified either by variable coefficients or by additional small terms. These theories give, for example, the spatial dependence of the slowly-varying soliton-amplitude, width, and speed in an inhomogeneous plasma. The soliton behavior is determined by carrying out a perturbation expansion based on the assumption that the soliton width is small compared with the scale length of the plasma inhomogeneity. Under this condition the soliton retains its identity, and its amplitude, width and speed are slowly-varying functions of position, hence, providing the *raison d'etre* for looking for a JWKB type solution for this problem.

Consider an ion-acoustic solitary wave propagating in an inhomogeneous plasma which is otherwise in a time-independent state. Assuming that the ions are cold and that the electrons follow a Boltzmann distribution, we have for the ions the following governing equations –

(i)   conservation of mass –

$$\frac{\partial n}{\partial t} + \frac{\partial}{\partial x}(nv) = v_1,$$ (6.8.21)

(ii) conservation of momentum –

$$\frac{\partial v}{\partial t} + v\frac{\partial v}{\partial x} = -\frac{\partial \phi}{\partial x} - v_1\frac{v}{n},$$ (6.8.22)

(iii) Gauss law –

$$\frac{\partial^2 \phi}{\partial x^2} = n_0 e^{\phi} - n_0,$$ (6.8.23)

where n is the number density of the ions, v is their velocity, ϕ is the electrostatic potential, and v_1 is the ionization rate per unit volume. n is normalized by a reference value of the unperturbed number density, v is normalized by the ion-acoustic speed C_s, ϕ is normalized by kT_e/e, and t^{-1} is normalized by the ion-plasma frequency ω_{p_i}. The subscript 0 refers to the undisturbed state.

Let us introduce two new independent variables –

$$\xi = \varepsilon^{1/2}\left[\int^x \frac{dx'}{\lambda_0(x')} - t\right], \quad \eta = \varepsilon^{3/2}x \tag{6.8.24}$$

where ε is a small parameter characterizing the typical amplitude of a wave, and seek solutions of the form –

$$\left.\begin{aligned} n &\sim n_0 + \varepsilon n_1 + \varepsilon^2 n_2 + \cdots \\ \phi &\sim \phi_0 + \varepsilon \phi_1 + \varepsilon^2 \phi_2 + \cdots \\ v &\sim v_0 + \varepsilon v_1 + \varepsilon^2 v_2 + \cdots \end{aligned}\right\} \tag{6.8.25a}$$

and take

$$v_1 = \varepsilon^{3/2} v. \tag{6.8.25b}$$

Since n_0 and λ_0 (to be determined below) are to be independent of t, we have

$$\frac{\partial n_0}{\partial \xi} = 0, \quad \frac{\partial \lambda_0}{\partial \xi} = 0. \tag{6.8.26}$$

Using (6.8.24)-(6.8.26), equations (6.8.21)-(6.8.23) give to $0(1)$:

$$\frac{\partial v_0}{\partial \xi} = 0, \quad \frac{\partial \phi_0}{\partial \xi} = 0. \tag{6.8.27}$$

Equations (6.8.26) and (6.8.27) imply that the unperturbed quantities depend only on the slow-space variable η. These are governed by –

$$\frac{\partial}{\partial \eta}(n_0 v_0) = v \tag{6.8.28}$$

$$\frac{\partial}{\partial \eta}\left(\frac{1}{2}v_0^2 + \phi_0\right) = -v\frac{v_0}{n_0} \tag{6.8.29}$$

$$n_0 e^{\phi_0} - n_0 = 0. \tag{6.8.30}$$

We have from equation (6.8.30) –

$$\phi_0 \equiv 0. \tag{6.8.31}$$

Using (6.8.31), equations (6.8.28) and (6.8.29) give –

$$n_0 \frac{\partial v_0}{\partial \eta} = -v \tag{6.8.32}$$

$$v_0 \frac{\partial n_0}{\partial \eta} = 2v. \tag{6.8.33}$$

We have from equations (6.8.32) and (6.8.33)

$$n_0 = \frac{1}{B}(\upsilon\eta + A)^2 \tag{6.8.34}$$

$$v_0 = \frac{B}{(\upsilon\eta + A)}. \tag{6.8.35}$$

A and B are constants to be determined using prescribed boundary conditions.

On using (6.8.25), to $O(\varepsilon)$, equations (6.8.21)-(6.8.23) give –

$$-\frac{\partial n_1}{\partial \xi} + \frac{1}{\lambda_0}\frac{\partial}{\partial \xi}(n_0 v_1 + n_1 v_0) = 0 \tag{6.8.36}$$

$$-\frac{\partial v_1}{\partial \xi} + \frac{v_0}{\lambda_0}\frac{\partial v_1}{\partial \xi} + \frac{1}{\lambda_0}\frac{\partial \phi_1}{\partial \xi} = 0 \tag{6.8.37}$$

$$-n_0\phi_1 + n_1 = 0 \tag{6.8.38}$$

from which, using equations (6.8.26) and (6.8.27), we may derive –

$$\frac{n_0}{\lambda_0}\left[1 - \lambda_0^2\left(1 - \frac{v_0}{\lambda_0}\right)^2\right]\frac{\partial v_1}{\partial \xi} = 0. \tag{6.8.39}$$

This leads to

$$1 - \lambda_0^2\left(1 - \frac{v_0}{\lambda_0}\right)^2 = 0$$

or

$$\lambda_0 = 1 + v_0. \tag{6.8.40}$$

Using (6.8.40), equation (6.8.36) then gives –

$$n_1 = n_0 v_1. \tag{6.8.41}$$

Next using (6.8.25), to $O(\varepsilon^2)$, equations (6.8.21)-(6.8.23) give –

$$-\frac{\partial n_2}{\partial \xi} + \frac{1}{\lambda_0}\frac{\partial}{\partial \xi}(n_0 v_2 + n_2 v_0 + n_1 v_1) + \frac{\partial}{\partial \eta}(n_0 v_1 + n_1 v_0) = 0 \tag{6.8.42}$$

$$-\frac{\partial v_2}{\partial \xi} + \frac{1}{\lambda_0}\left(v_0\frac{\partial v_2}{\partial \xi} + v_1\frac{\partial v_1}{\partial \xi}\right) + v_0\frac{\partial v_1}{\partial \eta} + v_1\frac{\partial v_0}{\partial \eta} + \frac{1}{\lambda_0}\frac{\partial \phi_2}{\partial \xi} + \frac{\partial \phi_1}{\partial \eta} = -\upsilon\phi_1\left(\frac{1 - v_0}{n_0}\right) \tag{6.8.43}$$

$$\frac{1}{\lambda_0^2}\frac{\partial^2 \phi_1}{\partial \xi^2} - n_0\phi_2 - \frac{1}{2}n_0\phi_1^2 + n_2 = 0. \tag{6.8.44}$$

Using (6.8.26)-(6.8.33), (6.8.38), (6.8.40), and (6.8.41), we may derive from equations (6.8.42)-(6.8.44):

$$\frac{\partial \phi_1}{\partial \eta} + \frac{1}{\lambda_0^2} \phi_1 \frac{\partial \phi_1}{\partial \xi} + \left(\frac{1}{2\lambda_0 n_0} \frac{\partial n_0}{\partial \eta} \right) \phi_1 + \frac{1}{2\lambda_0^4 n_0} \frac{\partial^3 \phi_1}{\partial \xi^3} = 0. \qquad (6.8.45)$$

Putting,

$$\phi_1 = g(\eta) \, \psi_1, \quad g(\eta) = e^{-\upsilon \int \frac{1}{2 n_0 \lambda_0} \left[\frac{2}{\upsilon_0} + (1 - \upsilon_0) \right] d\eta} \qquad (6.8.46)$$

equation (6.8.45) gives –

$$\frac{\partial \psi_1}{\partial \eta} + \frac{g}{(1 + \upsilon_0)^2} \, \psi_1 \frac{\partial \psi_1}{\partial \xi} + \frac{1}{2(1 + \upsilon_0)^4 n_0} \frac{\partial^3 \psi_1}{\partial \xi^3} = 0. \qquad (6.8.47)$$

In order to determine the properties of the solitary wave governed by equation (6.8.47), consider the equation –

$$\frac{\partial \psi_1}{\partial \eta} + \alpha(\eta) \, \psi_1 \frac{\partial \psi_1}{\partial \xi} + \beta(\eta) \frac{\partial^3 \psi_1}{\partial \xi^3} = 0. \qquad (6.8.48)$$

To lowest order in ε, equation (6.8.48) has the solution –

$$\psi_1 \approx a \, \text{sech}^2 \varphi \qquad (6.8.49)$$

where

$$\left. \begin{array}{c} \varphi = b\theta, \quad a = \dfrac{3\omega}{\alpha k}, \quad b = \sqrt{\dfrac{\omega}{4\beta k^3}} \\[4mm] k(\eta) = \dfrac{\partial \theta}{\partial \xi}, \quad \omega(\eta) = -\dfrac{\partial \theta}{\partial \eta} \end{array} \right\} . \qquad (6.8.50)$$

From

$$\frac{\partial k}{\partial \eta} = -\frac{\partial \omega}{\partial \xi} = 0, \qquad (6.8.51)$$

we have

$$k = \text{constant}. \qquad (6.8.52)$$

Equation (6.8.48) has an integral invariant to lowest order in ε given by –

$$\frac{\partial}{\partial \eta} \int_{-\infty}^{\infty} \psi_1^2 \, d\theta = 0. \qquad (6.8.53)$$

Using (6.8.49) and (6.8.50), equation (6.8.53) gives

$$\frac{\partial}{\partial \eta} \left(\frac{a^2}{b} \right) = 0$$

or

$$\omega = \omega_0 \left(\frac{\alpha}{\alpha_0} \right)^{\frac{4}{3}} \left(\frac{\beta}{\beta_0} \right)^{-\frac{1}{3}} \tag{6.8.54}$$

where the subscript 0 refers to some reference values. Using (6.8.54), (6.8.50) gives –

$$a \sim \left(\frac{\alpha}{\beta} \right)^{1/3}. \tag{6.8.55}$$

Returning to equation (6.8.47), the amplitude of the solitary wave is then given by –

$$\psi_{1_{max}} \sim \left(\frac{g/\lambda_0^2}{1/2 \, \lambda_0^4 \, n_0} \right)^{1/3}. \tag{6.8.56}$$

Thus, from (6.8.38) and (6.8.46), we have

$$n_{1_{max}} \sim n_0 \, g \left(2\lambda_0^2 \, n_0 \, g \right)^{1/3}$$

which on using (6.8.34), (6.8.35), (6.8.40), and (6.8.46), becomes

$$n_{1_{max}} \sim n_0 \left(\sqrt{n_0} + \sqrt{B} \right)^{2/3} e^{\left[-\frac{v}{3} \int \frac{2\sqrt{\frac{B}{n_0}} + \frac{B}{n_0} \left(1 - \sqrt{\frac{B}{n_0}} \right)}{2B \left(1 + \sqrt{\frac{B}{n_0}} \right)} d\eta \right]}$$

or

$$n_{1_{max}} \sim n_0 \left(\sqrt{n_0} + \sqrt{B} \right)^{2/3}. \tag{6.8.57}$$

On the other hand, the speed c of the solitary wave is given approximately from (6.8.40) to be

$$c \approx \lambda_0 = 1 + \sqrt{\frac{B}{n_0}}. \tag{6.8.58}$$

Equations (6.8.57) and (6.8.58) show that the amplitude of the solitary wave increases and the speed decreases as the solitary wave propagates into regions of increasing density.

6.9. Exercises

1. Solve by using the method of multiple scales,

$$y'' + 2\varepsilon y' + y = 0.$$

2. Solve by using the method of multiple scales,

$$y'' + \varepsilon\, xy' + y = 0, \quad x > 0,$$
$$y(0) = 0, \quad y'(0) = 1.$$

3. Solve using the method of multiple scales,

$$y'' + \omega_0^2 y = \varepsilon y^3 + \mu \cos \lambda t$$

and consider the cases $\omega_0 \neq \lambda$ and $\omega_0 \approx \lambda$ separately.

4. Solve the following initial-value problem using the method of multiple scales –

$$u_{tt} - u_{xx} + u = u^3,$$
$$t = 0 : u = \varepsilon \cos kx, \quad u_t = \varepsilon \sin kx.$$

5. Use Struble's method to solve Mathieu's equation

$$\ddot{u} + \left(\omega^2 - \varepsilon \cos t\right) u = 0$$

for cases $\omega \neq \dfrac{1}{2}$, $\omega \approx \dfrac{1}{2}$, $\omega \approx 1$ separately.

6. Solve by using the generalized multiple-scale method,

$$\varepsilon y'' + y' = a,$$
$$y(0) = 0, \quad y(1) = 1.$$

7. Solve by using the generalized multiple-scale method,

$$\varepsilon y'' + (1 + x)\, y' + x^2 y = 0,$$
$$y(0) = 0, \quad y(1) = 1.$$

8. Solve by using the generalized multiple-scale method,

$$\varepsilon y'' + xy' + x^2 y = 0,$$
$$y(0) = 0, \quad y(1) = 1.$$

9. Solve using the generalized method of multiple scales –

$$\varepsilon y'' - (2x + 1)\, y' + 2y = 0,$$
$$y(0) = \alpha, \quad y(1) = \beta.$$

10. Solve using the generalized method of multiple scales –

$$\varepsilon y'' + (1 + \alpha x) \, y' + \alpha y = 0, \quad \alpha > -1,$$
$$y(0) = 0, \quad y(1) = 1.$$

11. Investigate the evolution of the solitary-wave solution of the perturbed regularized long-wave equation (Shivamoggi and Rollins, 2002) –

$$u_t + u_x - u_{xxt} + uu_x = \varepsilon u \,.$$

Introduce a "tail", of which the near-tail portion remedies the "mass" defect and the far-tail portion exhibits a plateau structure.

Chapter 7

Miscellaneous Perturbation Methods

In addition to the standard perturbation methods discussed in Chapters 2-6, a variety of other perturbation methods have been developed for several applications. We will consider here only a couple of these methods since they have proved to be very useful in some applications.

7.1. A Quantum-Field-Theoretic Perturbative Procedure

Bender et al. (1989) gave a quantum-field-theoretic (QFT) perturbation procedure to solve nonlinear differential equations. In this procedure, one identifies a parameter δ as a measure of the nonlinearity in the differential equation in question. So, when $\delta = 0$, the equation becomes linear and can be solved analytically. As δ increases smoothly from zero, the nonlinear development gradually turns on. One assumes that the effect of changing δ from $\delta = 0$ to $\delta \neq 0$ is not associated with any sudden nonanalytic effects. The parameter δ is now treated as a small parameter and the solution is expanded as a formal perturbation series in powers of δ, even though the original equation actually corresponds to $\delta = 1$.

The validity of this procedure has been demonstrated by applying it to a number of differential equations of mathematical physics (Bender et al., 1989; Shivamoggi and Rollins, 1997 and 1999). The rapidity of convergence of this procedure has been found to be quite good because the δ-perturbation series has a finite (nonzero) radius of convergence.

7.1.1 Blasius Equation

Blasius equation (see equation (5.1.30)) is a third-order nonlinear differential equation that describes the velocity profile of the fluid in the boundary layer formed by a fluid flowing along a flat plate –

$$f''' + f'' f = O \tag{7.1.1}$$

along with the boundary conditions –

$$\eta = 0 : f = O, \ f' = O \tag{7.1.2a}$$

$$\eta \to \infty : f \approx \eta. \tag{7.1.2b}$$

Here, the primes denote differentiation with respect to η.

Equation (7.1.1) is now replaced by an equation that contains a parameter δ

$$f''' + f'' f^{\delta} = O. \tag{7.1.3}$$

Note that equation (7.1.1) is recovered when $\delta = 1$, and $\delta = 0$ corresponds to the linear zero-order approximation. By identifying δ as the perturbation parameter, the solution f is then expanded in a power series in δ –

$$f \sim f_0 + \delta f_1 + \delta^2 f_2 + \cdots. \tag{7.1.4}$$

This then leads to a set of linear equations for f_n:

$$O(1) : f_0''' + f_0'' = O \tag{7.1.5}$$

$$O(\delta) : f_1''' + f_1'' = -f_0'' \cdot \ln f_0 \tag{7.1.6}$$

etc.

Successive integrations of equation (7.1.5) along with use of the boundary conditions (7.1.2) lead to

$$f_0(\eta) = \eta - 1 + e^{-\eta}. \tag{7.1.7}$$

Equation (7.1.7) is compared with the exact numerical solution of equation (7.1.1) in Figure 7.1. The agreement is seen to be very good.

Next, we obtain from equation (7.1.6),

$$f_1''(0) = \int_0^{\infty} d\eta \cdot e^{-\eta} \cdot \ln(\eta - 1 + e^{-\eta}). \tag{7.1.8}$$

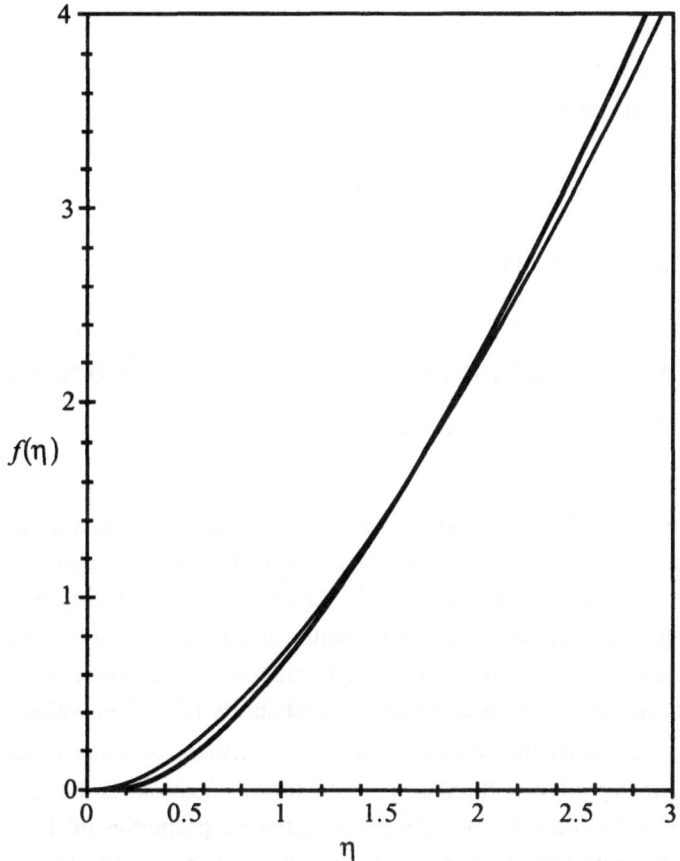

**Figure 7.1. Comparison of zeroth order approximate solution and
numerical solution (bold) for the Blasius equation (from Shivamoggi and
Rollins, 1999).**

Numerical integration (Bender et al., 1989) gives

$$f_1'''(0) = -2 \cdot 1332745. \tag{7.1.9}$$

Using (7.1.7) and (7.1.9), we have from (7.1.4) –

$$f''(0) \sim 1 - 2 \cdot 1332745 \ \delta + O(\delta^2). \tag{7.1.10}$$

In order to compare (7.1.10) with the exact numerical result (Schlichting,
1972) –

$$f''(0) = 0 \cdot 46960 \cdots \qquad (7.1.11)$$

following Bender et al (1989), one converts (7.1.10) to a (0,1) Padé approximation and evaluates the latter at $\delta = 1 -$

$$\left. \frac{1}{1 + 2 \cdot 1332745 \delta} \right|_{\delta = 1} = 0 \cdot 31915 \qquad (7.1.12)$$

which differs from (7.1.11) by 32%.

7.2. A Perturbation Method for Linear Stochastic Differential Equations

Wave propagation in a random medium[1] is usually described by a stochastic differential equation with the characteristics of the medium represented by the stochastic coefficients (Bourret, 1962, Keller, 1964, van Kampen, 1976, and Shivamoggi et al., 2000). A random medium is a family of media, each labeled by one value of a parameter α which ranges over a space A in which a probability density $p(\alpha)$ determines the probability of a given value of α, and therefore, represents the source of the waves which, in some cases, may be random. The objective of the theory of stochastic equations is the determination of the probability distribution of various statistical properties of the solution u, such as its expectation or mean value, its variance, and its higher-order moments.

Consider a linear operator $M = M(\alpha)$ in some space that depends on a parameter α, which ranges over a set A in which a probability density $p(\alpha)$ determines the probability of the given parameter α. In this case $M(\alpha)$ is a

---

[1] Propagation of waves in a random medium includes a number of applications, such as propagation of starlight through the turbulent atmosphere, propagation of radio waves through the ionosphere, and sound wave propagation in the ocean (Tatarskii, 1967, Chernov, 1969 and Andrews and Phillips, 1998). Although variations in the refractive index from its mean value in a turbulent medium are very small (on the order of 10^{-3}), the wave typically propagates through a large number of refractive-index inhomogencities and hence, the cumulative effect can be very significant.

stochastic operator. If g is a given element of the space and u is an unknown element, we may consider the linear stochastic equation for u given by

$$M(\alpha)\, u = g, \tag{7.2.1}$$

which is a family of equations depending on the parameter α. We assume equation (7.2.1) has a unique solution $u(\alpha)$ for each α. Moreover, since the probability density $p(\alpha)$ is known, it determines the probability density of the solution $u(\alpha)$ of the stochastic equation. We seek to find an equation for the mean value $\langle u \rangle$, defined by

$$\langle u \rangle = \int u(\alpha)\, p(\alpha)\, d\alpha. \tag{7.2.2}$$

Let us assume that $M(\alpha)$ depends on a small parameter ε, and that for $\varepsilon = 0$, $M(\alpha)$ reduces to a sure linear operator L_0. Upon expanding $M(\alpha; \varepsilon)$ in powers of ε, we may rewrite equation (7.2.1) in the form

$$\left[L_0 + \varepsilon L_1(\alpha) \right] u(\alpha; \varepsilon) = g. \tag{7.2.3}$$

The stochastic linear operator $L_1(\alpha)$ represents a stochastic perturbation of the sure linear operator L_0. We shall denote by u_0 a particular solution of the sure equation obtained by setting $\varepsilon = 0$ in equation (7.2.3), leading to

$$L_0 u_0 = g. \tag{7.2.4}$$

Now, assuming that the inverse operator L_0^{-1} is defined, we can rewrite equation (7.2.3) in the form

$$u = u_0 - L_0^{-1} \left[\varepsilon L_1 u \right]. \tag{7.2.5}$$

We may solve (7.2.5) for u iteratively as follows:

$$u \sim u_0 - \varepsilon L_0^{-1} \left[L_1 u_0 \right] + \varepsilon^2 L_0^{-1} \left\{ \left(L_1 L_0^{-1} L_1 \right) u_0 \right\} + \cdots. \tag{7.2.6}$$

On taking the expectation value, equation (7.2.6) gives

$$\langle u \rangle \sim u_0 - \varepsilon L_0^{-1} \left[\langle L_1 \rangle u_0 \right] + \varepsilon^2 L_0^{-1} \left\{ \langle L_1 L_0^{-1} L_1 \rangle u_0 \right\} + \cdots, \tag{7.2.7}$$

from which we obtain

$$\langle u \rangle \sim u_0 - \varepsilon L_0^{-1} \left[\langle L_1 \rangle \langle u \rangle \right] + \varepsilon^2 L_0^{-1} \left[\left(\langle L_1 L_0^{-1} L_1 \rangle - \langle L_1 \rangle L_0^{-1} \langle L_1 \rangle \right) \langle u \rangle \right] + \cdots. \tag{7.2.8}$$

Upon operating on (7.2.8) by L_0, and using equation (7.2.4), we find that

$$\left\{ L_0 + \varepsilon \langle L_1 \rangle + \varepsilon^2 \left[\langle L_1 \rangle L_0^{-1} \langle L_1 \rangle - \langle L_1 L_0^{-1} L_1 \rangle \right] \right\} \langle u \rangle + O(\varepsilon^3) = g. \tag{7.2.9}$$

Let $G(t,t')$ be Green's function associated with the unperturbed operator L_0 which is defined by

$$L_0^{-1} f(t) = \int G(t,t') f(t') dt'. \tag{7.2.10}$$

Hence, $G(t,t')$ satisfies

$$\left. \begin{array}{l} L_0 G(t,t') = \delta(t-t') \\ G(t,t') = 0, \quad \text{for} \quad t < t' \end{array} \right\}. \tag{7.2.11}$$

That solution of (7.2.11) which corresponds to outgoing waves at large distances should be chosen.

In order to give an explicit form of equation (7.2.9), let us write L_0 and L_1 as

$$L_0 = \frac{\partial}{\partial t} - A_0, \quad L_1(\alpha) = -A_1(t;\alpha). \tag{7.2.12}$$

Let us take A_0 to be independent of t and $\langle A_1 \rangle = 0$. The Green's function $G(t,t')$ associated with the unperturbed operator L_0 is then given by

$$G(t,t') = \begin{cases} e^{A_0(t-t')}, & t > t', \\ 0, & t < t'. \end{cases} \tag{7.2.13}$$

Using (7.2.10)-(7.2.13), equation (7.2.9) becomes

$$\left(\frac{\partial}{\partial t} - A_0 \right) \langle u(t) \rangle = \varepsilon^2 \int_0^t \langle A_1(t) e^{A_0(t-t')} A_1(t') \rangle \langle u(t') \rangle \, dt' + g, \tag{7.2.14}$$

which is valid if $\varepsilon t \ll 1$.

Noting that

$$\langle u(t) \rangle \sim e^{A_0(t-t')} \langle u(t') \rangle + O(\varepsilon^2), \tag{7.2.15}$$

equation (7.2.14) becomes

$$\left(\frac{\partial}{\partial t} - A_0 \right) \langle u(t) \rangle = \left[\varepsilon^2 \int_0^t \langle A_1(t) e^{A_0 \tau} A_1(t-\tau) \rangle e^{-A_0 \tau} \, d\tau \right] \langle u(t) \rangle + g. \tag{7.2.16}$$

If τ_c is the autocorrelation time of $A_1(t)$, i.e.,

$$\langle A_1(t) e^{A_0 \tau} A_1(t-\tau) \rangle = 0, \quad \text{if} \quad \tau > \tau_c, \tag{7.2.17}$$

and if $t > \tau_c$, we may write equation (7.2.16) as

$$\left[\frac{\partial}{\partial t} - A_0\right]\langle u(t)\rangle = \left[\varepsilon^2 \int_0^\infty \langle A_1(t) e^{A_0 \tau} A_1(t-\tau)\rangle e^{-A_0 \tau} d\tau\right]\langle u(t)\rangle + g, \quad (7.2.18)$$

which is valid if $\varepsilon\tau_c \ll 1$. Combining this restriction with the previous one, viz., $\varepsilon t \ll 1$, we have for the region of validity of equation (7.2.18), the interval

$$\tau_c \ll t \ll \varepsilon^{-1}. \quad (7.2.19)$$

7.2.1 Application to Wave Propagation in a Random Medium

Consider a linear wave propagating in a turbulent atmosphere, which has small random variations in the refractive index. The electric field of such a monochromatic wave satisfies

$$\frac{d^2 E}{dz^2} + k_0^2 \left[1 + \varepsilon\mu(z)\right] E = 0 \quad (7.2.20)$$

where k_0 is the wavenumber of the wave and the refractive index of the medium n is given by

$$n^2 = 1 + \varepsilon\mu(z). \quad (7.2.21)$$

$\mu(z)$ describes the small random variations.

Let us introduce the quantities

$$\xi = k_0 z, \quad F = \frac{dE}{d\xi}, \quad (7.2.22)$$

so that equation (7.2.20) can be written as the system of first-order equations

$$\frac{d}{d\xi}\begin{pmatrix} E \\ F \end{pmatrix} = \left[\begin{pmatrix} 0 & 1 \\ -1 & 0 \end{pmatrix} + \varepsilon\begin{pmatrix} 0 & 0 \\ -\mu(\xi/k_0) & 0 \end{pmatrix}\right]\begin{pmatrix} E \\ F \end{pmatrix}. \quad (7.2.23)$$

Identifying equation (7.2.23) with equations (7.2.3) and (7.2.12), and applying equation (7.2.18), we obtain

$$\frac{d}{d\xi}\begin{pmatrix} \langle E\rangle \\ \langle F\rangle \end{pmatrix} = \left[\begin{pmatrix} 0 & 1 \\ -1 & 0 \end{pmatrix} + \varepsilon^2 \int_0^\infty \langle\mu(\xi/k_0)\,\mu[(\xi-\eta)/k_0]\rangle\begin{pmatrix} 0 & 0 \\ -1 & 0 \end{pmatrix}\exp\left\{\begin{pmatrix} 0 & 1 \\ -1 & 0 \end{pmatrix}\eta\right\}\right.$$

$$\left. \times\begin{pmatrix} 0 & 0 \\ -1 & 0 \end{pmatrix}\exp\left\{\begin{pmatrix} 0 & -1 \\ 1 & 0 \end{pmatrix}\eta\right\} d\eta\right]\begin{pmatrix} \langle E\rangle \\ \langle F\rangle \end{pmatrix}. \quad (7.2.24)$$

Noting that

$$\exp\left[\begin{pmatrix} 0 & 1 \\ -1 & 0 \end{pmatrix}\eta\right] = \begin{pmatrix} \cos\eta & \sin\eta \\ -\sin\eta & \cos\eta \end{pmatrix},$$

equation (7.2.24) can be expressed as

$$\frac{d}{d\xi}\begin{pmatrix} \langle E \rangle \\ \langle F \rangle \end{pmatrix} = \left[\begin{pmatrix} 0 & 1 \\ -1 & 0 \end{pmatrix} + \frac{\varepsilon^2}{2}\begin{pmatrix} 0 & 0 \\ c_1 & -c_2 \end{pmatrix}\right]\begin{pmatrix} \langle E \rangle \\ \langle F \rangle \end{pmatrix}, \qquad (7.2.25)$$

where

$$\left.\begin{aligned} c_1 &\equiv \int_0^\infty \left\langle \mu(\xi/k_0)\, \mu\left[(\xi-\eta)/k_0\right]\right\rangle \sin 2\eta \; d\eta \\ c_2 &\equiv \int_0^\infty \left\langle \mu(\xi/k_0)\, \mu\left[(\xi-\eta)/k_0\right]\right\rangle (1-\cos 2\eta) \; d\eta \end{aligned}\right\}. \qquad (7.2.26)$$

If the wave propagation in a random medium is assumed to be stationary, Gaussian and a Markov process, then by Doob's Theorem (Doob, 1942) it is an Uhlenbeck-Ornstein process, so that

$$\left\langle \mu(\xi/k_0)\, \mu\left[(\xi-\eta)/k_0\right]\right\rangle = e^{-\eta/k_0}, \qquad (7.2.27)$$

then (7.2.26) gives

$$c_1 = \frac{2k_0^2}{1+4k_0^2}, \quad c_2 = \frac{4k_0^2}{1+4k_0^2}. \qquad (7.2.28)$$

If, on the other hand, we take the commonly used Gaussian form

$$\left\langle \mu(\xi/k_0)\, \mu\left[(\xi-\eta)/k_0\right]\right\rangle = e^{-\eta^2/2k_0^2}, \qquad (7.2.29)$$

then (7.2.26) gives

$$c_1 = 0, \quad c_2 = \sqrt{\frac{\pi}{2}}\, k_0\left(1 - \sqrt{2}\, k_0 e^{-2k_0^2}\right). \qquad (7.2.30)$$

From equation (7.2.25), it follows that the mean electric field $\langle E \rangle$ is governed by

$$\frac{d^2}{dx^2}\langle E \rangle + \frac{\varepsilon^2}{2}\, k_0 c_2 \frac{d}{dx}\langle E \rangle + k_0^2\left(1 - \frac{\varepsilon^2}{2}\, c_1\right)\langle E \rangle = 0. \qquad (7.2.31)$$

If the mean electric field $\langle E \rangle$ is assumed to vary in space, according to

$$\langle E \rangle = \frac{1}{2}\left(e^{ikz}\mathscr{E} + e^{-ikz}\mathscr{E}*\right), \qquad (7.2.32)$$

equation (7.2.31) gives

$$k^2 = k_0^2 \left(1 - \frac{\varepsilon^2}{2} c_1\right) + i \frac{\varepsilon^2 k_0^2}{2} c_2 + O(\varepsilon^3).$$ (7.2.33)

Taking the square root of (7.2.33) and substituting into (7.2.32), we obtain

$$\langle E \rangle = \frac{1}{2} \mathscr{E} \exp\left\{i k_0 \left[1 - \frac{\varepsilon^2}{4} c_1\right] z - \frac{\varepsilon^2}{4} c_2 k_0 z\right\} + \text{c.c.},$$ (7.2.34)

where c.c. denotes the complex conjugate of the preceding term.

Equation (7.2.34) shows the attenuation of the coherent wave due to random inhomogeneities in the medium. (This damping is due to phase interference among the individual solutions for different α, like the Landau damping of plasma oscillations (Landau, 1946), and has nothing to do with actual dissipation; thus, in this process, the medium does not gain energy.)

7.2.2 Renormalization Procedure

Application of perturbative procedures to random differential equations can sometimes run into qualitative problems. In order to see this, consider the following problem (Kraichnan, 1961 and Leslie, 1973):

$$\left.\begin{array}{l} \left[\dfrac{d}{dt} + \mu + ib(t)\right] u(t) = \delta(t) \\[2mm] t = 0 : u = 0 \end{array}\right\}$$ (7.2.35)

where μ is a determinate constant, and $b(t)$ is a real, centered stationary Gaussian random function of t having the auto-correlation function

$$\langle b(t') \, b(t'') \rangle = \sigma^2 \, e^{-\lambda(t'-t'')}$$ (7.2.36)

befitting an Uhlenbeck-Ornstein process, but otherwise unspecified.

7.2.2(a) Exact Solution

The initial-value problem (7.2.35) has the exact solution

$$u(t) = e^{-\mu t - i \int_0^t b(t')dt'}$$ (7.2.37)

Noting that $\int_0^t b(t')\,dt'$, being a linear functional of the centered Gaussian

random function $b(t)$, is a centered Gaussian random variable, we have

$$\langle u(t)\rangle = e^{-\mu t - \frac{1}{2}\int_0^t\int_0^t \langle b(t')\,b(t'')\rangle\,dt'\,dt''}.^2 \tag{7.2.38}$$

Using (7.2.36), (7.2.38) becomes

$$\langle u(t)\rangle = e^{-\mu t - \frac{\sigma^2}{\lambda^2}\left(\lambda t - 1 + e^{-\lambda t}\right)}.^3 \tag{7.2.39}$$

For long-range correlations (i.e., small λ) when the fluctuations are almost time-independent, (7.2.39) yields

$$\langle u(t)\rangle \approx e^{-\mu t - \frac{1}{2}\sigma^2 t^2}. \tag{7.2.40}$$

On the other hand, for short-range correlations (i.e., large λ) when the fluctuations differ very little from pure white noise, (7.2.40) yields

$$\langle u(t)\rangle \approx e^{-\mu t - \frac{\sigma^2}{\lambda} t}. \tag{7.2.41}$$

In this limit, the effect of randomness on the mean amplitude is only to cause a frequency shift.

7.2.2(b) Perturbative Solution

If the fluctuations $b(t)$ are taken to be very small, one may apply the perturbative procedure, namely, the result (7.2.9) to the initial-value problem (7.2.35), which leads to (Shivamoggi et al., 1999)

---

[2] This is based on the result (see van Kampen, 1992)

$$\langle e^x\rangle = e^{\left[\langle x\rangle + \frac{1}{2!}\langle\langle x^2\rangle\rangle + \frac{1}{3!}\langle\langle x^3\rangle\rangle + \cdots\right]}$$

where,

$$\langle\langle x^n\rangle\rangle \equiv \langle(x - \langle x\rangle)^n\rangle.$$

[3] In carrying out this integration, use has been made of the result

$$\int_0^T\int_0^T \rho(t' - t)\,dt\,dt' = 2\int_0^T\left(1 - \frac{\tau}{T}\right)\rho(\tau)\,d\tau.$$

$$\left.\begin{array}{l}\left(\dfrac{d}{dt}+\mu\right)\langle u(t)\rangle+\displaystyle\int_0^t\langle b(t)b(t')\rangle\,e^{-\mu(t-t')}\langle u(t')\rangle\,dt'=\delta(t)\\[4mm]\langle u(0)\rangle=1\end{array}\right\}. \qquad (7.2.42)$$

Putting

$$\langle u(t)\rangle=e^{-\mu t}\,v(t) \qquad (7.2.43)$$

and using (7.2.36), the initial-value problem (7.2.42) becomes

$$\left.\begin{array}{l}\dfrac{dv}{dt}+\sigma^2\displaystyle\int_0^t e^{-\lambda(t-t')}\,v(t')\,dt'=\delta(t)\\[4mm]v(0)=0\end{array}\right\}. \qquad (7.2.44)$$

Upon Laplace transforming according to

$$V(p)\equiv\int_0^\infty e^{-pt}\,v(t)\,dt \qquad (7.2.45)$$

the initial-value problem (7.2.44) yields

$$pV(p)+\sigma^2\,\frac{V(p)}{p+\lambda}=1, \qquad (7.2.46)$$

from which

$$V(p)=\frac{p+\lambda}{p(p+\lambda)+\sigma^2}. \qquad (7.2.47)$$

Upon inverting the Laplace transform, (7.2.47) leads to

$$v(t)=e^{-\frac{\lambda}{2}t}\left[\cos\sqrt{\sigma^2-\frac{\lambda^2}{4}}\;t+\frac{\lambda/2}{\sqrt{\sigma^2-\frac{\lambda^2}{4}}}\sin\sqrt{\sigma^2-\frac{\lambda^2}{4}}\;t\right]. \qquad (7.2.48)$$

Observe that, when $\lambda=0$, the perturbative solution (7.2.48) oscillates indefinitely, as $t\to\infty$, though the exact solution (7.2.40) vanishes in this limit (see Figure 7.2)!

7.2.2(c) Renormalized Solution

In an effort to remedy the qualitative difficulty mentioned above, Kraichnan (1961) proposed a renormalization of the perturbative procedure. (Kraichnan called this procedure the *direct interaction approximation*.) This involves replacing L_0^{-1} in the bracket in equation (7.2.9) by $u(t)$; we then obtain in place of the initial-value problem (7.2.42),

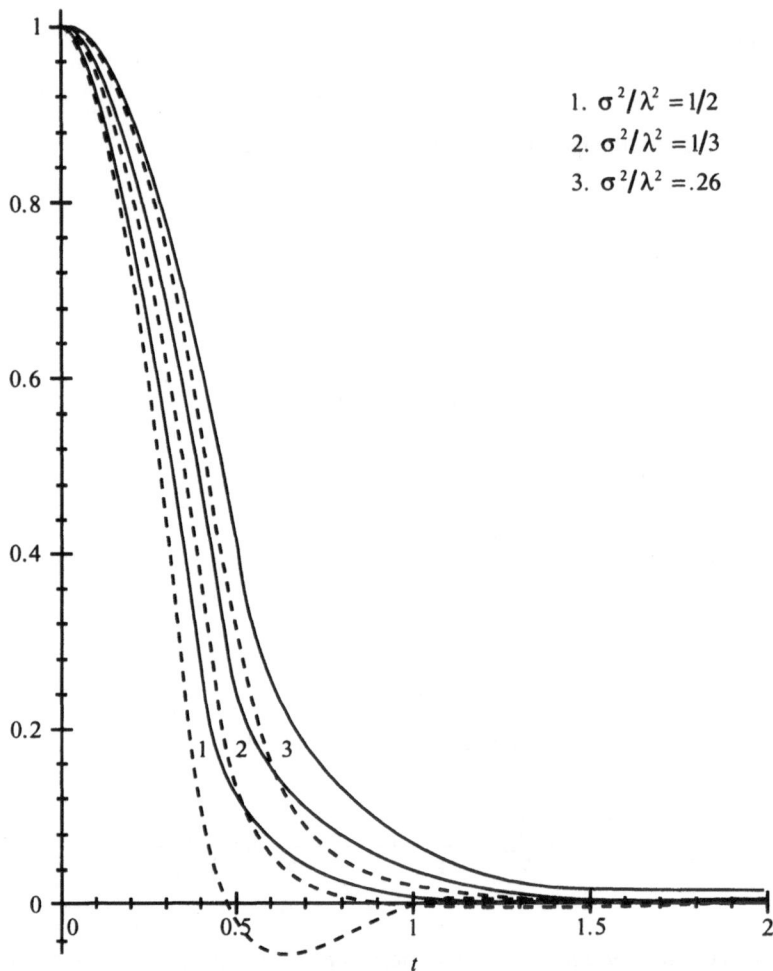

1. $\sigma^2/\lambda^2 = 1/2$
2. $\sigma^2/\lambda^2 = 1/3$
3. $\sigma^2/\lambda^2 = .26$

Figure 7.2. Comparison of the exact solution (solid line) with the perturbative solution (7.2.48) (dashed line) (from Shivamoggi et. al, 1999).

$$\left. \begin{array}{l} \left(\dfrac{d}{dt}+\mu\right)\langle u(t)\rangle+\int_0^t\langle b(t)\,b(t')\,u(t-t')\rangle\langle u(t')\rangle\,dt'=\delta(t) \\[2mm] \langle u(0)\rangle=1 \end{array}\right\} . \qquad (7.2.49)$$

Following Kraichnan (1961), we make the quasi-normality assumption[4]

$$\langle b(t)\,b(t')\,u(t-t')\rangle=\langle b(t)b(t')\rangle\langle u(t-t')\rangle \qquad (7.2.50)$$

and use (7.2.36) and (7.2.43); we then obtain from the initial-value problem (7.2.49),

$$\left. \begin{array}{l} \dfrac{dv}{dt}+\sigma^2\int_0^t e^{-\lambda(t-t')}\,v(t-t')\,v(t')\,dt'=\delta(t) \\[2mm] v(0)=0 \end{array}\right\} . \qquad (7.2.51)$$

Upon Laplace transforming with respect to t, the initial-value problem (7.2.51) yields the functional equation –

$$pV(p)+\sigma^2\,V(p+\lambda)\,V(p)=1, \qquad (7.2.52)$$

from which

$$V(p)=\frac{1}{p+\sigma^2\,V(p+\lambda)}. \qquad (7.2.53)$$

Equation (7.2.53), on iteration leads to the continued-fraction representation:

$$V(p)=\cfrac{1}{p+\cfrac{\sigma^2}{(p+\lambda)+\cfrac{\sigma^2}{(p+2\lambda)+\cdots}}}. \qquad (7.2.54)$$

Successive truncations of this continued fraction yield the following approximants –

$$V(p)=\frac{1}{p},\;\frac{p+\lambda}{p(p+\lambda)+\sigma^2},$$

$$\frac{(p+\lambda)(p+2\lambda)+\sigma^2}{p\left[(p+\lambda)(p+2\lambda)+\sigma^2\right]+\sigma^2\,(p+2\lambda)},\ldots . \qquad (7.2.55)$$

On inverting the Laplace transform, (7.2.55) leads to the following approximants:

---

[4] This amounts to expressing the triple correlations in terms of pair correlations while neglecting the residual fourth-order cumulants.

$$v(t) = 1, e^{-\frac{\lambda}{2}t} \left[\cos\sqrt{\sigma^2 - \frac{\lambda^2}{4}}\, t + \frac{\lambda/2}{\sqrt{\sigma^2 - \frac{\lambda^2}{4}}} \sin\sqrt{\sigma^2 - \frac{\lambda^2}{4}}\, t \right],$$

$$e^{-\lambda t} \left[\frac{\sigma^2}{2\sigma^2 - \lambda^2} + \frac{\sigma^2 - \lambda^2}{2\sigma^2 - \lambda^2} \cos\sqrt{2\sigma^2 - \lambda^2}\, t + \right.$$

$$\left. + \frac{\lambda}{\sqrt{2\sigma^2 - \lambda^2}} \sin\sqrt{2\sigma^2 - \lambda^2} \right], \cdots. \tag{7.2.56}$$

Comparison of (7.2.56) with the perturbative solution (7.2.48) shows that –

*   the first approximant corresponds to the *total* neglect of the random aspects in equation (7.2.35),

*   the second approximant corresponds to the incorporation of the random aspects in equation (7.2.35) in a *perturbative* way,

*   the third (and higher) approximant corresponds to the incorporation of the random aspects in equation (7.2.35) in a *non-perturbative (renormalized)* way.

Further comparison of (7.2.56) with the exact solution (7.2.39) shows that (see Figure 7.3) –

*   the third approximant is quite close to the exact solution,

*   the renormalized solution becomes more accurate as the ratio σ/λ becomes smaller.

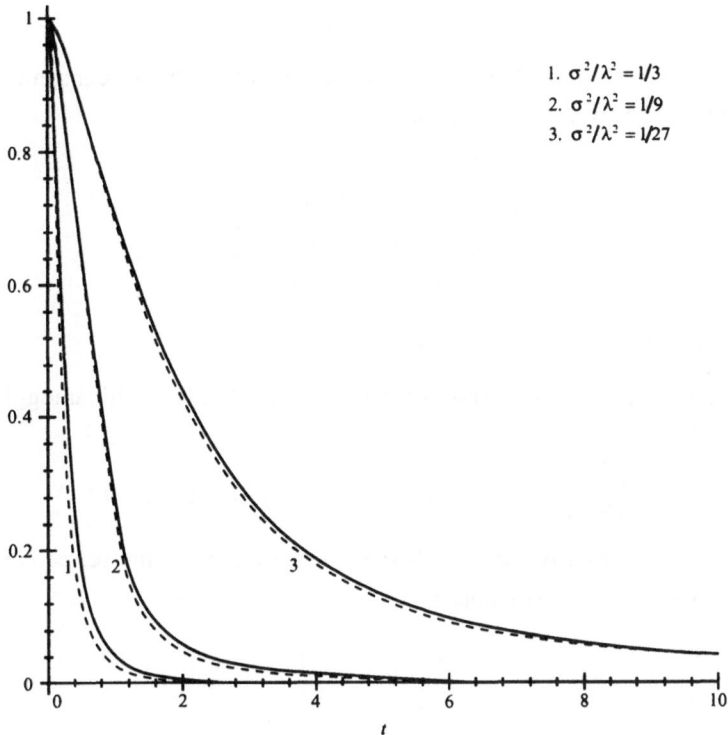

1. $\sigma^2/\lambda^2 = 1/3$
2. $\sigma^2/\lambda^2 = 1/9$
3. $\sigma^2/\lambda^2 = 1/27$

Figure 7.3. Comparison of the exact solution (solid line) with the third approximant (dashed line) in equation (7.2.56) (from Shivamoggi et. al, 1999).

7.3. Exercises

1. Use the QFT perturbation procedure to solve the Thomas-Fermi equation (Bender et al, 1989) –

$$\left. \begin{array}{l} \phi'' = \phi^{3/2} \, x^{-1/2} \\ \phi(0) = 1, \quad \phi(\infty) = 0 \end{array} \right\}.$$

2. Use the QFT perturbation procedure to solve the Kadomtsev equation (Shivamoggi and Rollins, (1997)) –

$$\left.\begin{array}{l}\phi'' = (x\phi)^{1/2} \\ \phi(0) = 1, \quad \phi(\infty) = 0\end{array}\right\}.$$

3. Use the QFT perturbation procedure to solve the Greenspan-Carrier equations (Shivamoggi and Rollins, (1999)) –

$$\left.\begin{array}{l}f''' + ff'' - \alpha\, gg'' = 0 \\ g'' + \beta(fg' - f'g) = 0 \\ \eta = 0 : f = 0, \; f' = 0, \; g = 0 \\ \eta \to \infty : f \approx \eta, \; g \approx \eta\end{array}\right\}.$$

Here, α and β are positive parameters.

4. Use the perturbation method of Section 7.2 to solve (Shivamoggi et al., (2000)) –

$$\frac{d^2 E}{dz^2} + k_0^2 \left[1 + \varepsilon\mu(z)\right] E + \varepsilon\left(aE^2\right) + \varepsilon^2\left(bE^3\right) = 0$$

where $\mu(z)$ describes the small random variations in the refractive index, and a,b are constant parameters.

References

Andrews, L.C. and Phillips, R.L. (1998), *Laser Beam Propagation through Random Media*, SPIE Press.

Bender, C.M., Milton, K.A., Pinsky, S.S., and Simmons, L.M. (1989), A new perturbative approach to nonlinear problems, *J. Math. Phys.* **30**, 1447.

Berman, V.S. (1978), On the asymptotic solution of a nonstationary problem on the propagation of a chemical reaction front, *Soviet Math. Dokl.* **19**, 1076.

Bogoliubov, N.N. and Mitropolski, Y.A. (1961), *Asymptotic Methods in the Theory of Nonlinear Oscillations*, Gordon and Breach.

Bolotin,V.V. (1964), *Dynamic Stability of Elastic Systems*, Holden Day.

Bourret, R.C. (1962), Stochastically perturbed fields with applications to wave propagation in random media, *Nuovo Cimento* **26**, 1.

Bretherton,F.P. (1964), Resonant interactions between waves: The case of discrete oscillations, *J. Fluid Mech.* **20**, 457.

Brillouin, L. (1926), Remarques sur la méchanique ondulatoire, *J. Phys. Radium* **7**, 353.

Burgers, J.M. (1948), A mathematical model illustrating the theory of turbulence, *Adv. Appl. Mech.* **1**, 171.

Cary, J.R. (1981), Lie transforms and their use in Hamiltonian perturbation theory.

Chandrasekhar, S. (1961), *Hydrodynamic and Hydromagnetic Stability*, Clarendon Press.

Chernov, L. (1969), *Wave Propagation in a Random Medium*, Dover.

Cole, J.D. (1968), *Perturbation Methods in Applied Mathematics*, Ginn-Blaisdell.

Curle, N. (1955), The influence of solid boundaries on aerodynamic sound, *Proc. Roy. Soc. (London)* A**231**, 505.

Doob, J.L. (1942), The Brownian movement and stochastic equations, *Ann. Math.* **43**, 351.

Drazin, P.G. and Johnson, R.S. (1989), *Solitons*, Cambridge University Press.

Eckaus, W. (1979), *Asymptotic Analysis of Singular Perturbations*, North-Holland.

Erdelyi, A. (1956), *Asymptotic Expansions*, Dover.

Ford, J. and Waters, J. (1963), Computer studies of energy sharing and ergodicity for nonlinear oscillator systems, *J. Math. Phys.* **4**, 1293.

Fox, P.A. (1955), Perturbation theory of wave propagation based on the method of characteristics, *J. Math. And Phys.* **34**, 133.

Fraenkel, L.E. (1969), On the method of matched asymptotic expansions, Parts I, II, and III, *Proc. Cambridge Phil. Soc.* **65**, 209, 233, and 263.

Friedrichs, K.O. (1955), Asymptotic phenomena in mathematical physics, *Bull. Am. Math. Soc.* **61**, 484.

Friedrichs, K.O. (1942), The mathematical structure of the boundary-layer problem, in *Fluid Mechanics*, Ed. von Mises, R. and Friedrichs, K.O., Springer-Verlag.

Gardner, C.S. (1959), Adiabatic invariants of periodic classical systems, *Phys. Rev.* **115**, 791.

Green, G. (1837), On the motion of waves in a variable canal of small depth and width, *Trans. Cambridge Phil. Soc.* **6**, 457.

Grimshaw, R.H.J. (1979), Slowly-varying solitary waves, I. Korteweg-deVries equation, *Proc. Roy. Soc. (London)* A**368**, 359.

Grimshaw, R.H.J. (1990), *Nonlinear Ordinary Differential Equations*, Blackwell.

Hasimoto, H. and Ono, H. (1972), Nonlinear modulation of gravity waves, *J. Phys. Soc. Japan* **33**, 805.

Hénon, M. and Heiles, C. (1964), The applicability of the third integral of motion: Some numerical experiments, *Astron. J.* **69**, 73.

Hinch, E.J. (1991), *Perturbation Methods*, Cambridge University Press.

Holmes, M.H. (1995), *Introduction to Perturbation Methods*, Springer-Verlag.

Infeld, E. and Rowlands, G. (1979), On the stability of electron-plasma waves, *J. Phys.* A**12**, 2255.

Janssen, P.A.E. (1981), Modulational instability and the Fermi-Pasta-Ulam recurrence, *Phys. Fluids* **24**, 23.

Jeffrey, A. and Kakutani, T. (1972), Weak nonlinear dispersive waves: A discussion centered around the Korteweg-deVries equation, *SIAM Rev.* **14**, 582.

Jeffreys, H. (1924), On certain approximate solutions of linear differential equations of the second order, *Proc. London Math. Soc.* **23**, 428.

Johnson, R.S. (1973), On an asymptotic solution of the Korteweg-deVries equation with slowly-varying coefficients, *J. Fluid Mech.* **60**, 813.

Kane, T.R. and Kahn, M.E. (1968), On a class of two-degree-of-freedom oscillations, *J. Appl. Mech.* **35**, 547.

Kaplun, S. (1957), Low Reynolds number flow past a circular cylinder, *J. Math. Mech.* **6**, 595.

Keller, J.B. (1964), Wave propagation in random medium, *Proc. Symp. Appl. Math.* **16**, 84.

Keller, J.B. (1968), *Perturbation Theory*, Lecture Notes, Department of Mathematics, Michigan State University.

Kevorkian, J. (1966), The two variable expansion procedure for the approximate solution of certain nonlinear differential equations, in *Space Mathematics*, Part 3, Ed. Rosser, J.B., American Mathematical Society.

Kevorkian, J. (1980), Resonance in a weakly-nonlinear system with slowly-varying parameters, *Stud. Appl. Math.* **62**, 23.

Kevorkian, J. and Cole, J.D. (1996), *Multiple Scale and Singular Perturbation Methods*, Springer-Verlag.

Knickerbocker, C.J. and Newell, A.C. (1980), Shelves and the Korteweg-deVries equation, *J. Fluid Mech.* **98**, 803.

Kraichnan, R.H. (1961), Dynamics of nonlinear systems, *J. Math. Phys.* **2**, 124.

Kramers, H.A. (1926), Wellenmechanik und halbzahlige quantisierung, *Z. Phys.* **38**, 828.

Kruskal, M.D. (1962), Asymptotic theory of Hamiltonian and other systems with all solutions nearly periodic, *J. Math. Phys.* **3**, 806.

Krylov, N. and Bogoliubov, N.N. (1947), *Introduction to Nonlinear Mechanics*, Princeton University Press.

Kuehl, H.H. and Imen, K. (1985), Finite-amplitude ion-acoustic solitons in weakly-inhomogeneous plasmas, *Phys. Fluids* **28**, 2375.

Kuzmak, G.E. (1959), Asymptotic solutions of nonlinear second-order differential equations with variable coefficients, *J. Appl. Math. Mech.* **23**, 730.

Lagerstrom, P.A. (1988), *Matched Asymptotic Expansions: Ideas and Techniques*, Springer-Verlag.

Lagerstrom, P.A. and Cole, J.D. (1955), Examples illustrating expansion procedures for the Navier-Stokes equations, *J. Rat. Mech. Anal.* **4**, 817.

Lake, B.M., Yuen, H.C., Rungaldier, H., and Ferguson, W.E. (1977), Nonlinear deep-water waves: Theory and experiment, Part 2. Evolution of a continuous wavetrain, *J. Fluid Mech.* **83**, 49.

Landau, L.D. (1945), On shock waves at large distances from the place of their origin, *Soviet J. Phys.* **9**, 496.

Landau, L.D. (1946), On the vibrations of the electronic plasma, *J. Phys. (USSR)* **10**, 25.

Langer, R.E. (1931), On the asymptotic solution of ordinary differential equations with an application to the Bessel functions of large order, *Trans. Am. Math. Soc.* **33**, 23.

Langer, R.E. (1935), On the asymptotic solutions of ordinary differential equations with reference to Stokes' phenomenon about a singular point, *Trans. Math. Soc.* **37**, 397.

Latta, G.E. (1951), *Singular Perturbation Problems*, Ph.D. Thesis, California Institute of Technology.

Leslie, D.C. (1973), *Developments in the Theory of Turbulence*, Clarendon Press.

Lighthill, M.J. (1949), A technique for rendering approximate solutions to physical problems uniformly valid, *Phil. Mag.* **40**, 1179.

Lighthill, M.J. (1952), On sound generated aerodynamically, I. General theory, *Proc. Roy. Soc. (London)* A**211**, 564.

Lighthill, M.J. (1967), Some special cases treated by the Whitham theory, *Proc. Roy. Soc. (London)* A**299**, 28.

Lin, C.C. (1954), On a perturbation theory based on the method of characteristics, *J. Math. And Phys.* **33**, 117.

Lindstedt, A. (1882), Über die integration einer für störungsthorie wichtigen differtialgleichung, *Astron. Nachr.* **103**, 211.

Liouville, J. (1837), Sur le développement des fonctions on partices de fonctions en séries, *J. Math. Pures Appl.* **2**, 16.

Luke, J.C. (1967), A variational principle for a fluid with a free surface, *J. Fluid Mech.* **27**, 395.

McGoldrick, L.F. (1970), On Wilton's ripples: A special case of resonant interactions, *J. Fluid Mech.* **42**, 193.

Nayfeh, A.H. (1964), *A Generalized Method for Treating Singular Perturbation Methods*, Ph.D. Thesis, Stanford University.

Nayfeh, A.H. (1969), On the nonlinear Lamb-Taylor instability, *J. Fluid Mech.* **38**, 619.

Nayfeh, A.H. and Hassan, S.D. (1971), The method of multiple scales and nonlinear dispersive waves, *J. Fluid Mech.* **48**, 463.

Nayfeh, A.H. (1973), *Perturbation Methods*, Wiley.

Nayfeh, A.H. (1981), *Introduction to Perturbation Techniques*, Wiley.

Nishikawa, K. and Kaw, P.K. (1975), Propagation of solitary ion-acoustic waves in inhomogeneous plasmas, *Phys. Lett.* A**50**, 455.

O'Malley, R.E. (1974), *Introduction to Singular Perturbations*, Academic Press.

Poincaré, H. (1967), *New Methods in Celestial Mechanics*, (1892), Translation NASA-TIF 450.

Prandtl, L. (1904), Uber Flüssigkeitsbewegung bei sehr kleiner Reibung, in Verhandlungen des III. Internationlen Mathematiker-Kongresses, Heidelberg.

Pritulo, M.F. (1962), On the determination of uniformly accurate solutions of differential equations by the method of perturbation of coordinates, *J. Appl. Math. Mech.* **26**, 661.

Rayleigh, J.W.S. (1912), On the propagation of waves through a stratified medium with special reference to the question of reflection, *Proc. Roy. Soc. (London)* A**86**, 208.

Schlichting, H. (1972), *Boundary Layer Theory*, McGraw-Hill.

Schrödinger, E. (1926), Quantisierung ais Eigenwertproblem, *Ann. Phys.* **80**, 437.

Sears, W.R. (1969), Aerodynamics, noise, and the sonic boom, *AIAA J.* **7**, 577.

Shivamoggi, B.K. (1977a), Nonlinear hyperbolic waves, *J. Sound and Vibr.* **54**, 603.

Shivamoggi, B.K. (1977b), Dynamic buckling of a thin elastic plate: Nonlinear theory, *J. Sound and Vibr.* **54**, 75.

Shivamoggi, B.K. (1977c), Uniformly-valid mach number expansion of the Navier-Stokes equations and mathematical formalization of Lighthill's theory of aerodynamically generated sound, *J. Sound and Vibr.* **51**, 303.

Shivamoggi, B.K. (1978a), Nonlinear hyperbolic waves, *J. Sound and Vibr.* **55**, 594.

Shivamoggi, B.K. (1978b), Method of matched asymptotic expansions: Asymptotic matching principle for higher approximations, *Z. Angew. Math. Mech.* **58**, 354.

Shivamoggi, B.K. (1978c), Wave propagation in an inhomogeneous medium, *J. Acoust. Soc. Am.* **63**, 1926.

Shivamoggi, B.K. (1979), Nonlinear theory of Rayleigh-Taylor instability of superposed fluids, *Acta Mech.* **31**, 301.

Shivamoggi, B.K. (1981a), Method of multiple scales and the problem of aerodynamically generated sound, *Arch. Mech.* **33**, 603.

Shivamoggi, B.K. (1981b), Effect of finite-ion temperature on ion-acoustic solitary waves in an inhomogeneous plasma, *Can. J. Phys.* **59**, 719.

Shivamoggi, B.K. (1982a), Rayleigh-Taylor instability of a plasma in a magnetic field: Nonlinear theory, *J. Plasma Phys.* **27**, 129 (erratum in **30**, 511, (1983)).

Shivamoggi, B.K. (1982b), Weakly-nonlinar dispersive waves: Group velocity, *J. Plasma Phys.* **27**, 507.

Shivamoggi, B.K. (1983), A variational principle for gravity wavetrains in magnetohydrodynamics, *Q. Appl. Math.* **41**, 31.

Shivamoggi, B.K. (1986), *Theory of Hydromagnetic Stability*, Gordon and Breach.

Shivamoggi, B.K. (1987), Internal resonances in weakly-nonlinearly coupled oscillators with slowly-varying parameters, *Z. Angew. Math. Mech.* **69**, 23.

Shivamoggi, B.K. (1988a), *Introduction to Nonlinear Fluid-Plasma Waves*, Kluwer.

Shivamoggi, B.K. (1988b), Ion-acoustic solitary waves in an inhomogeneous plasma, *J. Plasma Phys.* **40**, 579.

Shivamoggi, B.K. and Varma, R.K. (1988), Internal resonances in nonlinearly-coupled oscillators, *Acta Mech.* **72**, 111.

Shivamoggi, B.K. (1990), Nonlinear development of modulated gravity wavetrain in deep water, *J. Phys. A: Math. And Gen.* **23**, 4289.

Shivamoggi, B.K. and Muilenburg, L. (1991), On Lewis' exact invariant for a linear harmonic oscillator with time-dependent frequency, *Phys. Lett. A* **154**, 24.

Shivamoggi, B.K. (1997), *Nonlinear Dynamics and Chaotic Phenomena*, Kluwer.

Shivamoggi, B.K. and Rollins, D.K. (1997), An analytic perturbative solution for the Kadomtsev equation for a heavy atom in a very strong magnetic field, *J. Phys. A: Math. and Gen.* **30**, 3681.

Shivamoggi, B.K. (1998), *Theoretical Fluid Dynamics*, II Ed., Wiley.

Shivamoggi, B.K. and Rollins, D.K. (1999), Magnetohydrodynamic boundary layer on a flat plate: Further analytic results, *J. Math. Phys.* **40**, 3372.

Shivamoggi, B.K., Taylor, M.D., and Kida, S. (1999), On some mathematical aspects of the direct interaction approximation in turbulence theory, *J. Math. Anal. Appl.* **229**, 639.

Shivamoggi, B.K., Andrews, L.C., and Phillips, R.L. (2000), Nonlinear wave propagation in a turbulent medium, *Physica A* **275**, 86.

Shivamoggi, B.K. and Rollins, D.K. (2002), Evolution of solitary-wave solution of the perturbed regularized long-wave equation, *Chaos, Solitons, and Fractals*, **13**, 1129.

Simmonds, J.G. and Mann, J.E. (1986), *A First Look at Perturbation Theory*, Krieger.

Smyth, N.F. (1984), *Soliton on a Beach and Related Problems*, Ph.D. Thesis, California Institute of Technology.

Strauss, W.A. (1992), *Partial Differential Equations*, Wiley.

Struble, R.A. (1962), *Nonlinear Differential Equations*, McGraw Hill.

Tatarskii, V. (1967), *Wave Propagation in a Turbulent Medium*, Dover.

Timoshenko, S.P. and Woinowsky-Krieger, S. (1959), *Theory of Plates and Shells*, McGraw-Hill.

Timoshenko, S.P. (1959), *Elastic Stability*, McGraw Hill.

Tritton, D.J. (1988), *Physical Fluid Dynamics*, Clarendon Press.

Tsien, H.S. (1956), The Poincaré-Lighthill-Kuo method, *Adv. Appl. Mech.* **4**, 281.

Van Dyke, M.D. (1952), A study of second-order supersonic flow theory, *NACA Report* **1081**.

Van Dyke, M.D. (1975), *Perturbation Methods in Fluid Mechanics*, Parabolic Press.

van Kampen, N.G. (1976), Stochastic differential equations, *Phys. Rep.* **24**, 171.

van Kampen, N.G. (1992), *Stochastic Processes in Physics and Chemistry*, Elsevier.

von Karman, Th. (1910), Festigkeitsprobleme im Maschinenbau, *Enzyklopadie der Math. Wiss.* **4**, 349.

Wasow, W. (1965), *Asymptotic Expansions for Ordinary Differential Equations*, Wiley.

Wentzel, G. (1926), Eine verallgemeinerung der quantenbedingungen für die zwecke der wellenmechanik, *Z. Phys.* **38**, 518.

Whitham, G.B. (1952), The flow pattern of a supersonic projectile, *Comm. Pure and Appl. Math.* **5**, 301.

Whitham, G.B. (1965), A general approach to linear and nonlinear waves using a Lagrangian, *J. Fluid Mech.* **22**, 273.

Whitham, G.B. (1967), Variational methods and applications to water waves, *Proc. Roy. Soc. (London)* A**299**, 6.

Whittaker, E.T. (1914), On the general solution of Mathieu's equation, *Edinburgh Math. Soc. Proc.* **32**, 75.

Whittaker, E.T. (1964), *Analytical Dynamics of Particles and Rigid Bodies*, Cambridge University Press.

Wilton, J.R. (1915), On ripples, *Phil. Mag.* **29**, 688.

Yuen, H.C. and Lake, B.M. (1975), Nonlinear deep-water waves: Theory and experiment, *Phys. Fluids* **18**, 956.

Answers to Selected Problems

Chapter 1

4. $f(w) \sim 1 - \dfrac{1!}{w} + \dfrac{2!}{w^2} + \cdots + \dfrac{(-1)^n n!}{w^n} + (n+1)! \displaystyle\int_0^\infty \dfrac{(-1)^{n+1} w}{(w+x)^{n+2}} \, e^{-x} dx$

5. $E_n(x) \sim \dfrac{e^{-x}}{x} - n\dfrac{e^{-x}}{x^2} + n(n+1)\dfrac{e^{-x}}{x^3} + \cdots + (-1)^{r-1} n(n+1)\cdots(n+r-2)\dfrac{e^{-x}}{x^r}$

$$+ (-1)^r n(n+1)\cdots(n+r-1) \int_1^\infty t^{-(n+r)} x \dfrac{e^{-xt}}{x^r} \, dt$$

7. $y \sim a - \dfrac{e^{-2ax}}{2a} - \dfrac{1}{4a^2}\left(x + \dfrac{1}{4a}\right) e^{-4ax} + \cdots$

Chapter 2

2. $x_1 = 1 - \dfrac{\varepsilon}{z} + \dfrac{5\varepsilon^2}{8} + \cdots$

$x_2 = -1 - \dfrac{\varepsilon}{2} - \dfrac{5\varepsilon^2}{8} + \cdots$

$x_3 = -\dfrac{1}{\varepsilon} + \varepsilon - 2\varepsilon^3 + \cdots$

5. $y = -t + \varepsilon\left(1 - \dfrac{t^2}{2}\right) + \varepsilon^2\left(t - \dfrac{t^3}{6}\right) + \cdots$

6. $y = \sin t + \varepsilon\left[\dfrac{e^{-at}+1}{4+a^2}\sin t + \dfrac{2(e^{-at}-1)}{a(4+a^2)}\cos t\right] + \cdots$

7. $\quad y = a + bx + \int\limits_0^x (x-s) \, f(s) \, ds + \varepsilon \left(-\dfrac{1}{2} ax^2 - \dfrac{1}{6} bx^3 - \int\limits_0^x \dfrac{(x-t)^3}{6} f(t) \, dt \right)$

Chapter 3

2. $\quad x = \varepsilon a \cos(\omega t + \varphi) + O(\varepsilon^3)$

$$\omega = \sqrt{\dfrac{g}{p} - \omega^2} - \varepsilon^2 \dfrac{a^2}{4p^2} \sqrt{\dfrac{g}{p} - \omega^2} + \cdots$$

3. $\quad u = e^{\mu t} \, q(t)$ where

$$q = A_0 + \varepsilon \left(A_1 + \dfrac{1}{4} A_0 \cos 2t \right) + \cdots$$

$$\mu \approx \pm i \sqrt{\delta + \dfrac{1}{8} \varepsilon^2}$$

4. $\quad \varphi = \sqrt{2} \sin n\pi x - \dfrac{\varepsilon\sqrt{2}}{16 n^2 \pi^2} \sin 3n\pi x + \cdots$

$$\lambda = n^2 \pi^2 - \dfrac{3}{2} \varepsilon + \cdots$$

5. $\quad y = \dfrac{1}{x} + \varepsilon \left(\dfrac{x^2 - 1}{2x^3} \right) + \cdots$

Application of Lighthill's method leads to

$$y \approx \dfrac{2x + \varepsilon x}{2x^2 + \varepsilon}.$$

8. $\quad u = f\left\{ x - ct \left[1 + \dfrac{\varepsilon}{4} f'(x - ct) \right] \right\} + g\left\{ x + ct \left[1 + \dfrac{\varepsilon}{4} g'(x + ct) \right] \right\} + \cdots$

Chapter 4

2. $\quad u = a \cos \psi + \varepsilon \left(\dfrac{a^3}{32} \cos 3\psi - \dfrac{a^3}{32} \sin 3\psi \right) + \cdots$

where

$$\frac{da}{dt} = \varepsilon \frac{a}{2}\left(1 - \frac{a^2}{4}\right)$$

$$\frac{d\psi}{dt} = 1 - \varepsilon \frac{a^2}{8}$$

6. $\omega = 1 + \varepsilon \dfrac{3J}{4} + \cdots$

Chapter 5

1. $y = be^{x-1} - be^{-1} + a + \sqrt{\dfrac{2}{\pi}} \left(be^{-1} - a\right) \displaystyle\int_0^{x/\sqrt{\varepsilon}} e^{-\frac{\tau^2}{2}}\, d\tau$

2. $y = \dfrac{1+\alpha}{1+\alpha x} - (1+\alpha)\, e^{-\frac{x}{\varepsilon}} + \varepsilon\left[\left\{-\dfrac{\alpha}{(1+\alpha)(1+\alpha x)} + \dfrac{(1+\alpha)\alpha}{(1+\alpha x)^3}\right\}\right.$

$$\left. + \left\{\dfrac{1}{2}\alpha(1+\alpha)\dfrac{x^2}{\varepsilon^2} + \dfrac{\alpha}{1+\alpha} - \alpha(1+\alpha)\right\} e^{-\frac{x}{\varepsilon}}\right] + \cdots$$

5. $y \approx \begin{cases} \dfrac{x^2}{2} + 1 - \dfrac{5/2}{1 + e^{-\frac{5}{4\varepsilon}\left(x - \frac{1}{\sqrt{2}}\right)}}, & 0 \le x \le \dfrac{1}{\sqrt{2}} \\[4ex] \dfrac{x^2}{z} - \dfrac{3}{2} + \dfrac{5/2}{1 + e^{\frac{5}{4\varepsilon}\left(x - \frac{1}{\sqrt{2}}\right)}}, & \dfrac{1}{\sqrt{2}} < x \le 1 \end{cases}$

7. $y = a(2x+1) + (b - 3a)\, e^{-3\frac{1-x}{\varepsilon}} + \cdots$

$$y_c = a(2x+1) + \dfrac{9(b-3a)}{(2x+1)^2}\, e^{\frac{x^2+x-2}{\varepsilon}} + \cdots$$

Chapter 6

3. $u = a(t_1)\, e^{i\omega_0 t} + \text{c.c.}$

where

$$\frac{da}{dt_1} = \begin{cases} \dfrac{3i}{2\omega_0} |a|^2 a, & \omega_0 \neq \lambda \\[4mm] \dfrac{3i}{2\omega_0} |a|^2 a - \dfrac{\mu i}{4\omega_0}, & \omega_0 \approx \lambda \end{cases}$$

9. $y_c = a(2x+1) + \dfrac{9(b-3a)}{(2x+1)^2} e^{-\frac{2-x^2-x}{\varepsilon}} + \cdots$

10. $y = \dfrac{1+\alpha}{1+\alpha x} - (1+\alpha) e^{-\frac{1}{\varepsilon}\left(x + \frac{\alpha x^2}{2}\right)}$

$$+\varepsilon \left[\frac{\alpha(1+\alpha)}{(1+\alpha x)^3} - \frac{\alpha}{(1+\alpha)(1+\alpha x)} - \frac{\alpha^2(2+\alpha)}{1+\alpha} e^{-\frac{1}{\varepsilon}\left(x + \frac{\alpha x^2}{2}\right)} \right] + \cdots$$

Index

Permissions

Table 2.1 reprinted from Nayfeh, A.H., *Introduction to Perturbation Techniques*, 1981, with permission of John Wiley & Sons, Inc.

Figures 2.1, 5.14-5.16, 6.1, and 6.4 reprinted from Holmes, M.H., *Introduction to Perturbation Methods*, 1995, with permission of Springer-Verlag.

Figures 4.1 and 4.2 reprinted from Nayfeh, A.H., *Introduction to Perturbation Techniques*, 1981, with permission of John Wiley & Sons, Inc.

Figure 5.3-5.7 and 5.9 reprinted from Kevorkian, J. and Cole, J.D., *Multiple Scale and Singular Perturbation Methods*, 1996, with permission of Springer-Verlag.

Figure 5.12 reprinted from Lagerstrom, P.A., *Matched Asymptotic Expansions: Ideas and Techniques*, 1988, with permission of Springer-Verlag.

Figure 5.17 reprinted from Tritton, D.J., *Physical Fluid Dynamics*, 1988, with permission of Oxford University Press.

Figure 6.6 reprinted from Ford, J. and Waters, J., Computer Studies of Energy Sharing and Ergodicity for Nonlinear Oscillator Systems, *Journal of Mathematical Physics*, **4**, 1293, 1963, with permission of the American Institute of Physics.

Figure 6.9 adapted from Struble, R.A., *Partial Differential Equations*, 1962, with permission of McGraw-Hill Education.

Figure 6.11 reprinted from Shivamoggi, B.K., Dynamic Buckling of a Thin Elastic Plate: Nonlinear Theory, *Journal of Sound and Vibration*, **54**, 75, 1977, with permission of Elsevier.

Figure 6.16 reprinted from Lake, B.M., Yuen, H.C., Rungaldier, H., and Ferguson, W.E., Nonlinear Deep-Water Waves: Theory and Experiment. Part 2. Evaluation of a Continuous Wavetrain, *Journal of fluid Mechanics*, **83**, 49, 1977, with permission of Cambridge University Press.

Figure 7.1 reprinted from Shivamoggi, B.K. and Rollins, D.K., Magnetohydrodynamic Boundary Layer on a Flat Plate: Further Analytic Results, *Journal of Mathematical Physics*, **40**, 3372, 1999, with permission from the American Institute of Physics.

Figures 7.2 and 7.3 reprinted from Shivamoggi, B.K., Taylor, M.D., and Kida, S., On Some Mathematical Aspects of the Direct Interaction Approximation in Turbulence Theory, *Journal of Mathematical Analysis and Applications*, **229**, 639, 1999, with permission of Academic Press.